Praise for *SHOULD AMERICA PAY?*

"An extraordinary collection o⋯⋯⋯⋯⋯⋯ work to mainstream a topic that is sur⋯⋯⋯⋯⋯⋯'s social agenda." —*Library Journal*

"Compelling. . . . Beyond pro o⋯⋯⋯⋯⋯⋯ply concerned with having their reader⋯⋯⋯⋯⋯⋯ or accountability by looking at our collective past and present." —*Publishers Weekly*

"Whatever your opinions on the issue, *Should America Pay?* is . . . essential reading." —*Black Issues Book Review*

"*Should America Pay?* is a comprehensive look at a topic of increasing importance. The inclusion of dissenting voices and historical documents make the argument for reparations even more urgent and compelling. An indispensable text." —Herb Boyd, author of *Autobiography of a People: Three Centuries of African American History Told by Those Who Lived It* and co-author of *Brotherman: The Odyssey of Black Men in America*

"With the appearance of this scholarly and thought-provoking work, the reparations debate will undoubtedly be taken to new levels of intensity and sophistication." —Lewis V. Baldwin, Professor of Religious Studies at Vanderbilt University

"*Should America Pay?* is a beacon of light for an issue long tucked away in the dark. Winbush has assembled a group of essays that mix clear analysis with common sense and passion." —Chuck D, Public Enemy

"A collection of original and previously published essays addressing the legal, emotional, and practical issues surrounding the reparations debate from well-known commentators and noted scholars." —*Ebony*

"A comprehensive look at the controversial issue from such angles as the history, law, practical challenges, and global implications." —*Tennesseean*

"A complete and balanced look at a controversial topic that is gaining attention." —*Booklist*

RAYMOND A. WINBUSH, PH.D., is the Director of the Institute for Urban Research at Morgan State University. He received his undergraduate degree in psychology from Oakwood College in Alabama and received a fellowship to attend the University of Chicago, where he earned both his master's degree and Ph.D. in psychology. He has taught at Oakwood College, Alabama A&M, Vanderbilt University, and Fisk University. He is the recipient of numerous grants, including one from the Kellogg Foundation to establish a "National Dialogue on Race." He is the author of *The Warrior Method: A Parents' Guide to Rearing Healthy Black Boys*, the former treasurer and executive board member of the National Council of Black Studies, and is currently on the editorial board of the *Journal of Black Studies*. He lives in Baltimore, Maryland.

ALSO BY RAYMOND A. WINBUSH, PH.D.

The Warrior Method: A Parents' Guide to Rearing Healthy Black Boys

RAYMOND A. WINBUSH, PH.D.

SHOULD AMERICA PAY?

Slavery and the Raging

Debate on Reparations

Amistad

An Imprint of HarperCollinsPublishers

FIRST AMISTAD PAPERBACK EDITION PUBLISHED 2003.

Printed on acid-free paper

The Library of Congress has catalogued the hardcover as follows:

Should America pay? : slavery and the raging debate on reparations / [edited by]
Raymond A. Winbush.—1st ed.
p. cm.
Includes bibliographical references and index.
ISBN 0-06-008310-7 (acid-free paper)
1. African Americans—Reparations. I. Winbush, Raymond A. (Raymond Arnold),
date.
E185.89.R45 S56 2003
305.896'073—dc21   2002027927
ISBN 0-06-008311-5 (paperback)

HB 05.03.2021

To Africans both living and dead

who received no justice for being captured,

enslaved, colonized, and discriminated against.

Your struggles and sufferings

will never be forgotten.

ACKNOWLEDGMENTS

One of the challenges of editing a volume on an ongoing social movement is that it is a dynamic process that unfolds daily. It is like writing a history of the Civil Rights movement after the 1963 Birmingham demonstrations; events cry out for documentation but each scene is but one in a very lengthy play. This book represents an attempt to place key people and ideas in the reparations movement in one volume so that readers will have a comprehensive view of its history, legal issues, and the opinions of its supporters and detractors. I have asked persons to contribute whose names will be quickly recognized and some that may not. I was pleased with the eagerness of the authors amidst their extraordinarily busy schedules to write for this volume as a primer on reparations.

I want to thank several of the authors for supplying heretofore unpublished chapters about their involvement with and study of reparations. Many were busy, but provided materials that were undeniably essential to the reparations debate. I also want to thank publicly my staff at the Race Relations Institute at Fisk University, Nontombi Tutu, Hazel Joyner-Smith, and Bonolo Leburu, who allowed me to abandon them for long periods while the book was being compiled and edited. Morgan State University's support is always inestimable and continues to provide its faculty and students excellent opportunities to drink deeply of its rich tradition of providing a place where voices critical for understanding the African world can be heard.

I am especially indebted to Morgan State University president Dr. Earl Richardson and Dean of Graduate Studies Dr. Maurice Taylor, who made my transition from Fisk to Morgan smooth. They belong to that small but extremely dedicated group of African leaders who understand the importance of higher education for African students while others so easily compromise traditions that built the great Black colleges and universities of the United States.

My staff at the Institute for Urban Research at Morgan was essential in doing the tedious but necessary steps in editing this volume. Pat Thomas and Charlotte Stewart were particularly helpful in finding incredibly obscure historical information that was important for clarifying several points.

I also wish to thank my students Esther Jones, Sara Davis, Jeff Menzise, and Porcha Wafford, who convince me on a daily basis that young Black people have an African worldview larger than that of my generation and exhibit confidence that it will be the value system of the majority of our people in the very near future.

My brothers and sisters, Harold Jr., Myrna, LaVonne, and Ronald; my parents, Harold and Dorothy, who continue to inspire me along with other Ancestors, such as Bobby Wright, Ida B. Wells-Barnett, Sengbeh Pieh aka Joseph Cinque, Denmark Vesey, and Yaa Asantewa; friends Evelyn, Cheryl, Jerry, Jennifer Whatley-Williams, Opal Baker, Marimba Ani, Denise Wright, Ife Williams, Harry Allen, Soffiyah Elijah, Ramona Balbuena, Joyce, Yaa, Pat, best friends Chike and Carl Parker, and Anne all know the roles they played in bringing this book to fruition. Dawn Davis, Darah Smith, Tara Brown, Abbot Street, Rockelle Henderson, John Jusino, and David Koral were indispensable in their work at HarperCollins in making this book possible. Members of the Congressional Black Caucus, particularly John Conyers Jr., Cynthia McKinney, Eddie Bernice Johnson, and Barbara Lee, prove that elected Black officials need not compromise their principles on issues that impact their constituents. Dr. Jewel Crawford provided much-needed advice (and humor) and calls into question the stereotype that Black physicians are by nature apolitical.

Dr. Frances Cress Welsing and Neely Fuller have influenced my thinking more than anyone else regarding global white supremacy and its ways of distorting truth. Marimba Ani's extraordinary analysis of European thought contained in her remarkable book, *Yurugu*, keeps me centered in a world where leaders of governments, institutions, faith communities, and educational facilities continue their assault on people of color. I am still humbled by the people who surround me in my personal and professional life and love them dearly for their strong support in the work I do.

My children continue to be my primary inspiration in documenting the African world because they represent our future in a world that is drunk with white supremacy. Faraji, Sharifa, Omari, Beverly, and Jawanza, I love you all and simply want you to know that.

Ray Winbush, Baltimore, May 2003

CONTENTS

Foreword

When I was first approached to be a contributor for *Should America Pay?* I was unwilling. Not because I did not recognize the importance of the book but because I did. What finally convinced me was the opportunity to put down my thoughts and passions around this issue of justice for people of color. I was particularly passionate about the need to highlight the way in which issues of race and racism have tended to replace charity for justice when demands arise from people of color. This fact has been most strongly brought home to me in my participation since 2001 with the Jubilee movement and preparations for the World Conference Against Racism that took place in Durban, South Africa, in August 2001. For me both of these have been about opportunities for the world to look honestly at our economic and political history as a global community, and to change the system toward a more just one. However, it has become clear that there is real resistance to placing in a true historical context the present situation of so-called Third World people all over the world. The lengths to which Western governments have been willing to go to stop real discussions about both debt cancellation and reparations have been to me amazing and often heartbreaking. The resistance to the idea of even looking into or discussing responsibility for past atrocities and present inequities could be disheartening if it were not for the fact that it is equaled, and often surpassed, by the demands for justice that are beginning to be heard and litigated by historically oppressed groups from around the world. The honest discussion of reparations has come of age in the United States and the world. Maybe I should say that the world has come of age for the discussion of reparations. The truth is the discussion and claim for reparations from Africans and African descendants in the *Maafa* has been taking place since even before the Emancipation Proclamation. What has changed are not the *demands* from those due reparations but the realization by those in power that those demands will not simply go away if they are ignored.

In the last few years, the debate about reparations has moved steadily into the mainstream. The brave efforts of people like Congressman John Conyers (D-MI), who has introduced legislation calling for the study of the claim for reparations into every session since 1989, can no longer be regarded as quixotic. *The Debt*, the best-selling book by Randall Robinson, simply gave eloquent words and strong arguments to a sentiment that has been felt in the African American community for generations, and placed those arguments into the larger public forum. The very fact that right-wing commentator David Horowitz published "Ten Reasons Why African Americans Should Not Be Paid Reparations" speaks not to the weakness of the claim but to the need for the right to respond concretely to a legitimate demand. If there were no basis for the claim, there would be no need to attempt to repudiate it. Clearly the debate about reparations is now front and center on the United States and international public policy stage.

In fact the question that should be raised is not why has this happened now, but rather why has it taken so long? Why is it that the demands of African states for reparations from the colonial powers have been ignored for over forty years? Why is it that calls by African Americans for repayment for the years of unpaid labor as slaves and underpaid labor for generations and continuing to this day have been pushed aside? Part of the answer must be found in the world's willingness to ignore demands for justice by people of color. Rather, it has been easier to offer *charity*, which maintains a system where people of color are seen as *recipients* of the largesse of the West rather than *claimants* to what is rightfully theirs. African nations are given loans by Western governments, banks, and international financial institutions, with no recognition that the wealth that allows the latter to make those loans is partly built on wealth removed from the former during slavery and colonialism. African Americans are offered affirmative action to help them overcome years of injustice. The problem with charity, whether it be loans and grants or affirmative action, is that it is reliant on the goodwill of the benefactor. When the charitable feelings end, as has clearly happened in the United States with regard to affirmative action, there is no incentive to continue to give until the wrongs committed have been righted. A demand for justice, on the other hand, is not based on the willingness of those paying to pay, but on the right of the recipients to be compensated for crimes against them.

The response to two movements with a growing momentum in 2001–2002 shows how far the West is willing to go to forestall the discussion of justice in regards to demands by people of color. The first is the Jubilee movement calling for debt cancellation for the world's poorest countries. The initial response from the West and its international financial institutions was to talk about debt reduction. When this failed to stop the momentum, President Bush stepped in to suggest that the World Bank should offer grants to so-called Third World countries. However, the West is not the only one to realize that the momentum is growing, and these offers of continuing charity were swiftly denounced by organizations such as Jubilee South. The Jubilee South communiqué issued from Genoa in 2001 stated, "We do not accept the concept of grants from the North to South, because what is due us is, in fact, *reparations*"[1] (emphasis mine). The second is the inclusion of a call for compensatory remedies for the effects of racism in the Declaration and the Plan of Action for the United Nations World Conference Against Racism. This statement was one of the major obstacles in the view of the Western Group (Western Europe, the United States, Canada, Australia, and New Zealand) to the productive work of the conference. So much of an obstacle, indeed, that the United States threatened to withdraw from the conference if the issue remained on the table for discussion in Durban.[2] The fact that the West is unwilling to countenance even the discussion of compensatory remedies or the declaration of slavery to have been a crime against humanity indicates the level of anxiety that the demand for justice from people of color provokes.

We are told that these issues are in the past and we must focus on the present and the future, as though the past has no effect on either. We are told that it would be too difficult to find those who have been directly affected by slavery after such a passage of time. We are told that there is no one today who is directly responsible for slavery, that it was accepted practice at the time and that Africans sold other Africans into slavery. The authors in *Should America Pay?* more than adequately address the racial red herrings that are often the refuge of those resisting discussion on compensatory remedies. Such weak arguments are raised to convince the world that reparations cannot be seriously countenanced or even studied. All that these arguments prove is the way in which issues important to people of color in general and Africans and African descendants in particular are treated ahistorically. As though there were no similar claims made by others and accepted as legitimate by the very powers that denigrate the claim for reparations. For too long people of color have accepted the delegitimizing of their claims, little realizing that their unwillingness to press for justice would be used against them, such as the claim of the long passage of time.

Yet all the arguments against reparations are easily countered if the discussion is placed in context, historically and politically. The easiest question to answer is probably who should pay reparations. Those states that benefited from slavery and colonialism are clearly responsible. The fact that the responsibility of the state continues has been used against African states in the payment of debts accumulated by previous and illegitimate regimes. The clearest and most recent example is that of South Africa, where the majority government is now paying the debts for loans taken by the apartheid regime. The nationalist government of South Africa used many of the loans it received to bolster its ability to oppress black South Africans, yet today it is those who fought against apartheid who are called upon to repay those debts. If South Africans can be asked to pay for their oppression, how is it that Europeans and the United States cannot be asked to pay for the way in which their wealth was accumulated? It is a compelling argument that the felonious legacy of a nation outlasts the government under which crimes were committed against its citizens.

The argument that usually seems to leave those supporting reparations fumbling is the one about the passage of time. It is all too long ago now, let us move on. The argument is particularly strong in that it is used to bolster other parts of the argument against reparations, the difficulty in finding clear victims and clear beneficiaries of the oppression. We are told, for example, that the case of Jews and reparations for the European Holocaust is different in that there are survivors and the direct descendants of survivors. The establishment of the State of Israel is based on biblical claims to that land as the homeland given by God to the Jewish people. It is a claim based on something that happened over two thousand years ago. I am not here calling for a debate on the legitimacy of the State of Israel. I am merely arguing that the same Western governments that claim slavery happened too long ago for compensation by people of African descent accept as legitimate the right of European Jews to claim land based on theological interpretations that happened even longer ago. If one claim can be legitimately made, why not the other? These are the kinds of questions that the present-day reparations movement is asking. It is up to nations that have accepted some claims while ignoring others to explain such contradictions while listening to the arguments brought by plaintiffs for compensatory measures.

I do not imagine that the few words I have written here will convince those who are stubbornly opposed to reparations to change their minds; I do not even expect that the well-thought-out words and arguments of the authors of the different chapters in this book will do so. I do believe, though, that this book and the ongoing debate in national and international forums will at least make it clear that the issue of reparations is not one that will go away. No matter how many attempts are made to belittle this effort, the fact is the issue is here; the call for justice is being raised from Dakar to Detroit and will not be silenced until there is an honest discussion about it. The struggle for political and economic racial justice is one of the fundamental struggles facing us in this new millennium. The authors in this book discuss such issues head-on and are role models for governments who resist such discourses. How *they* address these matters will largely determine how successful we are in the attempts at building justice in a global community.

NONTOMBI TUTU *was raised in South Africa and is the daughter of Nobel Laureate Archbishop Emeritus Desmond Tutu. As Program Director for Fisk University's Race Relations Institute, she was instrumental in guiding Fisk students to elect the first African American mayor of Selma, Alabama. A delegate to the 2001 United Nations World Conference on Racism, she has spoken eloquently on women's issues, racism, and globalization.*

INTRODUCTION

> *Reparations—Payment of a debt owed; the act of repairing a wrong or injury; to atone for wrongdoings; to make amends; to make one whole again; the payment of damages; to repair a nation; compensation in money, land, or materials for damages.*
> —NATIONAL COALITION OF BLACKS FOR REPARATIONS IN AMERICA

A s this book goes to press the reparations movement, historically considered a fringe issue in the American Black nationalist community, is now firmly established among various constituencies in the United States as well as in African communities around the world. Its ascendancy as an important social movement—I would argue the most important since Civil Rights—is confirmed by the amount of print space and air time the media devote to it. Though the movement is picking up speed, compensatory measures for Africans have been elusive because of the entrenchment of white supremacy in world politics that provided legal sanction for this crime against humanity. Africans around the world have watched groups such as the Japanese, Jews, and others receive reparations for government-sanctioned crimes against them, while eyebrows are raised and arguments dismissed as nonsensical when similar justice for Africans and their descendants are made. Table 1 illustrates examples of payments made to various groups during the past sixty years. It is clear that the payment of reparations is not only a common occurrence but is firmly rooted in international law, which the United States recognizes. It is also important to note that while many view reparations as a radical solution to alleviate a historic wrong, conservative heads of state—President Ronald Reagan, for example—have endorsed them for victims of crimes against humanity.

TABLE 1: SOME EXAMPLES OF REPARATIONS PAYMENTS*

1952 Germany	$822 million	Holocaust Survivors
1971 United States	$1 billion + 44 Million acres of land	Alaska Natives Land Settlement
1980 United States	$81 million	Klamaths of Oregon
1985 United States	$105 million	Lakota of South Dakota
1985 United States	$12.3 million	Seminoles of Florida
1985 United States	$31 million	Chippewas of Wisconsin
1986 United States	$32 million, 1836 treaty violations	Ottawas of Michigan
1988 Canada	$230 million	Japanese Canadians
1988 Canada	250,000 square miles of land	Eskimos and Indigenous People
1990 Austria	$25 million	Jewish Claims on Austria
1990 United States	$1.2 billion	Japanese Americans

*Source: *Black Reparations Now! Part 1: 40 Acres, $50.00 and a Mule* by Dorothy Benton Lewis, 1998

Many people are unaware that the discussion of reparations for African people has a long history in the United States, going through three distinct stages with a nascent fourth beginning in 2002. *Stage I,* from 1865 to 1920, included the United States government's attempt to compensate its newly released three million enslaved Africans from bondage. This period also saw Callie House's heroic efforts at establishing the Ex-slave Mutual Relief, Bounty and Pension Association when she organized hundreds of thousands of ex-slaves for repayment from the government. *Stage II,* from 1920 to 1968, saw Marcus Garvey, Queen Mother Audley Moore, and numerous Black nationalists press for reparations by educating thousands of persons about the unpaid debt owed to Africans in America. This is the period during which the reparations movement was seen as a Black nationalist endeavor and civil rights organizations saw its goals as being unrealistic and extreme.

Stage III began in 1968 and continues today. The founding of several Black nationalist groups including the Republic of New Africa (1968), the National Coalition of Blacks for Reparations in America (1987), and James Forman's "Black Manifesto" (1969), which demanded $500 million from Christian churches and Jewish synagogues, served as catalysts for launching what some have called the modern reparations movement. Randall Robinson's 2000 book

The Debt aided in moving the discussion into even wider circles, as did the continuing attempts since 1989 by Congressman John Conyers Jr. (D-MI) to appoint a committee to study the effects of slavery upon the United States. I believe that *Stage IV* of the movement began in 2002 with the filing of several lawsuits against corporations and ultimately the government. This legal stage was temporarily discouraged by the Cato decision in 1995 in which a liberal federal court in California ruled that suing for reparations occurred too long after the incident (slavery) had happened.

A convergence of four groups provides a conceptual framework for understanding the current discussion of reparations: (1) grassroots organizers, (2) legislators, (3) attorneys, and (4) academics. A similar convergence of cooperation occurred during the late 1940s and resulted in what we now call the Civil Rights movement. A. Philip Randolph's Brotherhood of Sleeping Car Porters (grassroots) began conversations with Charles Hamilton Houston and Thurgood Marshall (legal), who consulted with politicians such as Hubert Humphrey of Minnesota (legislative) as well as psychologists Kenneth and Mamie Clark (academics). Together they formed national networks that led to the birth of the Civil Rights movement. Pioneering Black sociologist Charles S. Johnson of Fisk University provided research facilities and a place to discuss strategies for all of these groups with his establishment of the Race Relations Institute during the 1940s.

Reparations have a similar history. Grassroots organizations such as the December 12th Movement (D12), National Coalition of Blacks for Reparations in America (N'COBRA), and the National Black United Front (NBUF) worked closely with legislators in the mid-1980s—John Conyers Jr. (D-MI), for example—and collaborated with the Reparations Coordinating Committee (RCC), consisting of attorneys such as Willie Gary, Randall Robinson, and Johnnie Cochran and academics such as Manning Marable and Ron Walters. These groups conversed long and hard with each other, and as you will see, these discussions were often heated and difficult. What united them, however, was a common goal of pressing for reparations on a global level for African people.

The fertile ground for nourishing the movement came during the early 1990s when the December 12th Movement, along with other grassroots organizations, lobbied the United Nations to hold a World Conference Against Racism. This followed the tradition of Marcus Garvey during the 1920s, W. E. B. Du Bois during the 1940s and 1950s, and Malcolm X during the 1960s, who together encouraged bringing international attention to the struggle of Africans in America. The 2001 World Conference Against Racism (WCAR) presented an opportunity to press the issues of reparations at the global level. The three core issues adhered to consistently by the December 12th Movement that helped unify the struggle in the late 1990s were:

1. Declaration of the Transatlantic Slave Trade and slavery to be crimes against humanity.
2. Reparations for people on the African continent and victims of the *Maafa*.
3. The economic base of racism.

These were not haphazardly arrived at issues. Rather, the organizers had a steady eye on international law, as you will see in the various articles contained in this volume. Added to this list was the impact of colonialism on the first core issue—the Transatlantic Slave Trade—that allowed for even wider litigation efforts involving the former European colonial powers that divided Africa up at the 1884 Berlin Conference.

In retrospect, it appears that both Europe and the United States, though opposing every discussion in the international arena on reparations, underestimated how the movement would coalesce during the late 1990s. Similar to the Civil Rights movement in the United States and the antiapartheid movement in South Africa, the governments responsible for slavery and exploitation failed to understand how the reparations movement would be the first global dialogue on the past practices of slavery in the twenty-first century. The West, led by the United States, realized too late (2000) that the momentum of the discussion would accelerate at breakneck speed in 2001, as final plans for the WCAR began in earnest. A by-product of the meetings leading up to the WCAR was the extraordinary networking that took place among the global African community as they shared similar stories about patchwork programs provided as a panacea for sidetracking discussion on the continuing impact of slavery and colonialism on Africans and their descendants around the world. Both the United States and Europe failed in their strong-arm attempts at removing reparations from the agenda in Geneva and Durban, and the nongovernment organizations (NGOs) participating in the conference were even more encouraged to press the issue in the world arena.

Should America Pay? acquaints the reader with the many voices in a movement whose momentum is growing geometrically. Some compare it to the movement against *globalization,* and to be sure, they have much in common. The book will give you in-depth knowledge of where the movement has been, where it is now, and where it will be in the future. The contributors present facts, cajole, argue, and reveal their ideas about a volatile subject. Some will anger you, while others will leave you with a feeling of empowerment about what you ought to do next. Some readers will wish they'd embraced the movement earlier, while others will be dismissive of the assertions of those who are frontline soldiers in the debate. If any of these emotions come to you, and I am sure they will, we have done our job well in providing you with information that will be central to any dialogue about race during the first quarter of the twenty-first century.

The book is divided into seven parts. The foreword provided by Nontombi Tutu, daughter of the famous Nobel laureate, summarizes the global nature of reparations and how the struggle for justice involves both political understanding and reframing questions about justice, racism, and history.

The contributors in Part I give the historical context under which the reparations movement was born and grew. Congressman John Conyers, considered the legislative father of reparations, tells how it is important to understand that the reparations movement is attracting widespread support because it is an issue of justice and not charity. He also provides an overview of some of the major issues influencing the rise of the modern reparations movement. Professor Molefi Asante of Temple University gives the context in which Africans were stolen from their homeland and how the historical traumas of the Middle Passage and enslavement still manifest themselves in the life of contemporary Africans in America and form the basis of understanding the economic growth of the United States and Europe.

Deadria Farmer-Paellmann, whom some call the Rosa Parks of the Reparations Litigation Movement, provides personal insight into her journey to achieve reparations for Africans in America through the law and documents one of the most important yet neglected periods in reparations history, the ex-slave pension movement. Newly freed enslaved Africans in America tried to receive compensation for the labor performed for their slave masters and organized into one of the greatest mass movements in United States history. Its relevance to reparations history is monumental in that it shows how the federal government has a long and sordid history of derailing the struggle of Africans in America to receive justice.

Professor Haunani-Kay Trask's article, though focusing on the invasion of Hawaii, has direct relevance for how a movement in the United States seeking to right a historical injustice has fared. She shows that even when laws are enacted that support reparations, they may be a long time coming to the victims of the atrocities. My article provides emotional testimony from Africans in America whose land was stolen and humanizes the need for addressing gross

violations of the property rights of Africans in America. The issue of land takings from African Americans is the greatest unpunished crime in United
States history and will form the basis for many lawsuits involving reparations.

Part II presents the wide-ranging legal issues associated with reparations.
Any social movement seeking justice must eventually engage itself with whatever legal system governs it. The U.S. Civil Rights movement during the first
half of the twentieth century, though powerful in its symbolism and morality,
did not shake the status quo until it sought juridical remedies for *de jure* segregation in schools, housing, and voting. *Plessy v. Ferguson* and *Brown v. Board
of Education* are two legal decisions that changed the social contract between
African Americans and the United States forever. Legal redress is a necessary
step for achieving social justice, and the articles in this section look at the law
in both national and international arenas.

Jon Van Dyke of the University of Hawaii gives an overview of international reparations cases, how common they are and how they are directly connected with the struggle for reparations for Africans and African-descended
people. He sees reparations as a necessary step in achieving the elusive idea of
racial reconciliation in the United States between Black and white people. The
roundtable discussion reprinted from *Harper's* assembles some of the key
litigators from the Reparations Coordinating Committee to discuss legal
strategies and challenges that will unfold during the next decade. In a no-
holds-barred interview, they discuss the strengths and weaknesses of various
legal strategies in gaining reparations for African Americans. Following this
discussion is a letter from a former slave to his slave master and the critical
dates in the history of reparations, both a part of the original *Harper's* article.

Robert Westley of Tulane University gives legal arguments about how
reparations should be distributed and framed many of the opinions of Randall Robinson in his book *The Debt: What America Owes to Blacks.* He shows
how the idea for reparation litigation is firmly grounded in American jurisprudence. Kevin Outterson, a specialist on taxes and the law, takes on the
task of presenting actual numbers on the cost of American slavery as it related
to government revenue. The chapter presents a dispassionate view of how
slavery enriched the treasuries of governments at the local, state, and federal
levels. The article demonstrates how the early economy of the United States
was built on slave taxes collected over many years.

Omari Winbush and I provide a father-son dialogue to explain how *Plessy
v. Ferguson* was a watershed in the psychological relationships between Blacks
and whites in this country and marked differing views of progress, equality,
and reparations. Using a narrative technique developed by legal scholar Derrick Bell, we discuss a case that most persons know only by three words—

"separate but equal"—and show that it occupies a broader role in the history of Black/white relationships in the United States.

Part III presents those who debate reparations for Africans and African-descended people. Armstrong Williams describes the ominous implications of reparations, since it encourages minorities to believe that they are really lost souls. Christopher Hitchens's arguments around reparations were published in *Vanity Fair* during the summer of 2001 as a rebuttal to David Horowitz's ten oppositions to reparations point by point. John McWhorter, a linguist who often writes about race, gives a lengthy critique of Randall Robinson's *The Debt* and deplores the emphasis on a racial pessimism he feels characterizes the Black left. Finally, Shelby Steele echoes his long-stated claims that Blacks advocating reparations is another way of reinforcing the notion of victimization in the African American community.

Any book about social movements should discuss the organizations instrumental in bringing the issue to the forefront. *Brown v. Board* is better understood after reading about the internal workings of the NAACP during the late 1940s and early 1950s. Likewise, the strategies developed by Martin Luther King's Southern Christian Leadership Conference are critical in understanding how the Civil Rights movement focused the world's attention on Birmingham, Montgomery, and Chicago. The National Black United Front (NBUF), the National Coalition of Blacks for Reparations in America (N'COBRA), and the December 12th Movement (D12) are the primary organizations advocating for reparations in America. They were clearly ahead of the traditional civil rights establishment, who only recently saw the importance of reparations as a vital element in the human rights struggle for African Americans and is in the process of playing catch-up in aligning themselves with traditional Black nationalist organizations.

It is also important to note that Black nationalist organizations such as NBUF have no *white* benefactors and are not nearly as dependent on white financial support as are mainstream civil rights organizations. This allows for freedom in developing organizational strategies that do not have to please a white constituency that may have problems with a Black-only agenda. The civil rights organizations face a major dilemma in that the reparations movement is truly a grassroots movement, similar to the United Negro Improvement Association of Marcus Garvey during the early twentieth century, with very few whites involved. The civil rights establishment must decide how it will satisfy their white benefactors without alienating the masses of Africans in America. These groups faced the same challenge during early 1995 when Louis Farrakhan and the Nation of Islam began organizing the Million Man March that took place in October of that year. The boards of these organizations distanced themselves from Farrakhan because they feared a white backlash from donors with deep pockets who were frightened by Farrakhan. Unlike the Million Man March, however, the quest for reparations is not a onetime *event* but an ever-expanding *movement* that will be around for a very long time. The histories of the organizations critical to this movement are presented in Part IV so that readers can get an idea of how groups move from local agenda setting to organizing global social movements.

Conrad Worrill and Adjoa Aiyetoro discuss NBUF and N'COBRA's ascendancy in the reparations movement and the internal organizational struggles that led them to the forefront of the movement. Worrill shows how his organization, the National Black United Front, shaped its recent agenda to become active in the struggle for reparations while anchoring firmly in the historical quest for reparations. Roger Wareham, one of the lead attorneys in the historic Farmer-Paellmann case in which a class-action lawsuit was brought against three major corporations that benefited from slavery, describes the December 12th Movement's role in lobbying the United Nations for a World Conference Against Racism.

But reparations are not just the facts about crimes against humanity. The issue of reparations also generates powerful emotions associated with enslavement and colonialism and their still-lingering influence in the Western world. Teachers, doctors, psychologists, journalists, and lay people have strong opinions about reparations. Part V presents the voices of a variety of people—Black and white—who give prescriptive suggestions about healing the damage done by racism in the United States. Randall Robinson's *The Debt* began such a discourse and it shows how race relations in the United States cannot ignore the emotional aspect of these crimes.

Tim Wise draws a line from his own history as a Jew with Scottish ancestry to how he believes reparations will force an honest dialogue about race to occur in the United States.

Physician Jewel Crawford, psychologist Wade Nobles, and social worker Joy DeGruy Leary give their thoughts on the American health care system and the mental health establishment. Both see the historical thread of racism running through both systems and its terrible consequences on the physical and mental health of Africans in America. They too provide prescriptions through reparations for fixing these massive problems.

Author Haki Madhubuti in a single poem summarizes his feelings on the debt owed to Africans in America at all levels. For him, reparations include a host of crimes against Black people still being committed today.

Journalist Molly Secours discusses her personal experiences with white privilege and how it relates to reparations for African Americans. She presents the most common objections made by white people against reparations and tells why these arguments are unfounded. Yaa Asantewa Nzingha is a former Brooklyn teacher fired for teaching that Black students should refer to themselves as Africans rather than Americans. She argues that the psychological impact of enslavement is exemplified in the educational development of African American youth more than anywhere else and provides specific recommendations for repairing the United States' public education system.

The interview with the Hurdle brothers, the last living children of Andrew Jackson Hurdle, provides a coda, as their father was born a slave in 1845 in North Carolina. As of press time, they are suing several corporations for profiteering from the Transatlantic Slave Trade. They remind us how close slavery is to us, and of the influence it still exerts on all Americans.

Part VI presents several historical documents that are considered important in understanding reparations. The Thirteenth, Fourteenth, and Fifteenth Amendments to the United States Constitution, which many have not read, are presented in toto to see how citizenship and suffrage were given to Africans in America shortly after the Civil War. General William Sherman's 1865 Field Order No. 15 shows how the United States government felt that something should be done with their Black citizens, who had been enslaved for over two centuries. "Forty acres and a mule" have become synonymous with reparations, and the entire document is presented here. Thaddeus Stevens's bill came on the heels of Sherman's unfulfilled Field Order No. 15 and vividly shows how legislators understood this country's financial obligations to its formerly enslaved citizens.

Congressman John Conyers Jr.'s House Bill 40 has been introduced in every Congress since 1989 but has never left committee. It calls for a study of the impact of slavery on the United States and, for its opponents, represents the first step down a slippery slope of granting reparations to Africans in America. The Dakar Declaration supported by African governments in 2001 was one of the few times that African heads of state supported an international tribunal to assess how enslavement and colonialism impacted the development of Africa. The historic lawsuit filed by Deadria Farmer-Paellmann in 2002 marks the beginning of litigation against private corporations and their connection to the enslavement of Africans in America. Farmer-Paellmann's lawsuit followed lengthy personal research on how private corporations benefited from enslaving African Americans.

In 1967, Martin Luther King penned his last book, *Where Do We Go from Here: Chaos or Community?*. It served as a prophetic voice for the civil rights movement and located the status of the struggle for human rights in the United States. In Part VII, the most respected members of the reparations struggle, Dorothy Benton Lewis, Jill Soffiyah Elijah, Conrad Worrill, and Baba Hannibal Afrik, give opinions, strategies, and directions, including a summary of the legal aspects of the movement.

Finally, this book deliberately seeks to provoke debate and conversation over the greatest crime in world history, the Transatlantic Slave Trade. Until *all* people understand its causes and continuing impact on Africans throughout the world there will *never* be an honest dialogue about race and racism.

PART I

History and Reparations

MOLEFI KETE ASANTE

The African American Warrant for Reparations: The Crime of European Enslavement of Africans and Its Consequences

Until lions have historians, hunters will be heroes.
—KENYAN PROVERB

I n his 1993 monograph *Paying the Social Debt: What White America Owes Black America,* Richard America makes a forceful argument that reparations for Europe's enslavement of Africans in the United States is an idea whose time has arrived. Almost a decade before the powerful book *The Debt: What America Owes to Blacks,* written by Randall Robinson in 2000, America laid out the economic bases of the debt owed to African Americans. While the argument for reparations is a Pan-African one, we are most interested in this essay with the discourse surrounding the enslavement and its consequences in the American society. There are those who will immediately say that the people of the United States will never accede to reparations. I am of the opinion that the discussion and debate surrounding reparations has only recently occurred in any serious way; therefore, this essay is offered as an attempt to raise some of the philosophical ideas that might govern such a discourse.

Randall Robinson's *The Debt* has been one of the most popular and important books written on the subject so far because he has captured the warrants for reparations in very clear and accessible language. What he has demonstrated is that while a national *paralysis* of will may exist at the present time, there is no lack of national *guilt* and interest in this theme. There is every reason for the United States to shape and frame the culture of reparations that shall become an increasingly powerful moral and political issue in the twenty-first century. The highest form of law exhibits itself when a system of law is able to answer for its own crimes. Nothing should prevent men and women of moral and political insight from making an argument for an idea whose legitimacy is fundamental to our concept of justice. We must act based on our own sense of moral rightness.

When Raphael Lemkin started in 1933 to gain recognition of the term "genocide" as a crime of barbarity, few thought that it would soon become the language of international law. When genocide was adopted as a convention in 1948 with an international criminal court to serve as the home for judging genocide it was a victory for those who had fought to put genocide on the world agenda. My belief is that the current discussions about reparation undertaken by scholars, political activists, and the United Nations will advance our own plan to place reparations at the front end of the agenda for redress for African Americans.

THE GROUNDS OF THE ARGUMENT: MORAL, LEGAL, ECONOMIC, AND POLITICAL

The argument for reparations for the forced enslavement of Africans in the American colonies and the United States of America is based in moral, legal, economic, and political grounds. Taken together these ideas constitute an enormous warrant for the payment of reparations to the descendants of the Africans who worked under duress for nearly 250 years. The only remedy for such an immense deprivation of life and liberty is an enormous restitution.

When one examines the nature of the terms amassed for the argument for reparations it becomes clear that the basis for them is interwoven with the cultural fabric of the American nation. It is not un-American to seek the redress of wrongs through the use of some form of compensatory restitution. For example, the *moral* ideas of the argument are made from the concept of rightness as conceived in the religious literature of the American people. One assumes that morality based in the relationship between humans and the divine provides an incentive for correcting a wrong, if it is perceived to be a wrong, in most cases. Using *legal* ideas for the argument for reparations, one relies on the juridical heritage of the United States. Clearly, the ideas of justice and fair play, while often thwarted, distorted, and subverted, characterize the legal *ideal* in American jurisprudence. Therefore, the use of legal strategy in securing reparations is not only expected but required by Africans receiving compensatory redress for their enslavement. The Great Enslavement itself showed, however, how legal arguments could be twisted to defend an immoral and unjust system of oppression. Nevertheless, justice is a requirement for political solidarity within a nation and any attempt to bring it about must be looked upon as a valid effort to create national unity. Simply put, no justice, no peace.

Recognizing that justice may be both retributive and restorative, it seeks to punish those who have committed wrong and it concerns itself with restoring to the body politic a sense of reconciliation and harmony. *I believe that the idea of reparations, particularly as conceived in my own work, is a restorative justice issue.*

The *economic* case is a simple argument for the payment to the descendants of the enslaved for the work that was done and the deprivation that was experienced by our ancestors. To speak of an economic interest in the argument is typically American and an issue that should be well understood by most Americans. Simply put, Africans in America are owed back wages for nearly 250 years of uncompensated work by their ancestors and another 130 years for laws and behavior that continue to affect them economically.

Finally, the *political* aspect of reparations is wrapped in the clothes of the American political reality. In order to insure national unity reparations should be made to the descendants of Africans. It is my belief as well as that of others that the underlying fault in the American body politic is the unresolved issue of enslavement. Many of the contemporary problems in society can be thought of as deriving from the unsettled issues of enslavement. A concentration on the political term for reparations will lead to a useful argument for national unity.

WHY REPARATIONS?

One of the ironies of the discourse surrounding reparations for the enslavement of Africans is that the arguments against reparations for Africans are never placed in the same light as those about reparations in other cases. This introduces a racist element into the discourse. For example, one would rarely hear the question, Why should Germany pay reparations to the Jews? Or, Why should the United States pay reparations to the Japanese who were placed in concentration camps during World War II? If someone were to try to make arguments against those forms of reparations the entire corpus of arguments from morality, law, economics, and politics would be brought to bear on them. This is as it should be in a society where human beings respect the value of other humans. Only in societies where human beings are considered less than humans do we have the opportunity for enslavement, concentration camps, and gas chambers. It should be noted that when humans are considered the same as other humans no one questions whether compensatory measures should be given to an oppressed group. We expect all of the arguments for reparations to be used in such cases. This is why the recent rewarding of reparations to Jews for the Nazi atrocities is considered normal and natural. In Nazi Germany, Jews were considered inferior, and had Germany *won* the war, any thought of reparations to Jews would have been unthinkable. It is because Nazi Germany *lost* the war and other humans with different values had to make decisions about the nature of reparations that any were made at all. One can make the same argument for the Japanese Americans who lost their property and resources during World War II. A new reality in the political landscape made it possible for the Japanese to receive reparations for their losses. Eminent African and Caribbean scholars such as Ali Mazrui, author of *The Africans,* Jamaican ambassador Dudley Thompson, and others have argued for an international examination of the role the West played in the slave trade and the consequent underdevelopment of Africa. This is a laudable movement and I believe it will add to the intensity and seriousness of the internal discourse within the United States.

A strong sense of moral outrage has continued to activate the public in the interest of reparations. In early 2001 a lawsuit brought against the French national railroad in the Eastern District of New York Court charged the Société Nationale des Chemins de Fer with transporting 72,000 Jews to death camps in August 1944. The case was brought to the court on behalf of the survivors and heirs. In another case, a French court held that French banks that hoarded assets of Jews had to create a fund of $50 million for those individuals with evidence of previous accounts (*New York Times,* June 13, 2001, A-14). Similarly, on May 30, 2001, the German Parliament cleared the way for a $4.5 billion settlement by German companies and the government to survivors or heirs of more than one million forced laborers. This is in addition to much larger awards to Israel and the Jewish people for the Holocaust. The Swiss government has agreed to pay $1.25 billion to those Jewish persons who can establish claims on bank accounts appropriated during World War II.

Whenever people have been deprived of their labor, freedom, or life without cause, other than their race, ethnicity, or religion, as a matter of group or national policy, they should be compensated for their loss. In the case of Africans in the American colonies and the United States, the policy and practice of the ruling white majority in the country was to enslave only Africans after the 1640s. Prior to that time there had been some whites who had been indentured as servants and some native peoples who were pressed into slavery. However, from the middle of the seventeenth century to 1865, only Africans were enslaved as a matter of race and ethnic origin.[1]

A growing consensus suggests that some form of reparations for past injustice on a large scale should not be swept under the table. We have accepted the broad idea of justice and fair play in such massive cases of group deprivation and loss; we cannot change the language or the terms of our contemporary response to acts of past injustice. The recognition of reparations in numerous other cases, including the Rosewood, Florida, and the Tulsa, Oklahoma, burning and bombing of African American communities in the early 1920s, means that we must continue to right the wrongs of the past, so that our current relationships as citizens will improve through an appreciation of justice.

THE BRUTALITIES OF ENSLAVEMENT

Africans did not enslave themselves in the Americas. The European slave trade was not an African venture, it was preeminently a European enterprise in all of its dimensions: conception, insurance, outfitting of ships, sailors, factories, shackles, weapons, and the selling and buying of people in the Americas. Not one African can be named as an equal partner with Europeans in the slave trade. Indeed, no African person benefited to the degree that Europeans did from the commerce in African people. I think it is important to say that no African community used slavery as its principal mode of economic production. We have no example of a slave economy in West Africa. The closest any scholar has ever been able to arrive at a description of a slave society is the Dahomey kingdom of the nineteenth century that had become so debauched by slavery due to European influence that it was virtually a hostage of the nefarious enterprise. However, even in Dahomey we do not see the complete denial of the humanity of Africans as we see in the American colonies.

Slavery was not romantic; it was evil, ferocious, brutal, and corrupting in all of its aspects. It was developed in its greatest degree of degradation in the United States. The enslaved African was treated with utter disrespect. No laws protected the African from any cruelty the white master could conceive. The man, woman, or child was at the complete mercy of the most brutish of people. For looking a white man in the eye the enslaved person could have his or her eyes blinded with hot irons. For speaking up in defense of a wife or woman a man could have his right hand severed. For defending his right to speak against oppression, an African could have half his tongue cut out. For running away and being caught an enslaved African could have his or her Achilles tendon cut. For resisting the advances of her white master a woman could be given fifty lashes of the cowhide whip. A woman who physically fought against her master's sexual advances was courting death, and many died at the hands of their masters. The enslaved African was more often than not physically scarred, crippled, or injured because of some brutal act of the slave owner.

Among the punishments that were favored by the slave owners were whipping holes, wherein the enslaved was buried in the ground up to the neck; dragging blocks that were attached to the feet of men or women who had run away and been caught; mutilation of the toes and fingers; the pouring of hot wax onto the limbs; and passing a piece of hot wood on the buttocks of the enslaved. Death came to the enslaved in vile, crude ways when the angry, psychopathic slave owner wanted to teach other enslaved Africans a lesson. The enslaved person could be roasted over a slow-burning fire, left to die after having both legs and both arms broken, oiled and greased and then set afire while hanging from a tree's limb, or being killed slowly as the slave owner cut the enslaved person's phallus or breasts. A person could be placed on the ground, stomach first, stretched so that each hand was tied to a pole and each foot was tied to a pole. Then the slave master would beat the person's naked body until the flesh was torn off the buttocks and the blood ran down to the ground.

I have written this brief description to insure the reader that we are not talking about mint juleps and Sunday-afternoon teas with happy Africans running around the plantation while white people sang and danced. Africans on the plantations were often sullen, difficult as far as the whites were concerned, hypocritical because they would smile on command and frown when they left the white person's presence, and plotting.

SOME NUMBERS

It became popular in the 1980s to speculate over just how many Africans were captured, marched, shipped, and sold during European enslavement. Henry Louis Gates of Harvard has placed the number between 10 and 12 million, as has Philip Curtin. It is not my intention to enter the debate over these numbers, although I find the numbers quite conservative given the estimates made by other scholars and given the fact that Curtin particularly has demonstrated a penchant for minimizing African agency in the struggle against slavery and colonialism in his widely read book *The Atlantic Slave Trade*. The figure has reached as high as 100 million in the estimation of some scholars, such as W.E.B. Du Bois in his 1920 book *Darkwater*. I believe that the numbers are only important to ascertain just how deeply the Transatlantic Slave Trade affected the continental African economic, social, physical, and cultural character. However, for purposes of reparations the numbers are not necessary since there can be no adequate compensation for the enslavement and its consequences. The broad outline of the facts is clear and accepted by most historians. We know, for instance, that the numbers of Africans who landed in Jamaica and Brazil were different from those of Haiti and the United States. Furthermore, the establishment of concrete numbers of those captured and enslaved throughout the *Maafa** and in the United States, though difficult, will ultimately be achieved because of better data-gathering techniques and the lawsuits now emerging that will research such numbers. I believe it is necessary, however, to ascertain something more about the *nature* of the Africans' arrival in the American nation. At the end of the Civil War in 1865 there were about 4.5 million Africans in the United States, which means that there had been a steady flow of Africans into the American nation since the seventeenth century. *These Africans and their descendants constitute the proper plaintiffs in the reparation case.* Hundreds of thousands of Africans labored and died under the reign of enslavement without leaving any direct descendants. We cannot adequately account for these lost numbers, which include those who died resisting capture in Africa, those who died on the forced marches to the beach barracoons, those who died awaiting to enter the ships, those who died aboard ships, and those who continued to resist throughout 250 years of enslavement. We can, however, account for most of those who survived the Civil War and their heirs. In fact, some of the 187,000 who fought in the Civil War did not survive, but their descendants survived. These also constitute a body of individuals (class?) who must be brought into the discussion of reparations. Thus, two classes of people, those who survived the Civil War and their heirs and those who fought and died in the Civil War and their heirs, are legitimate candidates for reparations. Indeed, the consequences of the residual effects of the enslavement must be figured in any compensation.

THE NATURE OF THE LOSS

One of the issues that must be dealt with is, how is loss to be determined? Since millions of Africans were transported across the sea and enslaved in the

---

Maafa (pronounced Ma-AH-fa). A Kiswahili word meaning "disaster." It is used to describe over five hundred years of warfare and genocide experienced by African people under enslavement and colonialism and their continued impact on African people throughout the world. Until now words such as "diaspora" and "holocaust" had been appropriated from outside African culture and therefore do not embody the experiences of the African reality. The term was first popularized by the author Marimba Ani in her book *Yurugu: An African-Centered Critique of European Cultural Thought and Behavior* (1994), Africa World Press, and will be used throughout this book.

Caribbean and the Americas for more than two centuries, what method of calculating loss will be employed? It seems to me that loss must be determined using a multiplicity of measures suited to the variety of deprivations that were experienced by the African people. Yet the overarching principle for establishing loss might be determined by ascertaining the negative effects on the natural development of people, that is the physical, psychological, economic, and educational toll must be evaluated. What were the fundamental ways in which the enslavement of Africans undermined not only the contemporary lifestyles and chances of the people but also destroyed the potentialities for their posterity? I believe all of the issues of educational deficit, economic instability, poor health conditions, and the lack of estate wealth are directly related to the conditions of enslaved Africans both in the United States and throughout the world. What is called for is a national purpose to confront the historical abuse of Africans.

Given the fact that African Americans constitute the largest single ethnic-cultural grouping in the United States and will maintain this position into the future, reparations for the enslavement of Africans will have positive benefits on the American nation. African Americans number approximately 35 million people. Occasionally one reads in the newspaper that the Hispanic or Latino population will soon outstrip the African population in the United States. While it is true that taken together the number of Spanish-speaking Americans will soon outnumber the absolute number of African Americans, the Spanish-speaking population includes more than twenty different national origin groups, plus individuals who identify with African, Caucasian, and Native American heritages. One finds, for example, among the Spanish-speaking population, people from Mexico, Cuba, Puerto Rico, Dominican Republic, El Salvador, Honduras, Costa Rica, and numerous South American countries. Many of these people will self-identify as white; others will self-identify as Black or African.

Africans are an indispensable part of the American nation in terms of its history, culture, philosophies, mission, and potential. It is insane to speak of America without the African presence, and yet the deeper we get into the future the more important the nature of the relationship of Africans to the body politic will become. Reparations would insure: (1) recognition of the Africans' loss, (2) compensation for the loss, (3) psychological relief for both Blacks and whites in terms of guilt and anger, and (4) national unity based on a stronger political will. These are intrinsic values of reparations.

TOWARD A BASIS FOR REPARATIONS

Reparations are always based on real loss, not perceived loss. Take the case of the Japanese Americans who were taken from their homes in California and other Western states during World War II. They were removed against their wills from their homes, their property confiscated and their children taken out of schools. The Japanese Americans lost in real terms and were consequently able to make the case for reparations. Their case was legitimate and it was correct for America to respond to the injustice that had been done to the Japanese Americans.

The case of the Africans in America has some of the same characteristics, but in many ways is different and yet equally significant as far as real loss is concerned. What is similar is the uprooting of Africans against their wills. Also similar is the racism whites had against Africans, who they considered cultural and intellectual inferiors. From this standpoint it was easy to brutalize, humiliate, and enslave Africans since, as whites had argued, blacks were inferior in every way. What is different about the reparations case for African Americans is that it is much larger than the Japanese American situation, it has far more implications for historical transformation of the American society, and it is rooted in the legal foundations of the country.

It is possible to argue for reparations on the following grounds: (1) forced migration, (2) forced deprivation of culture, (3) forced labor, and (4) forced deprivation of wealth by segregation and racism. However, these four constituents of the argument for reparations are buttressed by several significant factors that emerged from the experience of the enslaved Africans. In the first place, Africans often lost their freedom because of their age. Most of the Africans who were robbed from the continent of Africa were between the ages of fifteen and twenty years—robbery of prime youth. A second factor is based on the loss of innocence where abuse—physical, psychological, and sexual—was the order of the day in the life of the enslaved African, as Trinidadian scholar Eric Williams detailed in his famous 1961 speech "Massa Day Done." Thirdly, one has to consider the loss in transit that derived from coffles and the long marches, the dreaded factories where Africans were held sometimes as long as seven months while the Europeans waited for a transport ship, and the severe loss of life in transit, where death on board the ships or in the sea further deprived a people. Fourthly, the factor of loss due to maimed limbs, that is, the deprivation of feet, Achilles tendons, and hands. Insurance companies such as Aetna understood the hazards of slavery and provided policies for slaveholders that allowed for premiums to be paid to them if death or maiming occurred to their slaves. Needless to say, none of the slaves received any benefits from this hideous business practice, but the effects lasted for generations, including sterility in women and castration in men.

Thus, to have freedom, will, culture, religion, and health controlled and denied is to create the most thorough conditions for loss of ancestral memory. The Africans who were enslaved in America were among the most deprived humans in history. It is no wonder that David Walker's 1829 "Appeal to the Colored Citizens of the World" argued that the enslaved Africans were the most abject people in the world. He also stated in his Appeal that "the White Christian Americans" were the most cruel and barbarous people who have ever lived.

THE REPARATIONS REMEDY

One way to approach the issue of reparations is to speak about *money* but not necessarily about *cash*. Reparations will cost, but it will not have to be the giving out of billions of dollars of cash to individuals, although it will cost billions of dollars. While the delivery of money for other than cash distributions is important, it is possible for reparations to be advanced in the United States by a number of other options. Among the potential options are educational grants, health care, land or property grants, and a combination of such grants. Any reparations remedy should deal with long-term issues in the African American community rather than be a onetime cash payout. What I have argued for is the establishment of some type of organization that would evaluate how reparations would be determined and distributed. For example, a National Commission of African Americans (NCAA) would be the overarching national organization to serve as the clearinghouse for reparations. The commission would investigate reparations as a more authentic way of bringing the national moral conscience to bear on the education of African Americans. Rather than begin in a vacuum, the NCAA would study various sectors of society, education, health and welfare, or economics and see how Africans were deprived by two and a half centuries of enslavement. For example, by the time Africans were freed from bondage in 1865, whites had claimed all land stretching from sea to sea, and had just about finished the systematic "cleansing" of indigenous people from the land, pushing thousands to Oklahoma in the Trail of Tears or, as in the case of the Oneida, to Wisconsin in a trail of sorrow. Furthermore, there were already five hundred colleges teaching white students, this during a time when it was a crime for Africans to learn and illegal for whites to teach Africans to read or to write. One likely answer to the reparations issue is free public and private education to all descendants of en-slaved Africans for the next 123 years, half the time Africans worked in this country for free. Students who qualified for college would be admitted and have all of their expenses covered by the government. Those who qualified for private schools would get government vouchers to cover the costs of their education.

The present educational deficit is not an individual but a collective and national deficit. This is not the same as saying that Kim Su or Ted Vaclav came to this country and could not read, but they made it by pulling themselves up by the proverbial bootstraps. Immigrants who choose to come to America are in no way enslaved and have different motivations for their immigration in-cluding economic, emotional, and political ones. Our coming was different and our struggle was epic because we were brought on slave ships and often worked nearly to death.

Despite the curious attempt to claim for all Americans the same heritage and the same history, the record of the country speaks for itself. From educa-tion to prison, the evidence of racial bias in interpretation of data as well as the data themselves show that African Americans have been treated unfairly due largely to the previous condition of servitude. Thus we have been under-developed by the very society that supposedly set us free.

At the head of America's race relations problems are the unresolved issues surrounding the institution of slavery. At the root of this irresolution is the be-lief that Africans are inferior to whites and therefore do not deserve compen-sation for labor or anything else. Indeed, it is this feeling that fuels the attacks on reparations for Africans as well. How whites feel about the condition of servitude forced on Blacks and how we feel about that condition or how we feel about the attachment of whites to the perpetuation of that condition are the central issues affecting race relations in this nation. Once we have come to terms with the basis for reparations we can begin to overcome the legacy of slavery.

MOLEFI KETE ASANTE *founded the first Ph.D. program in African American Studies in the United States, at Temple University, and is currently a member of the faculty in that department. He is recognized as a major contributor to the develop-ment of theory in African-centered thought and education.*

CONGRESSMAN JOHN CONYERS JR. WITH
JO ANN NICHOLS WATSON

Reparations: An Idea
Whose Time Has Come

*We all stand on many shoulders. Let us not forget that any
heights we may achieve are supplemented by much more than
our own accomplishments.*

As we address the topic of reparations in the United States, it is instructive to use the Reconstruction era as one of our backdrops. Let
us look specifically at George H. White, the last African American
Reconstruction congressman and the last African who had been enslaved to sit
in the House. Congressman White was born in Rosindale, North Carolina,
and was a graduate of Howard University. White studied law privately. He represented North Carolina's Second Congressional District and was elected in
1896 and reelected in 1898. Not surprisingly, Congressman White found it difficult to make his mark in Congress. He was able to obtain back pay for Black
Civil War veterans, for example, but his colleagues refused even to hear his
federal antilynching bill.

During his last speech, in January 1901, Congressman White said, "This,
Mr. Chairman, is perhaps the Negro's temporary farewell to the American
Congress. These parting words are on behalf of an outraged, heartbroken,
bruised and bleeding, but God-fearing people . . . full of potential force." It
would be more than twenty-five years before the next African American, Os-
car De Priest, of Chicago, Illinois, was elected to the United States House of
Representatives.

If Congressman White could offer testimony on the issue of reparations
today, he would certainly attest to the fact that Blacks never received forty acres

and a mule in the aftermath of the signing of the Emancipation Proclamation. On March 3, 1865, weeks before the end of the Civil War, and almost a year prior to the ratification of the Thirteenth Amendment, the Freedmen's Bureau was created by an act of Congress. According to Section 4 of the first Freedmen's Bureau Act, this agency "shall have authority to set apart for use of loyal refugees and Freedmen such tracts of land within the insurrectionary states as shall have been abandoned or to which the United States shall have acquired title by confiscation or sale, or otherwise; and to every male citizen, whether refugee or Freeman, as aforesaid there shall be assigned not more than forty acres of land." This portion of the Freedmen's Bureau Act (introduced by Congressman Thaddeus Stevens) was defeated by Congress on February 5, 1866, by a vote of 126 to 36 because many thought that it would disenfranchise *white* landowners who had been defeated in the Civil War. Land that had been distributed to Freedmen was reclaimed by the federal government and routed to the enslavers (who had lost the Civil War, fought for the Confederacy, and had already benefited unjustly from the unpaid labor of Africans).

In January 1865, General William Tecumseh Sherman had previously issued orders to General Rufus Saxton to divide land into forty-acre tracts and distribute them to freedmen after the creation of the Freedmen's Bureau in 1865. Just two months later, after the assassination of President Abraham Lincoln, President Andrew Johnson revoked the executive office's support for the Freedmen's Bureau and reneged on promises and commitments that had been negotiated by abolitionist/statesman Frederick Douglass in discussions with President Lincoln.

I believe that one of the best-kept secrets among Civil War historians is that the Union was losing to the Confederacy until enslaved Africans joined the Civil War to fight for the Union. As President Lincoln discussed the matter of introducing Africans who had been held in bondage to fight for the Union, Douglass strongly advocated on behalf of the Emancipation Proclamation, the Freedmen's Bureau, the provision of land to the newly freed Africans, and the adoption of the Thirteenth Amendment. Among the resources utilized to bring victory to the Union was Harriet Tubman, the renowned General of the Underground Railroad, who served as a scout during the Civil War conducting dangerous reconnaissance missions.

Upon learning that President Andrew Johnson had rescinded the order authorizing the Freedmen's Bureau Act and the distribution of land to freedmen, General Saxton wrote the following communique to the commissioner of the Freedmen's Bureau, Oliver O. Howard: "The lands which have been taken possession of by this bureau have been solemnly pledged to the Freedmen . . . it is of vital importance that our promises made to Freedmen should be faithfully kept . . . the Freedmen were promised the protection of the government, with the approval of the War Department . . . more than 40,000 Freedmen have been provided with homes under its promises . . . I cannot break faith with them now by recommending the restoration of any of these lands. In my opinion the order of General Sherman is as binding as a statute." Saxton's pleas were to no avail, however, as thousands of Freedmen were removed by force from land that had been granted by Congress and ordered by Sherman. This was done during the same period that witnessed the 1865 emergence of the Ku Klux Klan's unspeakable violent episodes targeting the newly freed Africans and President Johnson's removal of all federal protections guaranteeing the safety and protection of Africans in America.

The freedmen of the period included luminaries like Bishop Henry McNeal Turner, who had served as a chaplain in the Union Army. Bishop Turner was convinced that the U.S. federal government had betrayed African descendants. He was among many who publicly called for reparations, and he never forgave the nation for what he considered disgraceful ingratitude to Blacks who had built the wealth of the nation with unpaid labor and who had served the nation with courageous military valor during the Civil War. Years later, when he felt his last days were near, Bishop Turner transported himself to Canada, to assure that his remains would not be placed in American soil. (This was eerily prescient of W.E.B. Du Bois's decision, nearly a century later, to move to Accra, Ghana, and become a Ghanaian citizen, abandoning his lifelong work to assure that the United States would honor its ideals and constitutional protections to its citizens of African descent.)

As the ranking Democrat on the House Judiciary Committee, as the dean of the Congressional Black Caucus, and as the longest-serving African American and the second-most senior member of the House of Representatives, I believe it is vitally important that we look toward legislative remedies as a vehicle for addressing the critical issue of reparations for African Americans, just as legislative remedies have been approved for the redress of others. The United Nations World Conference Against Racism, held in Durban, South Africa, in August and September of 2001 declared that the Transatlantic Slave Trade was a crime against humanity, and should always have been so; which sets the proper stage for the timely consideration of H.R. 40, the Reparations Study Bill, which I have introduced every year since 1989. The UN World Conference Against Racism was also another tragic reminder of the deep moral flaws that have been etched into the fabric of America as the United States formally walked out of this historic gathering and days later walked into a terrorist attack on its own shores.

I believe it is vitally important that we look toward legislative remedies as a priority in the reparations movement not only to provide a level of redress for Africans who were enslaved but also to recognize the forces of legalized disparity that disenfranchised people of African descent, like Congressman White, *after* the signing of the Emancipation Proclamation and which continue to institutionalize racist policies and practices until this present day. We have gotten far too comfortable in accepting poverty, crime, and adolescent pregnancy as Black and their opposites as White. We have failed to trace the lineage of both of these economic conditions to slavery and its aftermath.

Why was a bill introduced to study reparations? H.R. 40—the Reparations Study Bill—was introduced in 1989, first and foremost, because of the request that I do so by Reparations Ray Jenkins, who is one of my constituents, a self-employed businessman, precinct delegate, and longtime community activist. Reparations Ray had been an advocate and proponent of reparations for African Americans for many years, and had become a fixture in community-based meetings, assemblies, church gatherings, and NAACP functions as a person who has been singularly committed to the priority of reparations as an issue for people of African descent.

After the introduction of the Civil Rights Redress Act, which paved the way for reparations awarded to Japanese Americans who had been illegally and immorally detained during World War II for three years, it seemed to be an appropriate juncture for the introduction of legislation to study reparations for African Americans, to address possible remedies and redress related to those victimized by the pandemic horrors of the Transatlantic Slave Trade and the long-term residual impact of institutional racism that has persisted among African descendants through Jim Crow segregation, hate crime terrors of lynching and cross burning, and the disparate practices and policies of the prison industry, which in many ways has begun to reenslave Africans, who are disproportionately incarcerated and performing slave labor under the oppressive structure of disparate sentences. Persons of African origin are 13 percent of America's population but account for more than 52 percent of America's 2 million prison population, notwithstanding the reality that Blacks are no more predisposed toward criminal behavior than any other population.

One of the other important factors for the introduction of H.R. 40 was the inescapable reality that legal precedence had long been established relative to the appropriateness of reparations by governmental entities in response to government-sanctioned human rights violations. For example, in

1990, the United States Congress and the President of the United States signed the Civil Rights Redress Act into Law, to lay the framework for $1.2 billion ($20,000 each) paid to Japanese Americans and a Letter of Apology as a federal redress to recognize the human, economic, and moral damage inflicted upon a class of people for a three-year period. Also in 1990, Austria paid $25 million to Jewish Holocaust survivors for its role in the genocidal Nazi regime during World War II; in 1988, Canada gave 250,000 square miles of land to Indians and Eskimos; in 1988, Canada gave $230 million to Japanese Americans; in 1986, the United States paid $32 million to honor the 1836 treaty with the Ottawas of Michigan; in 1985, the United States gave $105 million to the Sioux of South Dakota; in 1980, the United States gave $81 million to the Klamaths of Oregon; in 1971, the United States gave $1 billion plus 44 million acres of land to honor the Alaska Natives land settlement; in 1952, Germany paid $822 million to Jewish Holocaust survivors in the German Jewish Settlement—just to cite some historical backdrops of legal precedence that has been established.

Further, it should be noted that reparations for Africans has not only been an issue cited by Africans in America but also a significant point of discussion and action by Africans on the continent of Africa. James Dennis Akumu, former secretary-general of the Organization of African Trade Union Unity, states: "If you see the arguments the British are advancing in Zimbabwe and whites insisting on owning land and resources in Namibia, South Africa, and other parts of the continent, you can only come to the conclusion that in their minds, Africans should remain their slaves and should not own their own land and mineral resources." Akumu continues to press the point, "African labor and looted African wealth built these strong Western economies. Therefore, what we are claiming is what our people contributed to substantially, and is, therefore, rightfully ours."

Richard America, a Georgetown professor, estimates the U.S. government owes African Americans $5 trillion to $10 trillion for the enslavement of Africans: "Slavery produced benefits and enriched whites as a class at the expense of Blacks as a class . . . reparations is not about making up the past, but dealing with current problems."

These days Reparations Ray Jenkins is not the only voice articulating the importance of reparations in movement and activist circles. The National Coalition of Blacks for Reparations in America (N'COBRA), which was founded by Dr. Imari Obadele, attorney Adjoa Aiyetoro, attorney Chokwe Lumumba, attorney Nkechi Taifa, Brother Kalonji Olusegun, the late Queen Mother Moore, and recently deceased activist Charshee McIntyre, has been the premier organization promoting and organizing in support of reparations throughout the Diaspora. In addition, the publication of Randall Robinson's landmark book *The Debt* has served to broaden the support base and to sharpen the demands among people of African descent—and others who recognize the moral, ethical, and legal basis that has pushed reparations from marginal to mainstream and from the status of an extreme issue to the stature of an essential imperative.

The activism in support of reparations for Africans in America has also been greatly heightened by the advent of highly respected, highly regarded legal scholars who are central to the litigation that is currently being prepared on behalf of the reparations issue; consequently, litigation will join legislation and community education as key strategies to propel the reparations agenda and the reparations movement. Names like Professor Charles Ogletree of Harvard University, attorney Johnnie Cochran, attorney Willie Gary, among other giants in the field of law, have brought new levels of empowerment and widespread media coverage to reparations as revelations about corporate accountability and culpability in the Transatlantic Slave Trade are exposed to a public that now realizes that profits from the enslavement of Africans were not confined to plantations. Legal activist Deadria Farmer-Paellmann, who was featured in a February 21, 2001, reparations cover story in *USA Today,* and who was a keynote presenter during the N'COBRA 2001 National Convention in Baton Rouge, has devoted many years to extensive research that documents the linkages between corporate America to inhumane policies and profits during the 246 years of legalized chattel slavery in the United States. Based on her research as a law student, Farmer-Paellmann has told *USA Today* that she has identified about sixty companies that profited from slavery, and has taken her findings to nine of them (banks, insurers, a textile maker, and one estate) which, to date, have neither apologized nor offered reparations in recognition of their documented offenses. Another company, Aetna Insurance Agency, was confronted by her in 2000 with the irrefutable evidence that the company had insured the lives of Africans held in bondage for enslavers who reaped all of the profit for their unpaid labor. This revelation prompted the State of California to require other insurers to search their archives for slave policies, and Aetna issued a formal apology in March 2000.

In order to successfully move H.R. 40 out of the Judiciary Committee onto the House floor, strategic alignments must be crafted using the tried-and-true methodology of the political process, beginning with Africans in America and other reparations activists, holding every person they have supported in Congress accountable and responsible. Inexplicably, there are members of the Congressional Black Caucus who have not signed on in support of the Reparations Study Bill, and it is vitally important that their constituents let them know that they must be sponsors and adamant proponents of H.R. 40. Similar levels of accountability must be demanded for elected officials in the United States House of Representatives, the U.S. Senate, and other public posts by reparations proponents who demand successful passage of the legislation. Legislative remedies must always be advanced, notwithstanding the great potential of litigation redress and community education. One of the lessons learned from the successful reparations movement on behalf of Japanese Americans is that the litigation filed on their behalf was not successful but the *movement activities spearheaded by Japanese activists surrounding the litigation helped to pave the way for the ultimate success of the Civil Rights Redress legislation during the 1980s.*

CONCLUSION

From the perspective of Where do we go from here? with the reparations movement, the importance of grassroots organizing toward the development of a critical mass cannot be overstated. All of the elements necessary for the ultimate success of reparations for Africans and the descendants of Africa—the litigation process, the legislation efforts, and the community education process—are pieces of the movement that should be fueled by a critical mass of organizers, activists, proponents who exemplify the force and fire of the Reverend Nat Turner, General Harriet Tubman, the Singular Sojourner Truth, the Courageous John Brown, the fearless Ida B. Wells, the Genius of W.E.B. Du Bois, the self-determined Booker T. Washington, the Organizer Marcus Garvey, the By Any Means Necessary Malcolm X, the Majestic the Reverend Dr. Martin Luther King Jr., the Institution-Building of Dr. Mary McLeod Bethune, the Cooperative Economics of Madame C. J. Walker, the Quiet Regal Strength of the Incomparable Mother of the Civil Rights Movement, Mrs. Rosa Louise Parks, and the Indomitable Statesmanship of Frederick Douglass, who reminded us that Power concedes nothing without a demand. . . . It never has, and it never will; and that the limits of Tyrants are determined by the tolerance of those whom they oppress. All of these people flow and contribute to historian Vincent Harding's *There Is a River*. At the end of the day, we will look upon all of the struggles of Africans in America leading up to this final effort of achieving redress for the crimes committed against them and their ancestors. It will be a movement that will unite Africans in America with Africans from around the globe and show humanity that crimes against Africans in the form of enslavement and colonialism is an issue of justice that must be made right.

JOHN CONYERS JR. (D-MI) *is a founding member of the Congressional Black Caucus and author of H.R. 40, a bill introduced in the U.S. House of Representatives in 1989 to study the impact of slavery on the United States.*

JO ANN NICHOLS WATSON *has been the public liaison for Congressman John Conyers since 1997. She is president of the National Black Talk Show Association and president/CEO of Jo Ann Watson Systems.*

DEADRIA C. FARMER-PAELLMANN

Excerpt from *Black Exodus: The Ex-Slave Pension Movement Reader*

The corn is never declared innocent in the court of the chickens.
—EWE PROVERB

I discovered the ex-slave pension movement during my journey to satisfy my curiosity about forty acres and a mule. The elders in my family used to talk about the forty acres and a mule the government promised Blacks as compensation for slavery but never gave them. My grandfather, Willie Capers, whose grandparents were runaway slaves, planted the idea of pursuing reparations—payment for the repair of an injury or wrong done—in my head by occasionally punctuating complaints about how Blacks are mistreated in America with the statement, "And they still owe us our forty acres and a mule." His grandparents were enslaved on one of few rice plantations on St. Helena's Island in South Carolina.

As a child growing up poor in the predominantly Black East New York section of Brooklyn, New York, I clung to the "myth" that the government owed my family land. I lived in a two-bedroom apartment with my mother and five sisters, and would dream about how nice my life would be if the government made good on the promise my grandfather talked about. I thought, maybe if we could get our forty acres and a mule, we could afford to buy one of the old brownstone buildings that lined the streets in my neighborhood—that way my sisters and I could each have our own bedroom.

In the summer of 1973, when I was seven years old, we moved to a first-floor, three-bedroom apartment in Bensonhurst, Brooklyn, an area my mother referred to as an Italian ghetto with better schools than the ones in the Black ghetto we were leaving.

My first encounter with the neighborhood kids resulted in my earliest memorable experience with racism. A few of my sisters and I gathered excitedly in our bedroom window to greet our potential new friends upon their return from summer camp. Rather than matching our enthusiasm, one of them threw an empty soda can at the window as they all yelled "niggers!"

It seemed that in the seven years we lived in that neighborhood, not a day passed without some teenager reminding us that we were niggers. Very often, an object thrown through one of our windows would accompany the epithet.

My mom, Wilhelmena Farmer, taught us how to endure the terrorism of that community in order to achieve a better standing in life. Rather than respond to it with violence, she encouraged my sisters and me to fight their ignorance and hatred in the classroom through academic achievement. In other words, use negative experiences to fuel positive action. According to her, the worst thing we could do to a racist was get a good education. I suppose my mother's strategy for fighting racism played some role in my wanting to get what I thought was the best education—a legal education.

There was one other key lesson my mother taught me as a child that would help me on my journey to discovering the ex-slave pension movement. She emphasized the importance of learning about Black history and culture. We did not have much in our home, but there were plenty of magazines and books by and about Black people. One of my favorite books was *Before the Mayflower* by Lerone Bennett Jr., which highlighted a rich Black culture in Africa prior to the establishment of colonies in what became the United States of America. In addition, my mom made sure we watched documentaries and listened to radio shows on major figures from Black history like Dr. Martin Luther King Jr. and Malcolm X.

Information about Dr. King seemed plentiful on radio and television programs, but there was a mysterious shortage of information on Malcolm X; therefore, at the end of my undergraduate studies at Brooklyn College, in the late 1980s, I went on a Malcolm X binge. Four years of traditional political science studies left a void for progressive ideology that Malcolm X seemed to fill. I tried to read every book, speech, and interview he ever did. In the process, I was reminded of the idea of reparations for slavery.

In Malcolm's 1963 interview conducted by Alex Haley for *Playboy* magazine, he proposed that in return for the economic wealth that enslaved Africans brought the United States, that the government should give their descendants land and resources. This idea made immediate sense to me.

Learning that Malcolm X supported the idea that the government owed a debt to Black America for slavery gave even more weight to my grandfather's claim for forty acres and a mule. From that point on, I was on a mission to learn more about America's broken promise, and to determine how I could help encourage the nation to pay its debt to my people.

My first attempt to formulate a strategy for reparations was in 1988 while I was a lobbying student at the Graduate School for Political Management in New York City. My project for Lobbying 101 was to design a lobbying strategy for a slavery reparations organization. I wanted the strategy to have some basis in reality, so I searched for a reparations organization or information on modern demands for reparations. Outside of the 1973 Boris Bittker book *The Case for Black Reparations,* I could find no other information on reparations. Bittker's book was of limited help to me in developing an argument because it was filled with legalese that I was ill equipped to interpret. I abandoned that lobbying project and instead designed a less-interesting lobbying strategy for the League of Women Voters. I took a certificate in lobbying and moved on with my life.

My first long-term job after college was working as an administrator in the press office of the New York City Department of Health under the direction of Dr. Richard Atkins. From 1990 to 1994, I worked in this environment acquiring a basic education in press relations.

In 1992, my press skills were tested. I was invited to help organize a media event with a team of community activists engaged in preserving an eighteenth-century African burial ground located in lower Manhattan. The remains of nearly 400 of 20,000 enslaved Africans buried at this site were disinterred by archeological teams hired by the federal government in order to make room for a new federal office building and its parking pavilion.

We decided that the media event would be a twenty-six-hour drum vigil. While in the process of planning the vigil, I was given a tour of the site. I found archeologists kneeled over skeletal remains using toothbrushes to push away dirt that had lain undisturbed for two centuries. It was a bit surreal to me to be face-to-face with the actual remains of my African ancestors, many of whom were first-generation Africans in America. Many appeared to have died in pain, their mouths stretched wide open as if they were screaming. The chief archeologist reassured me that the haunted look resulted from the deterioration of jawbone muscle that normally holds the mouth shut.

I left the site thinking, my God, these people lived, worked, and died in this city to make this nation rich, but were never paid a dime. Compelled to find a way to do something about that injustice, during the spring of 1994 I quit my job with the city to go back to school to find a case for slavery reparations.

First, I finished my graduate studies in political campaign management at the George Washington University. I figured campaign skills would be crucial toward my goal.

Second, in August of 1995, I began classes at the New England School of Law with the intent of finding a case for slavery reparations against the federal government. I considered targeting the forty acres and a mule, but I was not sure whether the story was a myth. I soon discovered that the story was absolutely true!

In January of 1865, Major General W. T. Sherman issued Special Field Order No. 15, reserving land confiscated from rebellious landowners in various Southern states for settlement by Africans freed under President Abraham Lincoln's Emancipation Proclamation. Each family was to be given a plot of up to forty acres of tillable ground and the government would loan out mules to help work the land. By June of 1865, about 40,000 freedmen had been allocated 400,000 acres of land. This act of justice was short-lived. By September of 1865, President Andrew Johnson began the process of rescinding title to the land from the freedmen and returning it to its previous plantation owners.

The legality of Johnson's action was questionable at the time; however, subsequent legislative attempts to return the land allocations to the freedmen were unsuccessful.

I examined the outcome of a recent lawsuit, *Cato v. United States*, a case for government reparations filed by a woman and a small group of others against the federal government in a relatively liberal California federal court in 1995. The results were discouraging. The court dismissed the case holding, amongst other things, that the government had not given permission to the plaintiffs to sue, and that the case was being filed too long after slavery had ended. These legal hurdles are known respectively as *sovereign immunity* and *statutes of limitations*. I concluded that it was unlikely that the federal government would ever give African Americans permission to sue it for slavery reparations. As it stood, a piece of legislation introduced by Congressman John Conyers of Detroit, Michigan, simply asking for a congressional commission to study the effects of slavery on present-day African Americans remained stuck in committee for about eight years.

By 1997, I began to search for other parties that owed African Americans reparations. A footnote in a 1993 *Tulane Law Review* article entitled "If the Shoe Fits, Wear It: An Analysis of Reparations to African Americans" by Professor Vincene Verdun of Ohio State University College of Law led me to consider individuals as potential defendants in cases for reparations.

Verdun provided a brief hypothetical of a case for reparations filed by the children of former slaves against the children of former slave owners for assets acquired exploiting slave labor. This concept comes from the law of restitution that requires that a thing wrongfully acquired be restored to its rightful owner or the heirs of that owner. Sovereign immunity cannot be invoked since the government would not be a defendant; and statutes of limitations do not always apply in restitution actions.

I immediately felt her hypothetical case provided the key to winning reparations. Not only were there Americans today who owe their wealth to the exploitation of enslaved Africans, but there were also corporations similarly situated. I began to build a case for restitution from these private entities rather than reparations from the government. I also calculated that the government might ultimately consider paying reparations in an effort to protect corporations from exposure to restitution-based litigation.

In 1997, under the supervision of Professor Robert V. Ward Jr., I developed a strategy for restitution from corporations and private estates for a presentation in a class he team-taught with Professor Judith Greenberg at the New England School of Law. The class was entitled Race and the Law. My strategy included a legal action for restitution based on Professor Verdun's hypothetical, and a media campaign aimed at persuading tainted companies to apologize and share their tainted profits with the descendants of enslaved Africans.

The research I conducted uncovered many existing corporations that helped to maintain slavery. For example, I read Charles Blockson's book *Black Genealogy* and learned that Aetna Incorporated had a predecessor company that wrote life insurance policies on the lives of slaves with slave owners as beneficiaries, thereby helping to finance the institution of slavery.

In January 2000, a year after graduating from law school, I obtained copies of two of these slave policies from Aetna's archivist. Seeing documents bearing the words "Aetna Slave Policy," I wept, and thought to myself, the world needs to know what this multibillion-dollar corporation did. After all, the policies provided slaveholders with security to invest in human chattel with little risk. This practice helped to perpetuate slavery.

February, Black History Month 2000, was approaching. I thought reporters might be willing to write stories about my discovery. I contacted Aetna and asked that they apologize and pay restitution. They agreed to do both. By March of 2000, Aetna made an unprecedented apology for their role in slavery; however, they backed down on their promise to pay restitution.

I am still engaged in the process of targeting existing companies that profited from the enslavement of Africans in America. I am confident that in due time they will atone by apologizing and paying the slavery debts they owe into a trust fund to benefit the descendants of enslaved Africans. Furthermore, the tainted companies may even help to move the federal government to pay reparations as they participate in the electoral process seasonally.

My journey toward building a new approach to slavery reparations led me to the largest mass movement for slavery reparations in American history—the ex-slave pension movement. Between 1890 and 1917, over 600,000 of the 4 million emancipated Africans lobbied our government for pensions because they believed their uncompensated labor subsidized the building of the nation's wealth for two and a half centuries. Nashville, Tennessee, was the home of this effort, largely led by a Black woman, Callie D. House—the national spokesperson of the Ex-slave Mutual Relief, Bounty and Pension Association.

By 1917, the federal government effectively shut down the movement. Under the pretext of protecting the ex-slaves from the exploits of fraudulent organizers, the government pursued, prosecuted, and convicted Callie House and other leaders of the movement on fraud charges. Its position was essentially that the convicted leaders were acting fraudulently by collecting money to fund a lobbying effort that instilled the false hope in the hearts of ex-slaves that the government would give them a pension. This critical view of the ex-slaves' lobbying efforts was not only reflected in agency correspondence and court decisions on the matter, but it was also evident in written press of that time.

This movement is worth examining for several reasons. First, Walter R. Vaughan, a white Southern Democrat, initiated the request for slave pensions from the government. Vaughan drafted and secured the introduction of a slave pension bill that became the focus of the twenty-six-year lobbying effort. His contribution to the struggle shows that, in addition to ex-slaves, white humanitarians recognized that the government should pay this debt to African Americans.

Second, Callie House, a woman, was at the forefront of this massive human rights effort on behalf of ex-slaves. House served as the national spokesperson for the largest of all ex-slave pension organizations. Census records show that she was a widow raising five children alone. She lived and worked during the Jim Crow era—a period of time when lynching and other acts of terrorism against Blacks were on the rise. In spite of the dangers and challenges she faced as a Black person, a woman, and a mother, she traveled the nation for nearly two decades successfully organizing thousands of ex-slaves to demand that their humanity be recognized by the federal government. Although convicted on federal mail fraud charges, no proof was ever provided of her misappropriation of funds collected from ex-slaves.

Third, ex-slaves, a new group of American citizens, expressed their faith in the legislative process and belief in the Constitution by pursuing a twenty-six-year struggle for human rights. They expressed in their literature that it was their constitutional right to petition the government to redress their grievances by organizing and lobbying toward passage of ex-slave pension bills introduced into Congress. Conventions were held to educate toward this goal with participation of Black church, business, and political leaders from across the nation.

Fourth, today many people assume that ex-slaves did not want or expect reparations for their enslavement. The ex-slave pension movement is proof that the demand was made but the government actively refused to comply.

It is unclear why the ex-slave pension movement has remained relatively unknown. Perhaps established historians did not take the ex-slaves seriously enough to dedicate a book to the effort; or perhaps they did not want to document such a powerful showing of Black power in a book that could be used to encourage future generations of African Americans to take up the cause.

Maybe Black historians, believing that the movement's leaders were really stealing money from ex-slaves, avoided immortalizing an embarrassing episode from American history. Perhaps Booker T. Washington's bootstraps ideal, or Marcus Garvey's back-to-Africa approach, drew more credibility and eclipsed the ex-slave pension approach. Four groups of materials provide a balanced view of the movement and its key actors.

Walter Vaughan wrote a book in 1891 entitled *Vaughan's Freedman's Pension Bill*. It is the second edition of Vaughan's original appeal for pensions written in 1890, which marked the start of ex-slaves organizing themselves for pensions. He sets forth a wordy yet passionate and intelligent argument for ex-slave pensions using historical information that would strengthen arguments for reparations today.

Vaughan highlights the critical role that enslaved Africans played in winning the Civil War, emphasizing that without their military assistance, the Union would not have prevailed over the Confederacy. He reiterates the controversial position that President Lincoln was not really the Great Emancipator, but rather an opportunist that used emancipation only to the extent necessary to win the war. He states that emancipation resulted as a military necessity rather than as a political or social benefit conferred upon the recipients as a measure of justice and humanity. He sets forth thirteen reasons why ex-slaves should be pensioned that are applicable to today's argument for reparations. Included are the following:

> 1. *It will be a measure of recognition of the inhumanity practiced by the government in the holding, for a century, of men and women as slaves in defiance of human right.*

> 3. *It will afford foreign nations a complete refutation of the sentiment, often advanced, that American Freedom has been merely a disguised form of tyranny whereof human slavery was an exemplification.*

7. It will enable an impoverished race, reduced to penury through no fault of their own, to place themselves in a position of reasonable independence in their struggle for existence and recognition in general business affairs; and

12. It will remove the last barrier existing between the races which has made political solidarity an objectionable feature in the political affairs of any section of the Great American Republic.

Vaughan rounds out his argument for justice with examples of Black people who have been able to outgrow the conditions surrounding their birth, and become useful to the world in their day and generation. Included amongst these men of merit are legendary figures from the nineteenth century: Frederick Douglass, the well-known abolitionist and statesman; Robert Smalls, congressman from South Carolina; and Samuel R. Lowery, successful silk manufacturer.

Walter L. Fleming published an article in 1910 entitled "Ex-Slave Pension Frauds" in which he gives a critical examination of the ex-slave pension movement. He suggests that Walter Vaughan was ill balanced mentally with sincere intentions in his pursuit for ex-slave pensions. He credits Vaughan with having the only honest pension lobbying operation, and states that all other ex-slave pension groups were fraudulent, designed merely to secure money from the ignorant Blacks by the most barefaced misrepresentations.

Mary Frances Berry wrote an article in 1972 entitled "Reparations for the Freedmen, 1890–1916: Fraudulent Practices or Justice Denied?" She provides a political and legal analysis of the ex-slave pension movement, concluding that no evidence existed proving the movement's leaders acted fraudulently, and suggesting that more investigation needs to be done in the future.

There is a vast collection of original documents on the movement at the National Archives that give the reader an intimate look at the movement from the perspectives of the ex-slaves, government investigating agencies, courts, and the press from the enthusiastic start of the effort to the crushing end.

Included are documents in which ex-slaves articulate their demands for pensions from the government using historical analogies in recruitment literature and spirited songs in membership pamphlets. For example, I. H. Dickerson, a leader of the Ex-Slave Mutual Relief, Bounty and Pension Association, writes in one circular:

If the Government had the right to free us after 244 years of labor had been wrung from us, it had the right to give us something as other

Governments have heretofore. The Hebrews brought something from the land of Egypt; the captive Jews from Jerusalem, 70 years in bondage, were emancipated and compensated for their labor.

Adam Scales and J. W. Anderson, of the Ex-Slave Pension Association, include their powerful "Ex-Slave Pension Song" in their organization's membership pamphlet. The song chants:

> *We come not here for to ask an alm,*
> *We must have our money or never shall yield,*
> *But demand a debt from Uncle Sam,*
> *We must have our money or never shall yield.*
>
> CHORUS:
> *Why stay in the field,*
> *Camp on the grounds of the enemy.*
> *Stay in the field,*
> *We must have our money or never shall yield.*

A news clipping from the early 1900s cites a Judge Calhoun of an Atlanta, Georgia, court reprimanding I. H. Dickerson for organizing to lobby for slave pensions. While sentencing Dickerson in an action that accused him of defrauding ex-slaves who contributed money to lobby for passage of a pension bill, Calhoun stated: "This is the biggest swindling scheme I ever heard of. This scheme of getting money for the slaves is like pouring money into a rat hole and you know it."

Whatever one concludes about the motives of slave pension organizers and government officials around this movement, it cannot be denied that ex-slaves believed that they deserved reparations, and that they registered their demands with the United States for this form of justice over one hundred years ago. One of many petitions to Congress in support of slave pension legislation introduced in 1898 lists the names, ages, names of masters, and states of residence of a group of ex-slaves. It reads:

> *Whereas, It is a precedent established by patriots of this country to relieve its distressed citizens, both on land and sea, and millions of our deceased people, besides those who still survive, worked as slaves for the development of the great resources and wealth of this country, and Whereas, We believe it is just and right to grant the ex-slaves a pension.*

The ex-slaves' unfulfilled claim for pensions is remembered by review of these highlighted materials that serve as a foundation for the renewed journey toward reparations.

DEADRIA C. FARMER-PAELLMANN *is considered the Rosa Parks of the Reparations Litigation Movement. As an attorney-activist, she has studied the relationship of government and corporate policies that determined economic and social policies impacting African Americans. This chapter is reprinted from her forthcoming book,* Black Exodus: The Ex-Slave Pension Movement Reader.

HAUNANI-KAY TRASK

Restitution as a Precondition of Reconciliation: Native Hawaiians and Indigenous Human Rights

It is they who have not died in war, that start it.
—KIKIYU PROVERB

As one of the most violent and deadly centuries in the history of the world came to an end, questions of truth, reconciliation, and restitution began to appear in public forums, some with all the juridical structure of international tribunals. It is as if some deeply human yearning for peace has spawned an international quest for justice. In the South African case, a Truth and Reconciliation Commission brings perpetrators and their victims together where guilt is established and a resolution in the form of absolution and escape from punishment is granted to confessed criminals.

The stated purpose of these forums is more than the discovery of truth and deliverance of amnesty. Disclosure and exposure, in a public, communal sense, as well as a private, individual sense, are a part of South Africa's goal (and hope) of healing. Unquestionably, such an experiment is unique, indeed, daring. There is, of course, no meting out of retributive justice if one bares all. But what of reconciliation between peoples? As African novelist and Nobel laureate Wole Soyinka has argued in his elegant book *The Burden of Memory, the Muse of Forgiveness*, this risk-free parade of villains, calmly and occasionally with ill-concealed relish recounting their roles in kidnappings, tortures, murders, and mutilation, at the end of which absolution is granted without penalty or forfeit, is either a lesson in human ennoblement or a glorification of impunity.[1]

Soyinka fears that responsibility and justice are eroded by the price of truth and reconciliation. For him, and for many of apartheid's victims, neither peace nor justice is served by mere public disclosure. Soyinka is not alone. Political parties, such as the Inkatha Freedom Party, have encouraged efforts by whites to redress injustices through contributions to monetary funds established to assist apartheid victims. And Pan Africanist Congress secretary-general Thami Plaatjie has said, "It is high time that the broader white community embraced such an initiative. It serves no purpose to confess, if it is not accompanied by reparations. Mechanisms of atonement should be implemented."[2]

As might be expected, many native South Africans support criminal court proceedings with punishment for the guilty. For them, monetary reparations are separate from, and thus cannot substitute for, punishment. It may be that retributive justice, that is, punishment of criminals by tribunals and courts, is the only possible prerequisite to South African reconciliation.

In terms of the international community, the right to obtain financial compensation for human rights abuses and to have the perpetrator of such an abuse prosecuted and punished is itself a fundamental human right that cannot be taken from a victim or waived by a government. The Universal Declaration of Human Rights and the American Convention on Human Rights, among other international legal instruments, confirm that victims have rights to remedies, including compensation.[3]

It is apparent that courts appreciate, perhaps more than governments, that the yearning for justice is not soothed by the granting of amnesties. This is why Soyinka fears the South African experiment. He understands how the dangerous moral and political dilemma for South Africa is that confession with the promise of amnesty is an injustice to victims, and an impediment to the healing of the nation.

In Soyinka's words, "the problem with the South African choice is therefore its implicit, *a priori* exclusion of criminality and, thus, responsibility."[4] Truth, unfortunately, is clearly more central to the South African Commission than justice, which Soyinka astutely judges the first condition of humanity. Murderers who simply confess, knowing they will not suffer for their deeds, have no compunction against murdering again. Already the behavior of freed white supremacists suggests the Truth and Reconciliation process will not render enduring peace.[5]

Soyinka's troubled ruminations on the South African experiment are pertinent to any investigation of the current global move toward addressing historical injustices. Who will judge which *injuries of war*, for example, internment, forced marches, sex slavery, and which *injuries of imperialism*, for example, removal, plunder, and genocide, deserve reparation and restitution? And before there are any judgments of this magnitude, why are some peoples, in the words of Noam Chomsky, accorded the status of *worthy* victims, while others, like indigenous peoples and sex slaves, are treated as *unworthy* victims, whose suffering and historical subjugations are of lesser stature, and therefore of lesser consequence, than that of state-acknowledged *worthy* victims?[6]

Everywhere in the world, human rights violations against indigenous peoples and women are increasing but national and international investigations regarding them lag far behind. Even in the category of worthy victims, some are more worthy than others, say, Japanese internment victims over Aleutian internment victims during World War II, both in the sense of being newsworthy and in the sense of receiving equal treatment.

The case of Native Aleuts is particularly grievous. After Japan attacked the two westernmost Aleutian islands, Kiska and Attu, the American government ordered military evacuation of the Aleuts. They were removed to unheated, crowded barracks and vacant buildings, and left without adequate medical care. As a direct result of their ordeal, 10 percent or more of the Aleuts died in captivity. At war's end, the Aleuts were only haphazardly and slowly returned to their islands, where their houses had been looted by the Americans. Some of the stolen materials included irreplaceable sacred objects.

To address the issue of both internments, the United States Congress passed the Civil Liberties Act of 1988, in which a formal apology was made to the interned Japanese and Native Aleuts. But in the Japanese case, $20,000 was awarded to each internee, while only $12,000 was awarded to each Native Aleut.[7] More importantly for Native Aleuts, public recognition of their injury and American political acknowledgment of wrongdoing was but a minor sideshow to that received by the Japanese, including a monument on the Mall in Washington, D.C.

The Aleutian and Japanese internment cases reveal how discussions of reparation, restitution, and apology are framed more by political than ethical or moral considerations. Certainly, the successful reparation effort by interned Japanese is a testament to the dedicated and intelligent political organizing done by Japanese Americans. But such success is also a result of the colonial reality of Japanese Americans as settlers rather than indigenous peoples. Ideologically as well as politically, it is far easier for the United States government to address, in a public and official manner, the forcible internment of one of the most successful settler groups in America than it is for the same government to render equal acknowledgment of mistreatment of an indigenous nation. It is telling that partially repairing injury to Japanese Americans reinscribes the American ideology of equality among settlers, while recognition of harm done to indigenous peoples not only contradicts the dominant immigrant paradigm but raises prior issues of responsibility for genocide against native peoples.

As conquered nations, indigenous peoples were forced to become Americans. Our national status, then, is a result of subjugation, not choice. This indisputable history is critical because, as universally acknowledged, citizenship must be freely chosen or it is meaningless in terms of representing the interests and binding the loyalty of citizens. And while our lands of origin have been collectively renamed the United States of America, indigenous peoples are now classified by the federal American government as Native Americans, a nonsensical category that tells worlds about our contradictory status.

Given the primacy of homeland, of the place where native people understand an ancestral sense of belonging, identification as American has no correspondence to any cultural, familial, or tribal origins. The Lakota, for example, are native to the Black Hills; the Anishnabe are native to the northern woodlands of Minnesota; and my people, the Hawaiians, are native to the Pacific archipelago of Hawai`i. We are not Native to a recent creation called the United States of America. Rather, we are aboriginal to a specific land base that defines us linguistically, geographically, and historically. As indigenous peoples and nations, our ancestral attachments are prior to colonial categorization as Americans.

To assert national differences between native peoples of the Americas is no more contradictory than the assertion that Tahitians are not French, and Okinawans are not Japanese. If language is taken as but one marker of peoples, then it is telling to learn that some two thousand languages once characterized the Americas where, by conservative estimate, some 75 million to 100 million natives lived at contact.[8] In the same way that the French are both a people and a political entity defined by a particular land base, language, and culture, so too native nations are identified by different languages, lands, and cultural distinctiveness. In international human rights terms, indigenous peoples have claims to self-government, then, because we are the first nations of the land.

Given this aboriginal as opposed to settler history of native peoples, it should be obvious that no reconciliation between the American government and indigenous peoples can be achieved without a return to nationhood. *Restitution, in other words, must be a precondition for reconciliation.* And that restitution process must surely begin with self-government, then move on to land and water, and finally proceed to the repatriation of native artifacts, and compensatory educational, health, and other programs.

For native nations, the paramount issue is the return of native people to their native place. Unlike Japanese internment victims, land is the key to reconciliation if there is to be any reconciliation between native peoples, colonial governments, and the larger settler society.

INDIGENOUS HUMAN RIGHTS

When in conflict with colonizing powers, such as the governments of North and South America, native peoples have increasingly come to argue in the language of indigenous human rights. Individual civil rights of the kind common to modern constitutions are inadequate to enunciate indigenous claims to land, language, self-government, and religious practices such as protection of sacred sites. But these do appear as protected human rights in the United Nations Draft Declaration on the Rights of Indigenous Peoples. Framed by and for indigenous peoples, the Draft Declaration is the embodiment of the values and goals of native peoples. A product of twenty years' work in the international arena by indigenous peoples themselves, the Draft Declaration illustrates the widest scope of indigenous human rights.[9]

Most critically, indigenous peoples are defined in terms of collective aboriginal occupation prior to colonial settlement. Here, indigenous peoples are not to be confused with minorities or ethnic groups within states. Thus indigenous rights are strictly distinguished from minority rights. The numbers of indigenous peoples, therefore, does not constitute a criterion in their definition.

The difference between indigenous peoples and minorities turns, in part, on the critical identification of historical continuity such as occupation of ancestral lands; common ancestry with original occupants of these lands; culture, including such things as dress, religion, and membership in traditional communities; and finally, language. These distinctions are of first-order importance because, under international law, minorities, unlike indigenous peoples, do not have the right to self-determination.

Political self-determination is tied to land rights and restitution. The doctrine of discovery by which the Americas, the Pacific, and so many other parts of the world were allegedly discovered is repudiated. And the companion doctrine of *terra nullius* is identified as legally unacceptable.

A cursory overview of the Draft Declaration reveals the following. In Article 1: Indigenous peoples have the right to the full and effective enjoyment of all human rights and fundamental freedoms recognized in the Charter of the United Nations, and the Universal Declaration of Human Rights.

In Article 3: Indigenous peoples have the right of self-determination.

In Article 5: Every indigenous individual has the right to a nationality.

In Article 6: Indigenous peoples have the right to live in freedom, peace, and security.

In Article 7: Indigenous peoples have the collective and individual right not to be subjected to ethnocide and cultural genocide.

In Article 8: Indigenous peoples have the right to identify themselves as indigenous and to be recognized as such.

In Part V, Articles 21 and 23: Indigenous peoples have the right to maintain and develop their political, economic, and social systems, to be secure in the enjoyment of their own means of subsistence and development, and to engage freely in all their traditional and other economic activities.

Part VI of the declaration is often thought to be the most controversial to existing nation-states because land rights and restitution are addressed. The doctrine of *terra nullius,* that is, the vacant land argument used by Europeans who colonized the Americas, Australia, and other lands, is repudiated as an unacceptable legal doctrine.

It is in this part of the declaration that aboriginal peoples find a strong basis from which to argue that traditional lands should be restored to them.

In Article 26: Indigenous peoples have the right to own, develop, control and use the lands and territories they have traditionally owned.

In Article 27: Indigenous peoples have the right to restitution of the lands, territories and resources which they have traditionally owned, or otherwise occupied or used, and which have been confiscated, occupied, used or damaged without their informed consent.

In Part VII, Article 31, the declaration states: Indigenous peoples, as a specific form of exercising their right to self-determination, have the right to autonomy and self-government.

Tellingly, these rights, and others, are considered in Part IX to constitute the minimum standards for the survival, dignity, and well-being of the indigenous peoples of the world.

DOMINANT IDEOLOGIES AND REPARATIONS

The codification of human rights is part of a growing move to frame restitution issues in terms of an international morality. Human rights, and not simply civil rights, are becoming the standard for considerations of global justice. South Africa is but the most recent and best-known example of this evolution.

Regarding indigenous peoples, however, justice lags far behind legal codification. Another way to understand this, at least in the American context, is to ask why, beyond questions of effective organization and public relations, restitution appears more defensible and therefore much more likely when the injured are non-natives.

If we return to the issue of public morality, it becomes increasingly obvious that political systems move to compensate for injustices only when they are framed within a country's national ideology. Regarding the United States, that dominant ideology is a settler ideology: we are all immigrants. The notable exceptions to the immigrant paradigm are native peoples and African Americans. The latter did not voluntarily migrate but were forcibly transported from Africa to become slaves in the United States. Indigenous peoples became Americans through relentless, genocidal conquest. The historical and contemporary realities of both African Americans and native peoples, then, undercut the official ideology that the United States is a nation of immigrants. This explains, in part, increasing political resistance in the United States to Black reparations. Among other problems, acknowledging the debt owed to African Americans contradicts and therefore undermines the official ideology of the United States as a nation of immigrants.

In the case of reparations by Germany to the State of Israel for the Nazi Holocaust, the operative ideology is that modern Germany is once again an accepted member of Western civilization in good standing. This standing was bought, in part, through the reparation process. Israel is now, like the United States, a settler state and a member of Western civilization in the non-Western Middle East. From this perspective, payment to Israel ensures, in part, the return of Germany to the Western family of nations.

The Roma people, by contrast, are not considered by established Western governments to be family members in good standing. They are, in truth, a stateless people in a world of competing nation-states. As the much-maligned "vagabond, thieving Gypsies" of Europe, the Roma are "unworthy victims," in Chomsky's words.[10]

Given their little-known history, it is instructive to learn how similar to the Jews was the horrific suffering of the Romani during the Holocaust. In 1941, Reinhardt Heydrich's directive, which, according to scholar Ian Hancock, set the machinery of the Holocaust in motion, ordered the eradication of all Jews, Gypsies, and mental patients.[11]

The German effort to eradicate Gypsies began in 1721, 220 years before Hitler. In 1835, German scholar Teodor Tetzner described the Roma as the excrement of humanity. In 1886, according to Hancock, Otto von Bismarck called for especially severe treatment of Roma, and by 1905 a Gypsy Book compiled by the Germans listed all known Roma throughout Germany. Referred to as pests and a plague, the Roma were forbidden to enter public facilities such as parks, were required to be photographed and fingerprinted, and were incarcerated whenever the state deemed it necessary.[12]

From 1934 on, Roma were selected for sterilization by injection or castration and sent to death camps at Dachau and elsewhere. According to Hancock, the first mass genocidal action of the Holocaust took place at Buchenwald, where 250 Romani children were used as guinea pigs to test the Zyklon-B crystals later used in the gas chambers at Auschwitz-Birkenau. By 1945, Hancock reveals, between 1 million and 1.5 million Roma, perhaps half of all Roma in Nazi-controlled Europe, had perished in the Porrajmos, that is, the Devouring, as the Holocaust is called in Romani.[13]

If the magnitude of Gypsy suffering reveals that they, like the Jews, were part of the final solution, they have not been part of the effort at reparation, either by the Swiss or the Germans. This condition of nonrecognition undoubtedly stems from the relatively small number and lack of statehood of the Romani. Jews, by contrast, not only number in the millions but are well organized and well funded. Perhaps most importantly, in terms of ideology, Jews are perceived as worthy victims who, as testament to their worthiness, are no longer stateless, unlike the Romani.

The Romani case illustrates how political ideology—that thick layer of beliefs and justifications that bind citizens to nations—frames legitimacy. If, for example, the United States is a nation of immigrants, as the national ideology dictates, then an injury to one group of immigrants, say, interned Japanese, must be compensated. This compensation is not so much to repair damage to the injured as to reiterate the legitimacy of America's national existence. Restitution, however meager, protects and reinforces the ideology of a democratic republic. For the victims, restitution is the pittance that keeps them obedient and loyal. This explains what appears to be inexplicable, namely that apology and reparation make victims proud to be Americans.

Indigenous peoples, on the other hand, are *not* proud to be American since they are *not*, in truth, of America. Indeed, the very existence of native nations contradicts the dominant ideology of the United States as an immigrant nation founded in a vacant land. Restitution to indigenous peoples, not only in the United States but in the Americas, focuses on issues of conquest, of dispossession, of genocide. These historical realities make a mockery of the ideology that native people should be proud to be American.

Given the history of indigenous peoples, then, no monetary compensation, no apology, no effort to put the past behind us, is acceptable. The only acceptable reality is return of native lands and waters, a monetary indemnity, and recognition of native sovereignty.

THE HAWAIIAN CASE

Our human rights movement is now over three decades old. Beginning in the 1970s, our struggle evolved from anti-eviction actions and occupations of military reserves, including entire islands used as bombing targets, to civil and legal rights struggles, to the current demand for sovereignty. For twenty of these thirty years, I have been an advocate of my people's human right to self-determination, particularly self-government. Our focus has been on the injury our Native Hawaiian people continue to suffer at the hands of the American government. This injury began with those familiar American practices in the international arena: invasion, occupation, and takeover.

On January 17, 1893, the United States minister to Hawai`i ordered the landing of American marines in support of an all-white, all-male Committee of Safety, which had seized political power. Fearing the American military, our queen, Lili`uokalani, ceded her authority, not to the committee, but to the United States minister.

She wrote to Sanford B. Dole, descendant of white American missionaries and newly chosen head of the provisional government: "I yield to the superior force of the United States . . . Now, to avoid any collision of armed forces and perhaps the loss of life, I do under this protest, and impelled by said force, yield my authority until such time as the Government of the United States shall . . . undo the action of its representatives and reinstate me . . . as the constitutional sovereign of the Hawaiian Islands."[14]

On February 1, 1893, Minister Stevens proclaimed a United States protectorate and raised the American flag over Hawai`i. But his dream for swift annexation was short-lived. President Cleveland, a mere five days after his inauguration on March 4, 1893, withdrew the pending Hawai`i annexation treaty from Congress. On March 29, Commissioner James Blount, Cleveland's emissary, arrived in Hawai`i. After four months of investigation, Blount concluded that the overthrow, the landing of marines, and the subsequent recognition of the provisional government pointed to clear conspiracy between Minister Stevens and the alleged Committee of Safety.

After reading Blount's report, Cleveland explained to the Senate why he would never again submit the annexation treaty for ratification. In his concluding statement, he wrote: "By an act of war, committed with the participation of a diplomatic representative of the United States and without authority of Congress, the government of a friendly and confiding people has been overthrown. A substantial wrong has thus been done which a due regard for our national character as well as the rights of the injured people requires we should endeavor to repair."[15]

Thus was the issue of reparation, of undoing the harm and the injury to the Hawaiian people, first brought to the attention of the American government. It was an exquisite irony of history that an American president would be the first to argue the Hawaiian case for restitution and restoration.

In 1897, a protest petition against annexation to the United States was sent to Washington, D.C. Over 21,000 natives, representing the overwhelming majority of adult Hawaiians, had signed anti-annexation petitions.[16] Because Cleveland stalled annexation, the all-white oligarchy renamed themselves the Republic of Hawai`i. Our queen, meanwhile, was imprisoned in her own palace after a failed native counterrevolution. Final annexation in 1898 had to wait for a real imperialist, President William McKinley.

Union with the United States meant the transfer of 1.8 million acres of Hawaiian government lands, that is, nearly half the archipelago, to the all-white planter oligarchy. The Hawaiian language was officially banned from all public instruction, government business, and commerce.

HAWAIIAN HUMAN RIGHTS CLAIMS

In terms of international law, the American military invasion of our archipelago, overthrow of our native government, imprisonment of our queen, and immediate American diplomatic recognition of the hastily constructed, all-white, all-male provisional government, resulted in undeniable human rights violations, including our claim, as a native nation and people, to self-determination.

Under international law, these violations constitute:

1. an arbitrary deprivation of our nationality, and of our citizenship in an independent country;
2. an arbitrary deprivation of our national territory, including lands, waters, and other natural resources;
3. a denial of our human right to self-government as both an indigenous people and as a formerly independent country.

These deprivations, as a whole, comprise violations of Articles 15, 17, and 21 of the Universal Declaration of Human Rights. In addition, they are also violations of the American Convention on Human Rights. The fact that the overthrow and annexation occurred before international covenants went into effect does not invalidate the Hawaiian case. The ideal of universal self-determination is a settled principle of peremptory international law, superseding customary rules and bilateral treaties. This means that the principle of self-determination is of sufficient importance to be applied retroactively to relationships among states and peoples before the adoption of the 1948 United Nations Charter.

After a brief, five-year period as a republic, Hawai'i was annexed by congressional resolution, which required a mere majority vote rather than by treaty, which required a two-thirds vote. No popular vote was allowed in Hawai'i since Hawaiians were a majority of the population. Following annexation, Hawai'i became an American territory with all governors appointed by the American president.

At the creation of the United Nations in 1945, Hawai'i was listed as a Non-Self-Governing Territory under United States Administration. Such status was considered a trust relationship. Pursuant to Chapter XI of the UN Charter, the United States had a trust obligation to promote the political aspirations of the people of the territory, including self-government. But in 1959, when the United States allowed all citizens of Hawai'i, native and non-native, to vote, only two options were presented: territorial status and statehood. Neither commonwealth nor independence appeared as choices on the ballot.

Following the statehood vote, the United Nations, without inquiry or investigation, removed Hawai'i from the United Nations list of Non-Self-Governing Territories.

Today, the colonial policy of the United States toward Native Hawaiians continues to be one of state and federal wardship. Defined and divided by blood quantum, our native people are denied the following:

a. Our collective right to self-government;
b. legal control over our lands and waters;
c. economic and political power to develop our lands, fisheries, and cultural properties;
d. the ability to preserve and protect our entitlements for our *lahui*, or nation, including future generations.

Because of the colonial American policy of nonrecognition, Hawaiians are not allowed the civil or human rights due indigenous peoples. These include rights to control our trust assets, including nearly 2 million acres of lands and waters, and to sue the state and federal governments for land and water recovery.

The policies of the Reagan, Bush, and Clinton administrations have been, generally, ones of abandonment. The Department of the Interior under both Bush and Clinton administrations disavowed any trust responsibilities to Hawaiians either before or after statehood. Despite the presence of two federal trusts for Hawaiians, the federal government negotiates with the State of Hawai'i rather than with Hawaiians regarding our trust lands.[17]

In 1993, on the centenary of the American military invasion and occupation of Hawai`i, and the subsequent overthrow of our constitutional native government, the United States Congress passed Public Law 103–150, signed by President Clinton, and known as the Apology Bill. The bill states, in part, "the indigenous Hawaiian people never directly relinquished their claims to their inherent sovereignty as a people or over their national lands to the United States, either through their monarchy or through a plebiscite or referendum." In paragraph 4, Section 1, the Congress expresses its commitment to acknowledge the ramifications of the overthrow of the Kingdom of Hawai`i, in order to provide a proper foundation for reconciliation between the United States and the Native Hawaiian people. And in paragraph 5, the Congress "urges the President of the United States to also acknowledge the ramifications of the overthrow of the Kingdom of Hawai`i and to support the reconciliation efforts between the United States and the Native Hawaiian people."[18]

As an act of our collective right to self-determination and to self-governance, many Hawaiian sovereignty groups have proposed the following process for reconciliation as enunciated in the Apology Bill:

1. Final resolution of our historic claims relating to the overthrow, and to continued state and federal misuse of our resources, totaling nearly 2 million acres of land and close to $1 billion in trust assets. This resolution could come through negotiation, litigation, or some form of general settlement.

2. Express termination of the present U.S. policy of nonrecognition of Native Hawaiian self-determination. This would entail a repudiation of the continuing policy of wardship whereby the State of Hawai`i controls many of our entitlements against our interests and without our consent.

3. Recognition of our claim as a native nation to be included in the federal policy on recognized native nations, including American Indian nations. Once our claims are settled, lands and waters would be transferred to our control. Elected native leaders would then assume leadership. The manner and terms of their election and service would be formulated by our own native people.

As someone who has been working and organizing for twenty years for inclusion of our people in the federal policy on recognized native nations, I can say that at no time since the United States overthrew our government in 1893 has our nation been in greater danger. Every ameliorative federal program from educational grants, to diabetes treatment programs, to early childhood training systems is now in doubt because of federal funding deferments, indifference on the part of the U.S. Congress, and a persistent attitude that reparations payments would create a precedent that other U.S. citizens who have been victims of crimes against humanity could follow.

INDIGENOUS SOVEREIGNTY

For us, as native people, the return of native self-government is the only answer to total dispossession. In the context of the global move toward restitution and reparations, the guiding concern in every case, including that of my own Hawaiian nation, must be injury and human suffering. Apologies, no matter how authentically intended, carefully written, or legally encoded can never, ever substitute for restitution or reparations. The reasons are obvious. In the first place, apologies are no more than official statements. Costing nothing, they achieve what Christian theology calls cheap grace. It is the repairing of damage, of harm, which must be attempted even if, as in the case of native peoples, millions of native human beings will never be restored to life. Nevertheless, return of some lands, entitlements, and other negotiated rights go a long way toward helping natives to survive as peoples and nations, not only as individuals.

Here, the primary consideration guiding the reparations/reconciliation discussion must be the human relationship between victim and victimizer. Now that the South African case has given the world an example, a new kind of moral sensitivity is slowly appearing in discussions around the globe. The so-called North-South dialogue is part of this. Without doubt, Northern nations have benefited enormously from exploitation of the Southern nations. Therefore, one argument maintains, huge debts owed by nations of the South should be cancelled as partial reparations for centuries of plunder by nations of the North.

Questions about who is individually responsible and who is individually injured are secondary. In the case of American slavery, for example, it is true that no Black American alive today was a slave, and no white American alive today was a slaveholder. But these obvious statements miss the mark. The legacies of slavery continue through racial discrimination, ghettoization, exploitation, and a host of other evils such as high infant mortality and early death.[19] The legacies of white supremacy, meanwhile, continue through white socioeconomic and political dominance, widespread institutional racism, and increasing white violence such as the burning of Black churches and vicious hate crimes, including murder, against Black individuals. And these American legacies are separate from the monumental crippling and underdevelopment of Africa as a continent.

The question, then, is one of national injury and national responsibility. If we, as a community, and the United States as a country, are concerned about suffering, about citizenship, about the larger context of international relations, including human rights, then justice must be rendered before reconciliation can be considered. For it is justice, rather than specific amounts of money or carefully crafted apologies, that constitutes the first and primary obligation of nations and peoples.

HAUNANI-KAY TRASK *is considered a leader in the Hawaiian Nationalist Movement. She has been compared to Angela Davis for both her scholarship and activism in advocating for the rights of indigenous Hawaiians. She currently teaches and does research at the Center for Hawaiian Studies at the University of Hawai`i. She has published four books including the best-seller* From a Native Daughter: Colonialism and Sovereignty in Hawai`i.

RAYMOND A. WINBUSH

And the Earth Moved: Stealing
Black Land in the United States

If you steal an egg today, you will steal an ox tomorrow.
—HAITIAN PROVERB

TANGIBLE HORRORS

One of the historical challenges of the reparations movement has been to provide tangible evidence of crimes committed against Africans in America and appropriate measures of compensation for them. Even when there *is* documented evidence of such atrocities it has been difficult to receive reparations for the victims. The Tulsa Race Riot on June 1, 1921, and the Rosewood, Florida, massacre in January of 1923 are but two infamous examples of crimes toward Africans in America. These incidents alone included the murder of over 300 Africans in America, the destruction of two neighborhoods (W. E. B. Du Bois referred to the neighborhood of Greenwood, the site of the Okalahoma massacre, as "the finest example of Negro self-sufficiency in the United States") and their economic destabilization, from which neither fully recovered. In the Rosewood case, $150,000 was granted to nine survivors of the massacre and 143 descendants of the victims, and at the time of this writing only $750,000 had been approved by the governor of Oklahoma for a memorial to the 1921 slaughter. In both cases, these compensatory measures came after legislative struggles, heated discussions, and outright resistance to the demand of the victims for reparations.

These are but two of the thousands of examples of property being destroyed and lives taken from Africans in America and sanctioned by both individuals and governments. The resistance in *repairing* the damage done by the violence and fraud toward Africans in America is an intriguing but rarely discussed question. Why is it that Japanese Americans received an apology and compensatory measures for Franklin D. Roosevelt's Executive Order 9066 but Black Americans in all but a few instances have been unsuccessful in their efforts for remedies to the crimes inflicted upon them? Why do Jews continue to litigate successfully for and receive billions of dollars from nations and corporations nearly sixty years after the Holocaust, in what Norman Finkelstein calls the Holocaust Industry, yet Africans in America are subjected to paternalistic rejections of their movement for reparations for 350 years of enslavement and domestic apartheid? I believe it is because the history of Black/white relations in this country is so long and sordid that reparations for damages done to Africans in America would call for an enormous upheaval of the social fabric of the United States unmatched even by *Brown v. Board*. The examination of the treatment of Africans in America, unlike that of the Japanese or Jewish cases, forces the United States to question the very essence of values held by white Americans toward its Black citizens. The United States government and the vast majority of its white citizens resist examination, and both groups are nearly pathological in their denial of the continuing impact of enslavement on this country.

Africans in America were not culpable for what happened to them during enslavement, but its documentation and prosecution angers Americans who wish to maintain a mawkish view of the relationship between Black and white folks in America. It is no coincidence that two of the great American films—*Birth of a Nation* and *Gone With the Wind*—depict Africans in America as paradoxically dangerous and compliant. It is also no coincidence that two of the most powerful people in Hollywood, Steven Spielberg and Oprah Winfrey, both failed at the box office during the late 1990s to start a national dialogue on slavery and its aftermath in the United States with their respective films *Amistad* and *Beloved*. Americans reject the idea of confronting their racial past and will actively oppose even fictional attempts at retelling the relationships that existed during enslavement between Blacks and whites. Alice Randall's 2001 court battle over *The Wind Done Gone*, a parody of Margaret Mitchell's novel *Gone With the Wind*, illustrates how difficult it is to discuss lily-white icons without impunity. Even though Randall argued forcefully in court about it being a parody, the Mitchell Estate launched a full-frontal legal attack to block its publication. Though Randall won her case on appeal, the case, which drew national attention, indicated the antipathy toward anyone who deromanticizes racist assumptions about America's treatment of enslaved

Africans. Their obsession with preserving the image of the beloved Tara plantation in the novel at all costs is a powerful metaphor for how the United States was shaped by white obsession in stealing land from Black and indigenous people.

LAND THEFT AGAINST AFRICANS IN AMERICA

The primacy of land ownership is a core value in the European cosmology, particularly as it relates to the history of the United States. "Twenty-four dollars and some beads," "land rush," "Manifest Destiny," "the Louisiana Purchase," "Seward's Folly," and "squatter's rights" are but a few of the phrases deeply embedded in the American psyche connected with the idea of land ownership. The nebulous notion of the American Dream is rooted in the idea of entrepreneurship and home/land ownership. Historically, the reparations movement has faced two obstacles in seeking tangible information on crimes against African Americans: fraudulent production and destruction of public records that would provide historical proof of murders, theft, kidnapping, and rape. As we will see, fraud occurred primarily when there was theft associated with the property of Africans in America, sometimes "legally" and often violently. Perpetrators of this fraud often colluded with local, state, and even the federal government to defraud Africans in America of property. A similar obstacle has been the destruction of documents that proved illegal activities in the confiscation of Black property. Wholesale burning of courthouses, Black churches, and homes were common ways of destroying evidence of Black land ownership illegally obtained by white terrorists.

Whitecapping

The term "whitecapping" became popular between the years 1900 and 1929 and denoted the habit of night riders confiscating land from vulnerable Blacks during the era of Jim Crow. James R. Grossman, in his book *A Chance to Make Good: African Americans, 1900–1929*, describes this form of intimidation:

> During the last two decades of the nineteenth century and the first decade of the twentieth, the system stabilized but people moved. The South's black population churned as rural people moved from plantation to plantation, county to county, state to state in quest of the holy grail of land ownership. Some were lured by labor agents. Others, especially in hilly areas where whites already outnumbered blacks, were

pushed out in a process known as whitecapping, a term that referred to the practice of night riders pushing African Americans off their land through threat of violence. At least 239 instances of whitecapping were recorded during the two decades beginning in the late 1880s, with Mississippi the most common site. The term seems to have originated in Indiana, where night riders invading a small community or threatening an individual African-American resident would wear white caps as part of their disguise.

Despite its origins in the Midwest, whitecapping was mainly a Southern rural phenomenon. Successful Black farmers were the most likely targets of this kind of eviction because of the common assumption among whites that the success of some Blacks might unleash unrealistic (and dangerous) aspirations among the local Black population. A farmer in Alpharetta, Georgia, recognized that he "better not accumulate much, no matter how hard and honest you work for it, [since] you can't enjoy it." In some cases, whole communities were forced from their homes.

Whitecapping, as an institution, operated efficiently in disenfranchising Blacks from their land. It is a commonly held myth that the majority of Black men were lynched during the Jim Crow era because of spurious accusations of some impropriety toward whites, particularly white women. Tulsa and Rosewood reinforce this idea, but land theft was a frequent motivation in lynching Black families. It was a quick and easy way to illegally deprive Blacks of their land. It was not unusual for the grieving widow of the victim to abandon the land within a few days after the murder of her husband or other male relative. The abandoned property would then be placed in auction because it had been deserted and new deeds would be drawn and agreements struck with the new owners that nullified existing title papers of the murdered owner.

It is telling how the United States sexualized violence toward Black men rather than admit its complicity in depriving them of property. The castrations experienced by male victims who were, as James Allen's book title suggests, "without sanctuary," included not only their genitals, but their land, houses, places of worship, and farming equipment. Lynching because of imagined crimes against white women was far more acceptable psychologically as a rationale for murder than admitting guilt over land theft.

Race Relations Institute Land Theft Study

In April 1999, the Associated Press (AP) asked the Race Relations Institute of Fisk University (RRI), with which I was affiliated, along with dozens of other organizations and individuals, for example, the Penn Center in South Carolina and the Federation of Southern Cooperatives in Georgia, to provide leads in documenting land theft involving African Americans throughout the nation. Many of these institutions, including the Race Relations Institute, had both formally and informally collected information on Black land takings for many years and were eager to make this information known on a national basis. The magnitude of such research is daunting, since it requires hundreds of hours of document discovery, dozens of interviews, and collaboration among institutions covering the entire United States. When first approached by reporters from the AP, I was skeptical that we would get more than twenty or thirty responses from ads that were to be placed in Black newspapers around the country. In my reasoning I thought that half of those would not be qualified as true land takings, but since the burden of the analysis and discovery would fall squarely on the shoulders of a large news organization, we agreed to help out in any way we could. The RRI agreed to provide leads primarily because we felt that *any* case involving tangible evidence for reparations would be good for a potential lawsuit against either an individual, corporation, or government. Moreover, we saw this effort as a way of ultimately securing tangible evidence that would lead to lawsuits involving reparations. We drew upon the strategy of famous civil rights cases such as *Brown v. Board* that provided concrete evidence of the impact of enslavement and segregation on Africans in America *and* white America.

This latter point is often overlooked since an automatic connection is made between slavery and poverty among Africans in America, but rarely is it associated with wealth and privilege in white America. The economic impact of slavery had obvious extraordinary impact on Africans in America, but only recently have we seen serious research connecting slavery to what Peggy McIntosh refers to as white privilege. Understanding Black land takings exposes how these associations with race, wealth, and poverty were begun, nurtured, and maintained in a society that wants to remember Maines, Alamos, Tara, and the Confederate Battle Flag, but is pathological in forgetting its slaveholding past.

Since the reparations movement has lacked a cache of cases that would connect the historical dots between crime and compensatory remedies, the RRI saw this research as an opportunity to provide leads and gather information for violations against Black Americans in one of the most fundamental of Constitutional guarantees, the right to own property. It would also shed light on whether or not the federal government provided Black people with the equal protection stipulated under the Fourteenth Amendment to the Constitution.

Over two hundred ads were placed in Black newspapers around the nation as well as major dailies in population centers that had large Black populations.

The ads provided a telephone number, e-mail and mailing address, and fax numbers for any African Americans who were victims of land takings or knew other relatives and persons who were. The responses to the ad were overwhelming. Nearly 150 persons called the RRI, and these cases were forwarded to the AP, who did heroic work in locating, interviewing, fact checking, and investigating the stories of the respondents. The other unexpected occurrence was how readers of the ad passed this information on to persons whom they knew had been victims of such crimes. What is presented below, with the names changed, are excerpts from the letters and files sent to the RRI. In their own words, these people provide eyewitness testimony to the crimes committed against African Americans during the past one hundred years. At the time of this writing, the entire story was released in the AP's *Torn From the Land* series, and we are sure that after these cases are published, hundreds more will emerge as African Americans move toward receiving compensatory remedies for the crimes committed against them.

From Michigan:

My mother, _____, who is 85 years old, recalls giving large sums of money to her father, _____. Over the years, she too became disenchanted with the slowness of the process, the never ending requests for additional money, and the failure to generate documents that validate ownership of Williams's 100 acres to his son, Robert. My mother's sister Mary, who is in her 70s, recalls the ongoing years of court appearances, attorney fees, legal obstacles, and recurrent difficulties encountered by her father. According to my aunt, threats were made on her father's life, he was shot, and his family was subjected to many indignities (i.e., rape of his daughter and vicious beating of his son). My mother and her sister recall that their father was handcuffed and take [*sic*] to jail when he objected to efforts of the county to cut a road through his land. In addition, since he owned land, and his wife's family owned land, he and his family were not sharecroppers. It is my understanding that some of the people in the community respected my grandfather; however, I was told that he was deeply resented by others.

From Arkansas:
Dear Mr. Winbush:

I spoke to you Monday, March 20, _____ about some property in _____ AR that was taken from our family. I am sending a lot of information I have been trying to get our property back since 1990, but to no avail. Millie + John Robertson owned this heir property. Mr. Christianson illegally drew up a will for John Whitehall after Millie Robertson passed. John Whitehall is now deceased but before he died I, Glen Robertson, had a conversation with him and he said Millie Robertson never willed the property to him. He said Mr. Christianson and his attorney drew up all the illegal papers.

From Michigan:

We lost our land about 50 years ago. Someone forged (signed) all our names and sold the property. The person that forged the names is now dead. The property belonged to my Granddad who was deceased at that time.

From Ohio:

Dear Dr. Winbush:

I thank you for passing my letter on to Delores Barclay of A.P.I. She got in touch with me concerning how my father and uncle were illegally ran from our home in _____, Mississippi. She got in touch with me in the week of _____. She talked with my mother on _____. Enclosed is a letter for more information she needed from us. Thanks again for allowing our families story to be told. I was overly excited when Ms. Barclay called me. I am going to my father's grave today to vent out my final thoughts of a story that's been hid for over fifty years. Again, thanks.

From Ohio:

Dear Dr. Winbush,

I am part of two families that were forced to leave our homes and everything we had. My father, William Bolden, and his twin brother, Graham Bolden, were ran out of Midnight Mississippi in 1949, by white people that came by night. My mother who is 81 years old told me that it all started about the hauling cotton choppers and cotton pickers to the cotton fields. My father had two trucks, and my uncle had one. A white man from Midnight Mississippi wanted my father and uncle to haul the cotton pickers to his farm, and my father and uncle were in agreement with other cotton farms. These white people from Midnight Mississippi threatened my father and uncle. My mother said my father and uncle got real scared and left by night.

It was in January 1950 when my father and uncles sent for all of us. I remember coming to Cincinnati Ohio when I was nine years of age. It was a great shock to me. I remember waking late at night trying to jump out of [the] third floor window trying to escape back to Mississippi. I don't recall coming here affecting anyone the way it did me.

From California:

Dear Mr. Winbush:

First, I will say that I do not have documentation but I just want to tell my story as I told it to President Clinton when he first ran for Pres. And he answered me back and said that he was sorry.

As our parents told us, my grandfather owned land, two horses and a wagon back in Augusta Arkansas, when one morning there was a note on his gate telling him to leave or be killed. The note had $1.00 attached. He and my grandmother and eight (8) children left. Left land, wagon, horses and all.

In the late eightys, my now deceased brother who was a minister preached in our hometown of Palestine Ark and went to the courthouse in Little Rock and checked the records. There he found my grandfather's land in his name Hobart Gibson, sold for $1.00. The land at that time had been sold to the State of Arkansas where the light company sits. The state of Arkansas Light Co. is sitting on our land, and he said that the person that sold them the land was listed as owner of several, several parcels of land all around there.

I wrote Clinton and asked him if he was elected Pres., what would he do about that. A very nice letter said he was sorry. What an injustice!

There is nothing we can do, but thank you for reading this. At least I'm telling someone about it.

Mrs. Carolyn Gibson

PS: I wish those that have documentation luck.

This is a small sample of the thousands of Black victims of land takings and the violence associated with the crimes. Others who have kept these things quiet for fear of retaliation, ridicule, and shame will emerge. They represent tangible evidence that should be considered in the case of reparations. During the next ten years, the justice deferred in all of these crimes will form the basis of the expected lawsuits emerging from the push toward reparations for Africans in America. The dozens of lawsuits that will inevitably result from the reparations movement will reveal to the United States and the world the sordid history of property theft from Africans in America. The possession of land is so enshrined in the American psyche that it is part of the nebulous notion of the "American Dream." Depriving one of their land through illegal means is part of the bloody legacy of the United States, which began with the subjugation of indigenous people by European invaders of the Americas and continued unabated through the end of the twentieth century.

The reparations movement will be the greatest public education effort ever mounted on racism in this nation's history. The movement will begin the long-overdue unraveling of the nexus between race, poverty, and wealth in the United States. It will teach its students difficult lessons, such as the violence meted out to indigenous people by whites, and will strengthen the relationship between Africans in America and indigenous people relative to land and fraudulent transactions by whites toward their ancestors. This schooling will be painfully difficult but necessary if racial justice becomes a reality rather than a convenient sound bite in the mouths of public policy makers.

RAYMOND A. WINBUSH *is Director of the Institute for Urban Research at Morgan State University. He is a scholar-activist and is a member of the editorial board of the* Journal of Black Studies.

PART II

Reparations and the Law

JON M. VAN DYKE

Reparations for the Descendants of American Slaves Under International Law

> *The laborer is always in the sun, the landowner is always in the shade.*
>
> —YORUBA PROVERB

The world has long recognized the enslavement of one person by another as one of the grossest violations of the fundamental human rights of individuals. The international community has also reached a solid consensus that all violations of human rights must be investigated and documented, that the perpetrators of human rights abuses must be punished, and that the victims of human rights violations have a right to compensation. The obligation to investigate, punish, and provide compensation continues forward in time, and thus a new enlightened government that replaces a bad authoritarian government has a duty to punish the human rights abusers from the previous government and to compensate its victims.

All of these principles have been established and accepted, but challenging questions still remain about applying them to specific situations. How much compensation is appropriate to a victim who has suffered a grave human rights violation? Should the group punished include low-ranking individuals who carried out the orders, or only those at the top who made the decisions? Can amnesties or pardons be used in some situations to promote national healing? Is a genuine apology sufficient in some situations to bring closure to a violation of fundamental human rights? Will a careful investigation and documentation be adequate in some situations? If all sides are committed to reaching a true reconciliation, what will be needed to achieve that goal?

In the case of reparations for descendants of American slaves, additional questions need examination: Because those who were actual slaves are now dead, do their heirs have a continuing right to compensation? Is that right to compensation based on the wrong done to the ancestors, or to the continuing effects of that wrong that burden the descendants today, or both? How far back in time does the duty to address and correct human rights abuses extend? Who is responsible for paying the reparations?

Communities throughout the globe have been addressing and providing answers to these difficult questions, and we now have many actual case studies to analyze and evaluate. Although solutions will vary depending on the nature and extent of the abuses, it appears that the duty to address violations of fundamental human rights continues as long as the consequences of those violations continue to scar a community. As long as winners and losers from grave human rights violations can continue to be identified, it will be necessary to promote a reconciliation to heal the wounds and eliminate their consequences.

Legalized slavery occurred in the United States from 1619 to 1865, and was followed by another century of legalized racial discrimination. This overwhelming oppression of one group of humans by another allowed some to create wealth, while the oppressed group was first forbidden and then harshly constrained from accumulating any wealth. These uncontroverted facts meet the legal definition of unjust enrichment, and the claims of the victims must be viewed as a property right, recognizable in courts of law, and protected from infringement or elimination by the Fifth Amendment's Taking Clause.[1]

The word "reconciliation" plays a central role in any effort to bring closure to a grave violation of fundamental norms of human decency. This word is sometimes thought of as weak and meaningless, as a feel-good word requiring nothing more than admissions of sorrow and some hugs, handshakes, and smiles. But a true reconciliation requires much more, and this word has become powerful and dynamic. Reconciliation requires action to make amends, a settlement significant enough to heal the wound and to make right the wrong that has occurred, and it usually requires the transfer of something of value. In a traditional Pacific Island community, the family of the wrongdoer will offer animals and other products of value to the family of the victim. In our modern world, reconciliation requires a full and fair acknowledgment of the wrong, followed by a real settlement that will almost always include the conveyance or return of land, money, or other valuable property, as well as punishment and/or disgrace of those who committed the wrongs.

The purpose of this chapter is to put the claims for reparations brought by descendants of American slaves into an international context, to demonstrate that international law supports and even mandates the development of a reparations package for descendants of American slaves, and then to address some of the challenging questions that need to be answered to provide a fair reconciliation and thus to bring closure to this terrible act of human oppression that has dominated our country for most of its existence. The other chapters in this volume address many of the background issues, and writers elsewhere have also written eloquently about this topic.[2] Because others have set the stage for this analysis in substantial detail, the history is reviewed here only briefly, to illustrate why the claim by descendants of American slaves for reparations is credible under current international law.

SLAVERY IN THE UNITED STATES

Randall Robinson has made a compelling argument in his recent book *The Debt: What America Owes to Blacks*[3] that the enslavement of Blacks in America from 1619 to 1865 was "far and away the most heinous human rights crime visited upon any group of people in the world over the last five hundred years,"[4] and he has explained in detail why we must look at the claims of the descendants of slaves as a form of property. Their claim is a property right because the descendants of former slave owners (and other whites) have an enormous economic advantage over those who are descendants of former slaves, an advantage that can be linked directly to the wealth accumulated from the labors of the slaves and to the oppression that continued for a century after the end of slavery during the Jim Crow period. When it finally freed the slaves, the United States provided no meaningful compensation to the former slaves for the value of their labor, nor did it provide them with the economic assets needed to compete with the white community in the marketplace.[5] Robinson convincingly argues that a transfer of wealth from those who benefited to those who still suffer because their ancestors were oppressed is the only way to achieve a meaningful reconciliation: "Until America's white ruling class accepts the fact that the book never closes on massive unredressed social wrongs, America can have no future as one people."[6]

Since 1989, Representative John Conyers of Michigan has repeatedly introduced a bill to "acknowledge the fundamental injustice, cruelty, brutality, and inhumanity of slavery in the United States and the 13 American colonies between 1619 and 1865," and to establish a seven-member commission to determine whether the United States should make formal apology for slavery, examine whether African Americans still suffer from the lingering effects of slavery, decide whether compensation or other appropriate remedies should be provided to the descendants of slaves, and, if so, identify who should be eligible for such compensation.[7] He has been joined in this effort by Representative Tony Hall of Ohio, who has called for a formal apology, a commission to examine slavery's legacy, the development of educational materials to educate our youth about slavery, and the establishment of a scholarship fund.[8]

These bills have yet to make it out of committee, but they reflect a proper approach to the festering problem of racism in the United States. With the recent focus on affirmative action, those opposed to justice for African Americans have been able to dismiss these claims by arguing that African Americans are claiming preferential rights. In fact, African Americans are simply claiming a return of the property that was taken from their ancestors and that has been denied to them by virtue of the continuing impact of the mistreatment of their predecessors. If their claim can be recognized to be the property right that it surely is, it may be more understandable to the conservatives who currently dominate the federal judiciary.

SLAVERY UNDER INTERNATIONAL LAW

No principle of international law is more firmly established than the complete prohibition on slavery in all its forms. The prohibition is found in the 1948 Universal Declaration of Human Rights,[9] in the 1966 International Covenant on Civil and Political Rights,[10] and in the three regional human rights conventions in Europe,[11] the Western Hemisphere,[12] and Africa.[13] A separate treaty has also been drafted and widely ratified to define and prohibit explicitly all manifestations of slavery.[14] The definitive *Restatement of the Foreign Relations Law of the United States* lists slavery as one of the few universally condemned human rights violations, and says that persons engaging in slavery can be punished and sued by their victims for compensation wherever they are found.[15] The 1998 Rome Treaty Establishing the International Criminal Court includes enslavement as within the crime against humanity, which is within the jurisdiction of this new Court.[16]

Even before the U.S. Civil War and the addition of the Thirteenth Amendment to the United States Constitution, U.S. courts recognized that participation in slave trading constituted a fundamental violation of international law norms.[17] The enslavement of populations in World War II was recognized by all as a violation of fundamental principles of international law, and the Nazi perpetrators of such practices were punished at Nuremberg.[18] Recent United States cases have left no doubt that slavery is a universally condemned act, and that claims under international law can be brought by victims of slavery against the perpetrators of such acts.[19]

THE CONTINUING EFFECTS OF SLAVERY
IN THE UNITED STATES

Randall Robinson has called slavery a "human rights crime without parallel in the modern world," because "it produces its victims, ad infinitum, long after the active stage of the crime has ended."[20] The continuing effects of 246 years of slavery followed by 100 years of discriminatory Jim Crow laws can be seen everywhere in the United States, from the "yawning economic gap between Blacks and whites"[21] to the raw bitterness still separating the police and the Black community in many U.S. cities. Professor Chris K. Iijima has written that "the searing holocaust of American slavery has caused harm that is felt to this day. Not only is it felt in the social, economic, and political legacy and condition of the various peoples of this nation, but in the way we define our national character. It is ingrained in the very fabric of our institutions. It operates often unconsciously through unquestioned cultural assumptions about what is natural."[22]

White slave owners created wealth using Black slaves, who were, of course, forbidden from creating any wealth themselves. The economic differential that resulted from slavery continues today. Slaves, for the most part, were forbidden from establishing stable families, and the legacy from that deprivation is still seen in African American communities. Sociologist Glenn C. Loury has observed: "Who can say what the out-of-wedlock birth rate for blacks would be, absent chattel slavery? How does one calculate the cost of inner-city ghettos, of poor education, of the stigma of perceived racial inferiority? The severity of slavery's injury is far more profound than any cash transfer will be able to reverse."[23] In an attempt to put this injury in economic terms, Professor Adrienne Davis has explained, "[t]he slave-holding states did not confer legal status on black families; through inheritance, the family is one of the primary institutions of wealth transfer, but black slaves were excluded from inter-generational wealth transfer, one of the centerpieces of Anglo-American culture."[24]

Slaves were generally forbidden from learning even rudimentary educational skills, and the educational divide between whites and Blacks today is certainly linked to that deprivation, as well as to the long period of officially sanctioned segregated schools in the period after the Civil War. The Jim Crow laws penalized African Americans in many ways directly related to their economic status.

The systematic economic discrimination against the freed slaves began immediately after the Civil War was over. Arguments that former slaves be given at least forty acres and a mule were rejected even by opponents of slavery such as Horace Greeley, editor of the *New York Tribune,* who characterized the idea that the lands of former slave owners should be redistributed to freed slaves as "either knavery or madness."[25] Southern states then enacted the notorious Black Codes to regulate the Black labor force. Mississippi, for instance, "required that every January all Blacks be able to present written evidence of their employment for the next year, and also empowered all white persons to arrest any Blacks who left the service of their employers."[26]

This chapter is not the place to summarize the myriad abuses and discriminations enforced by local, state, and national governmental bodies and private organizations in the United States during the one hundred years of the Jim Crow period, but it is undisputed that Blacks were denied many of the most significant economic rights and opportunities enjoyed by the white community. Black baseball players, no matter how talented, could not play in the major leagues until Jackie Robinson finally broke the color barrier in 1947, and opportunities remained limited for a number of years thereafter. Randall Robinson has reminded us that the federal government included racially restrictive covenants in its mortgage loans to whites until 1950.[27] Banks participated in this systematic deprivation through the process of redlining neighborhoods. Racial profiling by law enforcement bodies continues at the present time.

Because this pervasive societal discrimination led inevitably to severe economic deprivation among African Americans, it is appropriate and necessary to characterize their claim for reparations based on the slave experience as a property claim, protected by the Fifth Amendment of the United States Constitution, and cognizable in courts of law.

THE RIGHT TO BRING A CLAIM FOR COMPENSATION FOR HUMAN RIGHTS INJURIES IS ITSELF A FUNDAMENTAL HUMAN RIGHT UNDER INTERNATIONAL LAW

This right includes the ability to recover monetary compensation, to insist on a proper and full investigation of the incident, and to see the perpetrators of the abuses brought to criminal justice. The obligation to meet these responsibilities extends to a successor government to the same degree that it applied to the former government that undertook or tolerated the human rights abuses.

The Universal Declaration of Human Rights says that "Everyone has the right to *an effective remedy* by the competent national tribunals for acts violating the fundamental rights granted him by the constitution or by law.[28] Similarly, the International Covenant on Civil and Political Rights,[29] which has been ratified by more than 140 countries, including the United States, says that "Each State Party to the present Covenant undertakes: (a) To ensure that any person whose rights or freedoms as herein recognized are violated shall have *an effective remedy,* notwithstanding that the violation has been committed by persons acting in an official capacity. . . ."

Regional human rights treaties also emphasize the right to redress for human rights violations. The European Convention on Human Rights says that "In the determination of his civil rights . . . , everyone is entitled to a fair and public hearing within a reasonable time by an independent and impartial tribunal established by law."[30] Similarly, the American Convention on Human Rights says that "Everyone has the right to simple and prompt recourse, or any other effective recourse, to a competent court or tribunal for protection against acts that violate his fundamental rights recognized by the constitution or laws of the state concerned or by this Convention, even though such violation may have been committed by persons acting in the course of their official duties."[31]

Decisions in the Inter-American system confirm that the right to an effective remedy is a continuing one that cannot be waived. The seminal case of the Inter-American Court of Human Rights is *The Velasquez Rodriguez Case,*[32] which holds that the American Convention on Human Rights imposes on each state party a "legal duty to . . . ensure the victim adequate compensation." The court explained that each country has the duty to protect the human rights listed in the Convention and articulated this responsibility as follows: "This obligation implies the duty of the States Parties to organize the governmental apparatus and, in general, all the structures through which public power is exercised, so that they are capable of juridically ensuring the free and full enjoyment of human rights. As a consequence of this obligation, the States must *prevent, investigate and punish* any violation of the rights recognized by the Convention. . . ."[33]

This decision is important, because it lists both the right to monetary compensation and the right to see the perpetrators of abuses punished as essential to the right to have one's human rights protected. Similarly, the European Court of Human Rights has ruled recently that Turkey violated the rights of citizens who were prevented from bringing a claim for the deliberate destruction of their houses and possession, noting that "the notion of an 'effective remedy' entails, in addition to the payment of compensation where appropriate, a thorough and effective investigation capable of leading to the identification and punishment of those responsible and including effective access for the complainant to the investigative procedure."[34]

The Human Rights Committee in Geneva, established by the International Covenant on Civil and Political Rights, has also gone on record opposing amnesties: "The Committee has noted that some States have granted amnesty in respect of acts of torture. *Amnesties are generally incompatible with the duty of States to investigate such acts;* to guarantee freedom from such acts within their jurisdiction; and to ensure that they do not occur in the future."[35] These uniform treaty provisions and opinions issued by international tribunals confirm that the right to an effective remedy is just as much a fundamental human right as the right to be free from murder, torture, slavery, and other grave human rights abuses.

Other international decisions that confirm this result include Report No. 36/96, Case No. 10.843, Inter-American Human Rights Commission, October 15, 1996, paragraphs 68, 105, 112 (ruling that Chile's 1978 Amnesty Decree Law violated Article 25 of the American Convention on Human Rights because the "[human rights] victims and their families were deprived of their right to effective recourse against the violations of their rights"); *Rodriguez v. Uruguay,* U.N. Doc. CCPR/C/51/D/322/1988, Annex (Human Rights Committee, 1994) (stating that "amnesties for gross violations of human rights . . . are incompatible with the obligations of the State party" under the International Covenant on Civil and Political Rights and that each country has a "responsibility to provide effective remedies to the victims of those abuses" to allow the victims to gain appropriate compensation for their injuries); *Chanfeau Orayce and Others v. Chile,* Cases 11.505 et al., Inter-American Commission on Human Rights 512, OEA/ser.L/V/II.98, doc. 7 rev. (1997) (stating that Chile's amnesty law violated Articles 1.1, 2, and 25 of the American Convention on Human Rights, that countries have a duty to "investigate the violations committed within its jurisdiction, identify those responsible and impose the pertinent sanctions on them, as well as ensure the adequate reparation of the consequences suffered by the victim").

U.S. decisions also support the conclusion that claims based on violations of international law cannot be waived or dismissed because of some other foreign policy goals, without violating the Takings Clause of the Fifth Amendment. The case of *Dames & Moore v. Regan*[36] involved the argument of a U.S. company that its claim for damages against Iran after the 1979 Iranian revolution had been unlawfully extinguished by the 1981 Algiers Accords that freed the U.S. hostages. In response to the argument made by Dames & Moore that its claim had been taken in violation of the Fifth Amendment to the U.S. Constitution, Justice William H. Rehnquist's opinion for the Court noted that claimants were not denied the right to pursue their claim, but rather were required to use an alternative forum, the Iran-U.S. Claims Tribunal in The Hague, Netherlands. The Court affirmed the requirement that Dames & Moore utilize this alternative forum, but also acknowledged that the company's claim for damages constituted a property right that could not be unilaterally eliminated without violating the Takings Clause, and indicated a willingness to take another look at the company's takings claim should the alternative tribunal not prove effective.[37]

Another seminal case is *Ware v. Hylton*,[38] where the U.S. Supreme Court ruled in 1796 that private British citizens were entitled to use the U.S. judicial system to collect pre–Revolutionary War debts despite language in the peace treaty that seemed to eliminate those debts. Justice Samuel Chase rejected the idea that a government can waive private claims without compensation to the claimants, even as part of a peace settlement, citing both the "immutable principles of justice" and the Fifth Amendment of the United States Constitution.[39]

Writing separately in the same case, Justice James Iredell said that "these rights [are] fully acquired by private persons during the war, more especially if derived from the laws of war . . . [and] against the enemy, and in that case the individual might be entitled to compensation."[40] He added that if Congress had given up the rights of private persons in a peace treaty "as the price of peace," the private individuals whose "rights were sacrificed" might "well have been entitled to compensation from the public" for their loss.[41]

State courts in the United States have also recognized the validity of such claims. In *Christian County Court v. Rankin*,[42] the court granted private compensation in an action against Confederate soldiers for burning the courthouse in violation of the laws of nations, saying that "[f]or every wrong the common law provides an adequate remedy . . . on international and common law principles."[43]

OPTIONS AND STRATEGIES THAT HAVE BEEN PURSUED ELSEWHERE

The strategies utilized in recent years to bring a sense of closure and reconciliation[44] can be categorized into the following four approaches, which are frequently utilized in combination: (1) an apology for the wrong, which can be general or specific; (2) an investigation and accounting; (3) compensation for the victims, either through a general class approach, or through individual determinations, or both; and (4) prosecution of the wrongdoers. These approaches are described below, with examples from recent history.

A. Apology.

A formal apology is a crucial element of any reconciliation process. Recent examples include President William Clinton's apology for the U.S. support of the military in Guatemala.[45] Secretary of State Madeleine Albright apologized for U.S. support for the 1953 coup that restored Shah Mohammed Reza Pahlevi to power in Iran and its backing of Iraq during the war with Iran in the 1980s.[46] Pope John Paul II issued a sweeping apology on March 12, 2000, for the errors of the Roman Catholic Church during the previous two thousand years, acknowledging intolerance and injustice toward Jews, women, indigenous peoples, immigrants, and the poor.[47] In 1993, the United States apologized for the participation by its military and diplomats in the illegal overthrow of the Kingdom of Hawai'i in 1893.[48] The United States has also apologized for the internment of Japanese Americans during World War II.[49] In 1999, Australia acknowledged the "most blemished chapter in our national history," and expressed its "deep and sincere regret that indigenous Australians suffered injustices under the practices of past generations, and for the hurt and trauma that many indigenous people continue to feel."[50] And, Benin and Ghana have apologized for their roles in facilitating the slave trade.[51]

B. Investigation and Accounting.

Documentation of the wrongdoing serves the important purposes of recognizing the suffering and acknowledging that wrongdoing occurred. Telling the victims' stories helps to restore a measure of humanity to those who have been deprived of life, liberty, or fundamental norms of human dignity. The two most significant accountings in recent years are those that took place in Chile and South Africa, but others have occurred as well.

- Chile's situation was unique in that General Augusto Pinochet allowed elections to take place in the late 1980s, but retained firm control over the military and kept a watchful eye on the new government. The new president, Patricio Aylwin, was effectively blocked from prosecuting Pinochet and his military associates, but he wanted nonetheless to acknowledge and honor the victims, and so appointed a Commission of Truth and Reconciliation, which prepared a comprehensive report documenting two thousand human rights abuses.[52]

- In South Africa, a Truth and Reconciliation Commission met for two and a half years to document as many of the human rights abuses as possible and issued a report blaming both sides for abuses. Persons who came forward with truthful accounts of their participation in violent acts linked to a political objective were pardoned as part of the national healing effort, but others have been prosecuted for their role in these atrocities.[53] As of January 2000, when the Commission completed the bulk of its work, some 7,000 individuals had applied for political amnesty, but only 849 requests had been granted.[54] More than half of those who were pardoned were members of the African National Congress, who had been working to overthrow the apartheid government, and a smaller number were members of the apartheid security forces or were in the Zulu-based Inkatha Freedom Party. Only one pardoned individual (Adriaan Vlok, who had been minister of law and order from 1986 to 1994 and confessed to ordering a bomb attack in 1987) was a member of the governing apartheid National Party.[55] Although the commission has recommended that compensation be provided to 20,000 victims who suffered particularly from the apartheid period, the government has been slow to decide how to handle this issue.[56]

- In February 1999, an independent United Nations–sponsored Historical Clarification Commission concluded an eighteen-month investigation and reported that the Guatemalan military—with United States money and training—committed acts of genocide against the indigenous Mayan community in Guatemala during the country's long civil war and were responsible for 42,000 human rights violations, including 29,000 deaths or disappearances.[57] The next month, President Clinton apologized for the United States' participation saying that "support for military forces and intelligence units which engaged in violence and widespread repression was wrong, and the United States must not repeat that mistake."[58]

- Brazil is finally addressing the abuses that occurred during the military dictatorship that lasted there from 1964 to 1985. A new investigation began in the spring of 2000 to determine, for instance, Brazil's role in Operation Condor, a program in which South American dictators helped each other hunt down political enemies hiding abroad; to determine what really happened on April 30, 1981, when two military personnel were killed by a bomb in the parking lot outside an arena containing 20,000 supporters of left-wing causes, and learn whether they were *agents provocateurs* trying to disrupt the event; and to determine whether former president Joao Goulart was murdered in his sleep while in exile in Argentina in 1976.[59]

- The city of Tulsa, Oklahoma, established the eleven-member Tulsa 1921 Race Riot Commission to investigate and document the rampage of May

31–June 1, 1921, during which whites—acting under the authority or acquiescence of the police and sheriff's deputies—destroyed more than one thousand homes, thirty-five stores, eight doctors' offices, and five motels, and killed some three hundred African Americans in the thriving Black Greenwood section of Tulsa, also called the Negro Wall Street. The Commission spent two years interviewing hundreds of witnesses and victims, and in February 2000 recommended that the State of Oklahoma pay reparations to the eighty survivors—now in their eighties and nineties—whose relatives were killed or whose property was destroyed and also fund a scholarship program, establish an enterprise zone for Black businesses, and erect a monument to the victims, a package that might cost as much as $33 million.[60]

• In 2001, Peru established a Truth Commission to investigate the killings and forced disappearances of some 29,000 Peruvians during the previous twenty years.[61] Comprised of two Catholic priests, an evangelical pastor, and six representatives of civil society, the commission's responsibility was not to impose sanctions, but to "act with the purpose of purifying the national conscience, and to bring to light what was hidden."[62]

C. Compensation for the Victims.

International law has always been clear that reparations are essential whenever damages result from violations of international law. This principle is securely rooted in the 1928 decision of the Permanent Court of International Justice in the *Chorzow Factory Case*,[63] and it was reaffirmed in 1999 by the International Tribunal for the Law of the Sea in the *M/V Saiga Case*.[64] Reparations are just as important and just as mandatory in cases of human rights abuses as in any other cases. The requirement of appropriate compensation is being recognized increasingly in a wide variety of contexts.

• In 1992, after more than two thousand human rights abuses were documented by a Chilean commission, the Chilean Legislature enacted a law providing a wide range of economic benefits for the victims and their families.[65]

• The Japanese Americans interred in World War II received $20,000 each,[66] and those persons of Japanese ancestry brought to camps in the United States from Latin America have received $5,000 each.[67]

- Canada has provided a reparations package for the First Nation children who were taken from their families and transferred to boarding schools, where they were denied access to their culture and frequently physically mistreated.[68]

- New Zealand established a process to address the wrongs committed by the British against the Maori people in the late 1800s, systematically taking their lands and resources, and has returned lands and transferred factories, fishing vessels, and fishing rights to the Maori groups to compensate them for their losses.[69]

- In Puerto Rico, Governor Pedro J. Rossello publicly apologized and offered restitution of up to $6,000 each to thousands of *independentistas* and others who were spied on by a police intelligence unit starting in the late 1940s.[70]

- In 1994, Florida governor Lawton Chiles signed into law a bill providing for the payment of $2.1 million in reparations to the descendants of the Black victims of the 1923 Rosewood massacre, in which white lynch mobs killed six Blacks and drove others from their homes to destroy a prosperous Black community.[71] Under this bill, Florida provided $150,000 to each of nine survivors, and other monetary awards to 143 descendants of victims.[72]

- Lawsuits have been filed in the United States by victims of human rights abuses for compensation. One of the prominent cases was brought by 9,500 victims of human rights abuses in the Philippines against former president Ferdinand E. Marcos after he fled in exile to Hawai'i in 1986, and continued against his estate after he died in 1989. A federal court jury ruled that Marcos was liable for the torture, murder, and disappearances that these victims suffered, and ordered his estate to pay $775 million in compensatory damages and $1.2 billion in exemplary damages.[73]

- The German government has funded various compensation programs to pay victims of the World War II Holocaust, and to make payments directly to the State of Israel as well.[74] More recently, lawsuits were filed in U.S. courts by the victims of slave and forced labor during World War II against the German banks and companies that profited from such abuses,[75] and on July 17, 2000, an agreement was reached to provide

$5 billion to the 250,000 members of this victimized class.[76] Payments of about $7,500 to each person forced to work as a slave laborer and about $2,500 to each person obliged to work in forced-labor situations began to be distributed in the summer of 2001.[77]

• Austria has similarly established a $400 million fund to provide compensation to slave and forced laborers sent to Austria by Hitler after the 1938 annexation.[78]

D. Prosecution of the Wrongdoers.

The trials at Nuremberg and in the Far East after World War II still stand as models for systematic and conscientious prosecutions of those who have violated the laws of war and fundamental human rights principles. But for almost half a century after those trials, no other international trials took place.

Then in the early 1990s, the United Nations Security Council established tribunals to prosecute those who violated fundamental norms during the fighting in the former Yugoslavia and Rwanda.[79] These tribunals were slow in getting started, but have recently been proceeding steadily through their caseload. As of April 2002, the Rwandan Tribunal had delivered seven verdicts, was holding twenty-one individuals who were then on trial, and was holding another twenty-nine who were awaiting trial in Arusha, Tanzania.[80] As of that same time, the Yugoslav Tribunal had completed sixteen trials, sometimes with multiple defendants, and had forty-four matters in various stages of disposition, again frequently with multiple defendants.[81] In June 2001, the Serbian government finally extradited Slobodan Milosovic, and flew him to the Netherlands to face trial. His trial began in early 2002 and was expected to last for two years. Other examples of prosecutions have emerged in the last few years:

• A unique war crimes tribunal was established in early 2002 to try those responsible for the horrific atrocities committed during the ten-year civil war in Sierra Leone, which was notorious for the systematic raping, enslavement, and hacking off of limbs of civilians, including children. This tribunal will be established in Freetown, Sierra Leone, and will include UN and local judges sitting side by side to consider the evidence[82] brought by the foreign prosecutor, an experienced U.S. Army lawyer named David Crane.[83]

• General Augusto Pinochet, the dictator of Chile from 1973 to 1989, was held under house arrest in England for sixteen months, fighting his extradition to Spain to be prosecuted for the torture and murder of Chileans, but he was finally returned to Chile in February 2000 after British officials concluded that he was medically unfit to stand trial. Although this protracted episode did not lead to an international trial of Pinochet, the British House of Lords reached a significant decision during the period of house arrest, ruling that Pinochet's status as a former head of state did not give him an immunity from prosecution and that prosecution for his egregious universal crimes would be appropriate in any country.[84] Since his return, Chile's courts have concluded that General Pinochet is not entitled to immunity, and are also pursuing cases against military officers who served in the Pinochet government.

• In November 1999, Judge Baltasar Garzon, the same Spanish magistrate who has been pursuing General Pinochet, charged ninety-eight former Argentine officers with genocide, terrorism, and torture in connection with the atrocities perpetrated by the military dictatorship that controlled Argentina from 1976 to 1983, when between 9,000 and 30,000 persons died or disappeared.[85] Previously, Judge Garzon ordered the arrest of Adolfo Scilingo, an Argentine officer who testified in the Spanish court that he had thrown dissidents from planes during the Argentine "dirty war."[86] Jorge Rafael Videla, the Argentine dictator during this period, was rearrested in June 1998 for his participation in the systematic kidnapping of children, even though he had previously been pardoned (in 1990) after his life sentence (in 1985) for his role in the death squads.[87] In February 2000, Argentina's newly inaugurated president, Fernando de la Rua, ordered a purge from the government payroll of some fifteen hundred military personnel and civilians connected with the dirty war from the 1976 to 1983 period.[88]

• In February 2002, the International Court of Justice ruled that Belgium could not prosecute the foreign minister of the Congo (Abdulaye Yerodia Ndombasi) for war crimes and crimes against humanity while he held that office, because of the importance of the principles underlying diplomatic immunity, but the Court also went out of its way to say that "immunity does not mean impunity."[89] The Court carefully explained that prosecution could take place after Mr. Yerodia ceased to hold his office, that the Congo could waive the immunity or bring a prosecution in their own courts, and that prosecution would be permissible in an international criminal court that had jurisdiction over the crimes.[90]

These many situations illustrate the complexity of these issues. International law has confirmed that every victim of a human rights abuse has the right to an effective remedy, and that such a remedy will generally include the right to a thorough investigation, compensation for the injuries suffered, and prosecution of the perpetrator. But the different factual situations listed above demonstrate that no one approach works for every historical event. Just as prosecutors exercise discretion to refrain from prosecuting in certain situations, and to accept plea agreements for reduced charges in many other situations, some historical episodes seem to justify a more merciful approach, with reduced penalties or simply a full description of what actually happened. In some situations, pardons may be justified after part of a sentence has been served to foster national healing.

But in each situation, as an irreducible minimum, a full investigation and disclosure of what occurred is essential to ensure that the culprits' deeds are known by all and to prevent them from ever exercising power again. And for a true reconciliation, the transfer of property from those who have benefited to those who have suffered seems essential to bring the matter to a just and acceptable resolution.

WHO SHOULD PROVIDE THE REPARATIONS FOR THE DESCENDANTS OF AMERICAN SLAVES?

How do these international precedents apply to the claims of African Americans for slavery reparations? Once the wrongs done to African Americans are recognized as constituting a claim that creates a property right cognizable under our legal system, what remedy is appropriate? The wrongs occurred long ago. Every person who owned and abused slaves is now dead. Many argue that current generations have no responsibility, because none of us ever held slaves. Many Americans have immigrated recently to the United States and many others are descendants of immigrants who arrived after the Civil War. Still others are descendants of soldiers who fought bravely in the Union Army in the effort that helped to free the slaves.[91]

It is true that it would be inappropriate to pursue individual guilt at this point, but the collective responsibility of our government cannot be denied, nor can it be denied that white people, speaking generally, continue to benefit from the burdens that were imposed for hundreds of years on Blacks. Even those whites whose ancestors immigrated to the United States after the Civil War benefit from the white skin privilege that still brings substantial benefits in our country. Whites have long tried to distance themselves from responsibility for racial wrongs, but it is time to face up to the situation, and to deal with it in economic terms.

But who or what institutions should provide the reparations? Because slavery was so pervasive in the Southern states of the United States, and because Jim Crow laws continued both in the South and in many parts of the North long after the Civil War, identifying the source of reparations payments today is complicated. Obvious targets are the federal government and the governments of the Southern states, because they provided the legal regime that permitted people to be enslaved and to be treated as property, and then collected tax revenues from the slave owners for their property.[92] Some private companies, such as Aetna Insurance, FleetBoston Financial, and CSX have already been identified as having profited from the institution of slavery or benefited from the labor of slaves, or being the successors to companies that did,[93] and they should certainly contribute to the reparations fund. Universities may also be liable. The first endowed chair in law at Harvard University was, for instance, funded by Isaac Royall from lands he obtained by selling slaves he owned in Antigua.[94]

And what about the heirs of those families that owned slaves and amassed wealth from the work of their slaves? Randall Robinson has rejected that approach, saying, "No one is suggesting personal culpability. We are talking about crimes by governments against people, crimes that should not be touched by statutes of limitations, because when governments commit such crimes, they 'have a certain immortality.'"[95]

Do we need a new approach, thinking outside the typical legal boxes, to fashion a remedy for such a systemic wrong done pervasively by a myriad of institutions and individuals over centuries? One approach might be to draw upon the strategy utilized in defective product litigation, when it is impossible to be sure which company made the specific IUD, for instance, that caused the damage to any one specific woman. In those situations, the market share of the company is used as the basis for the amount that should be put into a fund, which is then distributed to the victims. Another example can be drawn from the slave-labor claims against German companies, which led in 2001 to the establishment of a $5 billion fund for the victims.[96] All German companies (and non-German companies doing business in Germany) implicated in the slave- and forced-labor practices during the Nazi period contributed to this fund in relationship to the scope of their business activities, as did the German government. Utilizing this approach, the Southern states and the federal government, plus those companies that benefited or profited from slavery, should contribute to the reparation fund.

As mentioned earlier, Benin and Ghana have apologized for the role they played in facilitating the slave trade.[97] Is an apology enough, or should the African nations whose residents rounded others up and brought them to the waiting slave ships on the coast bear some of the cost as well? That is a significant question, and the answer must be that those who shared the blame, and profited from the slave trade, should share in providing appropriate reparations to those who suffered. It may be hard to sort out responsibility with regard to the role of Africans, however, particularly since much of the slave trade was conducted at a time when much of Africa was controlled by European countries.

How large should the reparations fund be? Others have written in detail on this topic, and it obviously should be the product of careful thought and negotiations. The level of unjust enrichment should provide the benchmark for the amount, and the figure should take into account all the economic consequences of the centuries of lost wages, plus the Jim Crow period of home equity discrimination, restrictive covenants, redlining, and so on. The final figure should indeed be very large, to give everyone involved a sense of honest closure, but, as in the German slave-labor settlement, it will never be large enough to provide a true level of compensation for the losses that were suffered by so many generations of African Americans.

WHO SHOULD RECEIVE THE PAYMENTS?

Because of the passage of time, it may seem improper just to hand money to all the descendants of slaves, some of whom have been able to prosper economically despite the obstacles in their path. The approach of funding a foundation to address specific needs of the descendants of slaves has been suggested by Deadria Farmer-Paellmann, a New York attorney whose great-great-grandmother was a slave in South Carolina. She has demanded that Aetna Inc. set up a $1 billion foundation to benefit minority education and business because it profited from slavery, selling policies in the 1850s that reimbursed slave owners for financial losses when their slaves died.[98]

A foundation could direct the distribution of payments to address needs linked to the lingering effects of slavery. Education, health, and economic development are obvious targets, and because of the property losses that are so evident, economic development should probably be highest on the list of targets. Low-interest loans, sources of venture capital, mortgage money, and the other infrastructure revenues that are so central to economic growth are examples of appropriate destinations for the reparations assembled for the descendants of African American slaves. As Deadria Farmer-Paellmann has said: "The idea is to become a part of the business world in America. I'm looking for us to not have to rely on programs and to become more economically independent."[99]

WHAT ABOUT THE DESCENDANTS OF THE WHITE PEOPLE WHO TRIED TO HELP THE SLAVES?

Many whites participated in the abolition movement and the underground railway, taking risks to help free slaves, and many fought and died in the Civil War, which was undertaken in part to end the morally unjustifiable practice of slavery. During the following century, whites continued to discriminate against Blacks, but some whites joined in efforts to confront and end officially sanctioned discrimination, frequently taking significant risks and sometimes suffering injury or death because of their actions. These efforts must be acknowledged, and they have been somewhat successful in turning our country away from the slavery and apartheid that has divided us for most of our history.

Those whose ancestors participated in these noble initiatives, as well as those whose ancestors were not slave owners, may ask why they should now have to participate in reparations payments. They should not have to pay as individuals, of course, but it is still their country that authorized and sustained, through laws, courts, and police enforcement, an oppressive system that continues to stain our communities with discriminatory perceptions, beliefs, and practices. Because all white Americans benefit by being white, we all must participate in providing the economic compensation necessary to restore to the African Americans the property that was taken from them through the institution of slavery and its aftermath.

STATUTE OF LIMITATIONS

This doctrine is utilized in all jurisdictions to discourage stale claims and ensure that credible evidence still exists to allow for a proper determination of a controversy. Exceptions exist for those wrongs that we view as most serious—such as murder and crimes against humanity—and slavery at the scale it was practiced in the United States certainly falls within any modern definition of a crime against humanity.[100] Another exception is the doctrine of equitable tolling, which allows plaintiffs to sue despite the passage of time if they had been prevented from suing earlier, or if an earlier suit would have been futile. This doctrine might apply in the slavery situation, because it certainly would have been unrealistic for African Americans to have brought a suit for reparations at any time prior to the present period, since they were subject to active, officially sanctioned discrimination until recently. Another possible approach is the continuing wrong theory, which would emphasize the continuing property deprivation resulting from slavery and its aftermath. This issue could be eliminated by Congress, if it were to enact a statute permitting a claim to be brought for reparations despite the statute of limitations that would otherwise apply.

SOVEREIGN IMMUNITY

The doctrine of sovereign immunity prevents suits against governments unless they consent to the suit. It will present another formidable procedural barrier to any claim against the federal or state governments, unless Congress enacts a statute waiving immunity in this case.[101] The U.S. Supreme Court has recently breathed new life into the Eleventh Amendment to protect states from being sued in federal court,[102] but if Congress prepares a proper legislative record and relies upon the Thirteenth and Fourteenth Amendments as the source of its power, it should be able to waive the immunity of the states in this situation.

WHAT ABOUT OTHER DESERVING GROUPS?

Some fear that providing reparations for African Americans will only serve to invite a never-ending Pandora's box of claimants to make similar claims. Other groups have, of course, suffered injuries in the United States, and other such claims may well be raised. The damage done to indigenous communities stands out as a particularly disgraceful chapter in United States history, and discrimination has been pervasive against many immigrant groups.

Nonetheless, the claim of African Americans based on slavery and its aftermath stands out from all the others, because it is based on the most outrageous wrong committed during our country's history and because the racial divide it created continues as a national calamity. Although terrible wrongs were done to the indigenous people in the Americas, our country has tried to address these wrongs, and has in recent years provided claims commissions and settlement agreements for most of the Native American claimants.[103] Discrimination has been systematic and harmful toward Asian immigrants and other groups, and the contract labor programs through which many immigrants came to the United States resembled slavery to a disturbing degree. But one still has to rank the 1619–1865 period of actual slavery and the Jim Crow century that followed as the most egregious human rights abuse in our country's history. Today's African Americans are unique in the suffering they continue to experience because of that unforgivable history.

WHAT SHOULD BE DONE?

The other chapters in this volume provide a spectrum of ideas about how to address and resolve this national nightmare. As Representative John Conyers has repeatedly suggested, a full investigation and accounting must be the first step, to document the entire experience and reduce our national denial of this historical tragedy. The claim must then be developed in a forum willing to listen and make recommendations. An analogy can be found in the Waitangi Tribunal in New Zealand (Aotearoa), which was established to adjudicate the claims of the Maori under the 1840 Waitangi Treaty. It held years of hearings on the claims of each Maori tribe and then made careful recommendations for solutions. Although the recommendations were not binding, they were taken seriously by the New Zealand government, and led to settlements with each group. The Tulsa 1921 Riot Commission mentioned above[104] provides another example. The full documentation of slavery and analysis of its continuing impacts called for by Representative Conyers is long overdue.

Once the commission's work is done, a good-faith solution must be found. Because the claim of the slave descendants is a property claim, it is protected under the Fifth Amendment of the U.S. Constitution and cognizable and enforceable in our judicial system. But because the actual wrongdoers are no longer alive, a proper solution must be somewhat creative and flexible. Most participants in the reparations movement agree that it would be wasteful and inappropriate to hand out money or resources equally to individual African Americans, and instead favor distribution targeted at the specific problems caused by slavery and its aftermath. Health and education needs would be high on this list, but the most pressing category would be economic development. Low-interest loans, venture capital, assistance with mortgages, and other financial assistance directly helping African American entrepreneurs are urgently needed to reduce the economic gap between the races in our country.

In New Zealand, in the Waitangi Treaty settlements mentioned above, the Maori have recovered land and money, but they have also obtained given factories, ships, fishing rights (which are private property in New Zealand), and other economic resources. As a result, they are now strong players in the New Zealand economy, and are deeply involved whenever important national decisions are made.

Reparations for African Americans should be aimed to achieve those same goals. This process must be designed to provide a measure of honorable reconciliation for the centuries of oppression, to return to African Americans the property that was denied to them during the long periods of slavery and systematic Jim Crow discrimination, and thus to enable them to be economic participants and to be at the table when important decisions in our country are made.

JON M. VAN DYKE *has been named Outstanding Law Professor four times at the University of Hawai`i, where he teaches, and is an expert on international law in the area of human rights violations and their consequences.*

Does America Owe a Debt to the Descendants of Its Slaves?

HARPER'S EDITOR NOTE: *Hardly a week goes by that we don't read of another gigantic lawsuit with thousands of plaintiffs and billions in damages. Once an esoteric legal device, the class-action lawsuit has become the dominant form of litigation to resolve bitter disputes over collective guilt and innocence that not so long ago played out in Congress. Indeed, our preening national legislature, besotted with special-interest money, seems rivaled by the big budgets and major issues that now thrive in the class-action courtroom.*

At the same time, one hears rumblings among historians and philosophers to consider a lawsuit for slave reparations. After all, class-action lawyers have ridden to the rescue of those forced into slave labor in Germany and prostitution in Korea. The academics discuss such a slavery suit in moral, historical, or metaphysical terms. That's nice. But in this, the land of show-me-the-money, the thinking quickly becomes practical: Who gets sued? For how much? What's the legal argument? How do you get a case into court?

To answer these questions, Harper's magazine invited four of the country's most successful class-action lawyers to strategize about how to bring America's most peculiar sorrow into a court of law.

CAUSE OF ACTION

JACK HITT: We're here today to talk about how: how, that is, to repay blacks for what they suffered under slavery and what they've suffered since because of it. But first let's talk about why, because when many people hear the term "slave reparations," they go nuts. "Oh, that was so long ago," they say. Can't we just leave this alone? Everybody's got gripes. Blacks should just get over it. To a lot of people, the very idea of a lawsuit seems unreasonable.

DENNIS C. SWEET III: That's because people think slavery ended in 1865. And it did, but the aftermath of slavery is still with us.

ALEXANDER J. PIRES JR.: Every great lawsuit tries to tell a story of injustice in a way that will resonate. There's a lot to work with here. Slavery's the most unacknowledged story in America's history.

HITT: Unacknowledged?

SWEET: This is what Randall Robinson says in his new book, *The Debt*.

HITT: What's unacknowledged?

SWEET: Oh, just about everything. Take our nation's capital. Nearly every brick, every dab of mortar, was put there by slaves. There's not a plaque in all of Washington acknowledging that slaves built the Rome of the New World. This is how it is with slavery. We've heard of it, but we don't really know anything about it.

WILLIE E. GARY: Think about this. In 1865 the federal government of this country freed 4 million blacks. Without a dime, with no property, nearly all illiterate, they were let loose upon the land to wander. That's what begins the aftermath of slavery.

SWEET: How many Americans know that 25 million blacks died in slavery? And how many know that virtual slavery was perpetuated for nearly a century after emancipation? Peonage laws made unpaid workers out of debtors. There were sharecropping schemes. Then Jim Crow laws. And even after that, there were other entrenched policies that have kept African Americans living in ghettos.

HITT: Robinson points out that until 1950 the federal government included in mortgage loans restrictive covenants preventing blacks and only blacks, no other group, from buying houses in white neighborhoods. So blacks could not make their equity work for them. They couldn't move up.

RICHARD F. SCRUGGS: A house is the largest single investment and asset most people have.

PIRES: And it's how every immigrant first got into the middle class. So that policy effectively delayed the arrival of the black middle class by half a century.

GARY: And banks kept it up denying loans to blacks, often by redlining, by which they literally would draw lines on a map around a neighborhood and not give loans to even creditworthy people living there. That happened until almost last week.

HITT: These are all compelling examples. And so you wonder: If Koreans can sue the Japanese about heinous acts carried out in the 1930s, and Jews are suing over prewar slave-labor camps, and American POWs imprisoned in Pacific camps are even suing the Japanese for slave-labor wages, why can't blacks also sue for similar recompense?

SWEET: It can be done. In fact, Alex won one of the great reparations lawsuits in the last few decades. He filed on behalf of how many was it, Alex?

PIRES: Twenty-four thousand black families.

SWEET: Until 1997 the United States Department of Agriculture had an almost zero rate of granting black farmers loans. Until 1997, okay?

PIRES: Ninety-five percent of all farm loans went to white farmers. And until the 1960s, the United States D.A. had a special section called Negro Loans, which ensured that black applicants were rejected. It's amazing.

SWEET: It's amazing how young this country is and how close in time, when you come to think about it, all our history is. No part of our history is that far off. The effects of slavery are still with us, we all know that: single parents, black men wandering off from their families, a tradition of not going to school, distrust of the future. This is not black culture. It's slave culture.

GARY: My children and I have talked about your case, Alex. They've never been on a farm, but for them and a lot of black people you changed our whole thinking about what we can be in this country. If you're black, I mean, you're thinking, it's not going to happen, the government isn't going to give you a fair shake. Then all of a sudden you see this happen. I mean, the country stepped up to the plate.

PIRES: In the end, it did.

GARY: Man, it was such a big message. You know, as an African American I felt really good about it.

SWEET: So how do we make a case?

LEGAL STRATEGY I: BREACH OF CONTRACT

GARY: I think this could be a tort, a simple lawsuit where one party sues another.

HITT: So a variation of the classic you-done-me-wrong lawsuit. Just really big.

GARY: Specifically, it could be a breach-of-contract suit, too. After the war, former slaves were promised forty acres and a mule, and we never got it. That was a contract. It was a promise. We just have to stand up and tackle this wrong or try to make it right. So it could be a tort, it could be a breach of contract. You almost have to start a lawsuit to see where it can go. And I don't think that the fact that it's 135 years later should be a hindrance to people waking up, realizing that it was a grave injustice. And until America accounts for its actions, this friction is always going to be there.

SCRUGGS: Breach of contract after 135 years? You do have a statute of limitations problem.

GARY: Not if Congress steps in.

HITT: But can you count on that?

PIRES: I don't think the legislature's going to help until the lawsuit goes forward. You have to file that suit, and you have to go forward yourself.

SWEET: Al, it's just like your black farmers' suit. I joined Al on that case. They had studies showing rank discrimination, years and years and years of it, and the courts never did anything about it.

PIRES: This is how these lawsuits go. Everybody said the statute of limitations is against you, as well as other legal problems too. But we just kept marching, marching, marching. Getting more folk, going around the country. And there were motions to dismiss. Then motions against class certification, the process by which you define who is suing. But as the facts started getting out, people started to say, Hey, this argument makes sense. These people were wronged. Something's not right. The judges began to think, are we going to let this great injustice go unanswered because of a technicality? I think this situation is like that.

SWEET: We still have to get specific here. We have to get past Cato.

HITT: Cato?

PIRES: *Cato v. United States* was a slave-reparations decision issued in 1995 by the Ninth Circuit Court of Appeals. This case was easily dismissed, because the judges said, We can't find a theory with which to move the case forward. Since the Ninth Circuit is basically liberal, this decision sort of took the wind out of reparations thinking. You either find a theory under statutes or you're going to have to find it under constitutional law.

SCRUGGS: We have to find a legal theory, then.

PIRES: And people have tried. They considered the Thirteenth Amendment, didn't work. They've considered a tort claims act, didn't work. They've considered the civil rights acts, didn't work. The Cato case is useful, because the Ninth Circuit said, We've looked real hard at it. We can't find a way.

GARY: Let's think about the breach of contract: forty acres and a mule.

PIRES: Well, how can you sue?

HITT: The promise of forty acres and a mule was an executive order sanctioned by Congress.

GARY: It's in writing! Breach of contract.

HITT: Well, that contract was voided by President Andrew Johnson in 1865. Not only did he reverse the very first stab at reparations, but the few thousand blacks in Florida and South Carolina who actually did get the forty acres and a mule had them taken away.

GARY: Well, that's it right there. The government can take property only under the eminent-domain clause. But they didn't deprive the former slaves of their property for any national purpose. This was theft. If anything, the takings clause applies.

HITT: Takings?

GARY: Takings is part of the Fifth Amendment: "nor shall private property be taken for public use without just compensation."

SCRUGGS: But you still have the statute of limitations problem. Look, I've been wrestling with this thing. I mean, I love big stuff. We all do. We love to think of elegant solutions to major national social problems, and this is the biggest one there is.

GARY: I think either way we look at it we're going to need help politically, because we don't have the law squarely on our side. We don't have the statute of limitations on our side. We don't have any of that stuff. So it's going to require more than just a simple, single legal theory.

PIRES: In a federal case you've got six years before the statute of limitations runs out.

SWEET: Unless you can prove fraudulent concealment.

SCRUGGS: But there wasn't any fraudulent concealment here.

PIRES: Wait. In recent cases involving World War II slave-labor victims, the statute of limitations doesn't apply if there is a war crime or if there's a crime upon humanity. So I say: If there ever was a crime upon humanity, what white folks did to black people is the worst that ever happened in this country. We would argue that it's not fair to apply the statute of limitations upon us. But Dickie is right. It's the main problem. Well, actually there are two. The second is sovereign immunity—you can't sue the government without its consent.

HITT: But to take Willie's breach-of-contract argument, no former slave could have sued in, say, 1870 and expected to get a hearing. Is the clock not suspended for the century or so it took us to recognize that the courts should be open to African Americans?

SWEET: If not, you have to ask yourself: Is an injustice no longer an injustice so long as you get away with it for a long period of time?

HITT: Well, that gets us to the next logical question. Who is suing here, who are the plaintiffs?

SWEET: No, no, Jack. That's never the first question.

PIRES: The first question is: Who are the defendants?

SWEET: That's better.

HITT: Who are the defendants, i.e., who pays?

GARY: That's it.

THE DEFENDANTS

SWEET: I think you have two defendants here. The government and private individuals.

HITT: Private individuals?

SWEET: I mean private companies.

GARY: And private individuals. There are huge, wealthy families in the South today that once owned a lot of slaves. You can trace all their wealth to the free labor of black folks. So when you identify the defendants, there are a vast number of individuals.

HITT: Descendants of former slave owners? You're making me nervous.

GARY: Well, like Dennis said, you've got those families that owned slaves, had the plantations, worked the slaves, and because of the sweat and suffering of the slaves those families are major players now in the United States. I think you just track them down. You have to go into North Carolina, South Carolina—

SWEET: Mississippi, Alabama, all over.

HITT: As the descendant of—I'm not making this up—Martin Van Buren Hitt, a slave owner, I think I speak for a lot of people when I ask, "How do you do that?" But you're saying it's possible?

GARY: It's possible. Look, nobody ever said it was going to be easy. You know, if you're not ready to get in the trenches and fight, if you're not ready to get knocked down, kicked out of court, and everything else then it's not the type of issue you want to pursue, because it will be a struggle.

PIRES: Let's talk about the corporate defendants. You have in America, from the 1830s, 1840s, 1850s, the beginning of the greatest accumulation of our wealth. The early oil industry, for example, predates the Civil War. You've got to look back and find out, because many of those companies still exist, under other names. Standard Oil, a.k.a. Esso, a.k.a. Exxon, is still here. They're all still here.

GARY: Aetna Inc., which has been around since 1853, just apologized to blacks for underwriting slave insurance policies. And the *Hartford Courant*, which is still publishing, also apologized for running ads that assisted in the capture of slaves running away, making a break for freedom.

PIRES: You look at a banker like J. P. Morgan and you look at the other trusts, like the railroads. Fleet Bank used to be the Providence Bank, whose original wealth dates back to the family of John Brown, whose descendants underwrote Brown University enough to cover up the embarrassment of where he made his money.

HITT: Embarrassment?

PIRES: The Browns made much of their money as slave traders in the late eighteenth century.

GARY: So we've got the federal government.

PIRES: And we've got the states.

GARY: And we've got the private profiteers.

SWEET: Looking good so far.

THE PLAINTIFFS

PIRES: Since we've agreed that our case would extend beyond the end of slavery, it seems to me that the issue of plaintiffs would be best dealt with if you think of the suit as falling into three time brackets. There's slavery up until 1865. Then you have government-approved segregation for, what, seventy or eighty years? And then we have this kind of fuzzy land we lived in until the 1960s, and I don't even know what you call that.

SWEET: Denial.

PIRES: It's easy to go back to the 1940s, the 1950s, because we have precedent—the Japanese internment, the case of the Jews in Germany, we have lots of cases. So if I say, We're going to go back to the 1940s, which is sixty years ago, and pick up claims, that's an easy case. But to go back to the second time bracket, which is to the 1860s, is a little tougher, because of sovereign immunity and the statute of limitations.

HITT: Well, that's one of Robinson's points in his book. Why not just sue for the more recent cases, which stem from slavery? Their proximity in time gives us two things: living victims and a quantifiable economic case. As was mentioned, the federal government sanctioned mortgage covenants that essentially restricted blacks from entering the middle class. That is an economic damage. We can measure it and put a numerical figure on it, a precise amount of money. And then you can sue. You have a class of African Americans who were struggling to join the middle class—very appealing. Your average jury would support it, easily.

PIRES: And even a case as good as that one leads us to another problem. Willie, let's suppose you are the lead lawyer. We have a plaintiffs' meeting and people say, How far back are we going? Are we going to go back 250 years, 150 years, or 50 years? Isn't that the question?

GARY: Obviously as a lawyer you want to make your job as easy as you can and also put yourself in the best position to win. And to do that, you want to put your hands on those damages that you can quantify so that you can develop them. You've got evidence still available, and you've got people. The only problem with it, though, is that this kind of lawsuit is going to appear in the court of public opinion, and you're going to need the support of the people.

PIRES: The white people.

GARY: And black people.

PIRES: Black people you got, right?

GARY: No, not necessarily.

PIRES: Black people aren't going to be happy with such a suit?

GARY: What about those people who, for whatever reason, maybe were excluded because you started in the 1940s? Would we pick up everybody? One-third, two-thirds, of all living blacks? I don't know. If you leave a substantial number of people outside . . . I mean, it's got to be like we are not leaving anyone behind. Because if we pick the best, most recent case, some people are going to complain. Then we've got to be prepared to meet the fight within the fight. And that could be a major problem.

PIRES: What's more important, to tell the real story of American slavery or to win specific damages from 1940 onward?

GARY: It's something we'd have to think about. You would have some people saying—

PIRES: —you lawyers didn't do your job.

GARY: No, that we're taking the easiest way out. And that we're leaving people behind. No one should be left behind, and if you're going to do it, you should do it right, and it's not fair just for a few and not for all.

SWEET: Let me say this to you: All black folks are not going to be happy. You can go back 200 years, include every damage in the world, and you still will have blacks who are not happy.

PIRES: Why?

SWEET: I mean, Clarence Thomas'll probably write an opinion saying it stigmatizes black folks to bring any action.

PIRES: Why?

SWEET: You have some self-hating black folks. You'll even have some black folks who feel like they don't want to have the issue brought to the forefront.

SCRUGGS: There are so many different parts here that you have to think about. Let's get to this one. Who are your plaintiffs, first of all?

PIRES: Black people.

SCRUGGS: Well, okay. Does that include Tiger Woods? What are his damages? What are Denzel Washington's damages?

GARY: Well, one thing about a class-action lawsuit, you don't have to try to figure out the damages. You can do that on a grid.

HITT: Do you use some kind of damages formula when you're dealing with a large class like this?

GARY: Yes. But look here, we've got more non–Tiger Woodses than we have Tiger Woodses. For every Tiger Woods, we have 100,000 non-Tigers.

PIRES: That's true.

SWEET: Well, the thing about it is, in a class action, you have people with different degrees of damage. In my fen-phen case, some people were hurt. But some weren't. And, hell, some people even lost weight. I mean, they were better off, I guess. Some people died, of course.

GARY: But they were all still members of the class. They each experienced the risk.

SWEET: They were all members of the class, and their damages vary along with the degree of impact.

SCRUGGS: I think you'd need to really define the class pretty well, so you keep the Tiger Woodses and people like that out. You can't have people say, "Well, damn, you're going to give all this money to Tiger Woods or Denzel Washington."

SWEET: When you say "the class," you've got to remember that you're still covering a large majority of the black people in the United States of America. Then you have certain representatives who'll be in court with you.

SCRUGGS: Right, but you've got to come up with an appealing plaintiff, just like you would in an individual personal-injury case.

HITT: Who would you want to be representatives?

SWEET: You go and you pick them.

SCRUGGS: You interview a lot of people to pick somebody who's articulate, who's got an appealing personal case, and who is typical of the class that he's going to represent.

HITT: In the Japanese-American case, they brought forward as class representatives only the Japanese Americans who in the 1940s said, "I acquiesced to this because it was my patriotic duty." They did not bring forward the ones who rioted or resisted the draft.

SCRUGGS: Right. You don't want to trot Mike Tyson out there.

HITT: You want to trot out the black guys who fought in World War II and came back to freedom's home only to be told they couldn't sit at a lunch counter.

SCRUGGS: That's right. You carefully pick them.

PIRES: All our famous plaintiffs are selected. Rosa Parks was selected.

SCRUGGS: Yeah, Rosa Parks. Perfect example.

PIRES: They're all selected. I mean, the history of American plaintiffs, Jack, is that they are all selected. Remember Darrow's famous case, the Scopes trial? Well, the lawyers found Scopes by taking an ad out in the paper.

DAMAGES

PIRES: I have this theory about big lawsuits. Their chance of success is not really a matter of the plaintiff or the defendant, nor of legal theory, nor of arguments of liability. It is a matter of the damages to the plaintiff. People react to damages in a visceral way. Take Dickie's tobacco case. Why did people warm to it? Because the average person knows what medical costs are and how much he or she is spending on them. Poor folks have no medical insurance, and middle-class folks are gagging on paying for it. Then someone like Dickie—

SCRUGGS: —comes along and says, "We're going to sue on behalf of the attorneys general because the tobacco people are responsible for a lot of our medical costs, and they're not paying their fair share!" The average person says, "That's right, yeah. Screw them! It's hundreds of billions of dollars. Good! GET them!" No one cares about the technicalities.

GARY: That case also changed our whole attitude toward tobacco.

PIRES: In my black farmers case, people finally said, "Hey, blacks are farming without access to loans? I've had trouble getting loans. Give them a hearing!" People relate to it. They react to those damages, and they say, "We're going to pay these black folks for what we did to them. It wasn't liability-based, it was damages-based. Every decade has its case. In the seventies it was IBM. In the eighties it was AT&T. And in the nineties it was tobacco. People reacted to those damages too. Break up this big monopoly. Yeah, that's wrong. So how do we make our damages appealing?

SCRUGGS: First, by making it clear that the damages are not just about money. You know that old saying: If you catch a man a fish, you've fed him for a day; if you teach him how to fish, you've fed him for a lifetime. And that's what I'm talking about here. Regardless of whether the defendant is the federal government or a corporate institution that profited from the inhuman treatment of blacks, like German corporations that used Jewish slave labor to make money and are still reaping the benefits, you've got to describe the damages in such a way that makes sense to the public.

GARY: But a lawsuit is also about the money.

SCRUGGS: I worked with a tribal corporation established under the Alaska Native Claims Settlement Act. And the way the federal government settled similar claims was that it vested the Indian tribes with large sums of money and land and resources. And I'm afraid that it gave too much wealth too soon to people who were not sophisticated enough to do anything with it, and they were victimized.

PIRES: They lost it all.

SCRUGGS: Many did. In one generation.

PIRES: People will worry about that.

SCRUGGS: I'm not saying that blacks are less sophisticated than whites. If you gave money to a bunch of WASP Harvard graduates, they'd blow through it, too. It's human nature.

SWEET: But Dickie, the better part of the solution would be the victory itself, the benefit that comes when it's recognized by this country that reparations are in order.

SCRUGGS: That's a different goal.

GARY: But there's got to be money, because it goes a long way toward achieving the very goals you're talking about.

PIRES: Let me put my question this way: Say the government finally admitted, "You're right. It's the worst injustice in the history of our country. We're the most successful nation on Earth ever. You win! What do you want for damages?"

GARY: That's the big issue. You want healing, because you can change the thinking of generations unborn, the future of race relations in America.

HITT: Let's stick with the law.

GARY: No, this is important. It would say that America stepped up to the plate and acknowledged its wrongdoing and reached out to the people and said there is justice for all. It would change things the way you and I see each other. It would be nice, you know, sometime to sit down together, and you say, "I'm sorry" and I say "I'm sorry," and then we could just break bread together. We can go forward, we can do greater things than we ever anticipated.

HITT: But when I was asking a black woman once about this, I said, "You know, maybe in the end the money should all be directed toward the poorest of African Americans, because they are the real heirs of slavery's worst tragedy, the people in the ghettos. And we'll aim that money at them." And she goes, "No, no. I want some of the money." Wouldn't you want just a little bit of that money, Willie? For the symbolism of it, if nothing else? Just for the satisfaction?

GARY: There is the money.

SCRUGGS: Money is not the solution. It's setting in place institutions and programs.

GARY: Education.

SCRUGGS: Exactly. It's the difference between giving them the fish and teaching them how to fish.

SWEET: I think a small part of it is going to be the money and the remedies. But the message that will be sent is so important. By having the whole country come forward and say, "This situation has gone on too long," that's a huge step.

GARY: And for every dollar paid, the government would get a $100 return.

HITT: Charles Krauthammer, a conservative columnist, is very much in favor of black reparations.

GARY: Really?

HITT: He says that black reparations make sense. We've done something wrong, we need to pay for it, right? He sees it in pure economic terms, as if it were the nation's biggest tort claim in history. They were done wrong. Let's figure out an adequate sum of compensation and pay. Then he adds: And affirmative action doesn't make sense, because we're unfairly putting one person ahead of another person. So let's eliminate all the minority-preference programs, and then let's move forward with the pay schedule.

GARY: It makes a lot of sense. No doubt about it.

PIRES: Wait a minute. What fundamentally separates black folks from white folks? Not money, but education?

SWEET: I say if you're forced to go through a trial, and you're forced to stand up there and talk about damages, the only thing to do is quantitatively ask for damages, for money.

PIRES: You can't think that money will be enough. What do you really want, Dennis? Huh?

SWEET: I'm saying that's all you can ask for.

PIRES: Suppose the judge says, "What do you want me to give you? You want money? You want education? You want access to housing? You want health care?" Dennis, what do you want?

SWEET: Al, Al, Al, hold on, hold on a second. Let me tell you this: If you're in a situation with a jury, then you can only ask for monetary relief. If you have a judge, you can say, "Judge, I need you to create these programs," or other more nuanced solutions.

SCRUGGS: That's right. All a jury can do is award money.

PIRES: What's that going to fix when the only major difference between black folks and white folks today is education.

SWEET: Oh, no, no, no, no. Noooo.

PIRES: It's the level of education. That's the biggest difference.

SCRUGGS: Well, that attitude's pretty rough.

SWEET: It's not just education. It's like, you know, Chris Rock, the comedian, said it best. He has a bit in his act where he's talking to just a normal white guy and says, "Despite all the changes in society, you wouldn't switch places with me, a black man." Then he pauses. "And I'm rich!" The thing is, there are a lot of benefits to being white. A lot.

HITT: But that's the nice thing about arguing about damages. You get specific. What do we need to fix this? Alex is right to ask, If the judge said, "Okay, you win," then what would you ask for?

SCRUGGS: That was the very toughest thing we faced in our tobacco case. We asked ourselves hypothetically: If the chairmen of the boards of these major tobacco companies walked in here today and said, "Okay, we're ready to do a deal," what did we want? It took us a year to come to some general consensus among the attorneys general and some of the public-health advocates as to what we really wanted if we had these guys by the throat. What we found out later was once we got what everybody had said he or she wanted, that wasn't enough.

PIRES: Not enough money, $368 billion?

SCRUGGS: No, no. It wasn't enough money. There wasn't enough money in the world to satisfy some people. And it wouldn't have mattered. The problem is that there are people invested in the fight, okay? I mean, like, some people in Palestine or Northern Ireland don't want the wars to end.

GARY: That's the other fight within the fight. Some black people are not going to want you to file this case and then win.

SCRUGGS: Exactly. This was the biggest mistake we made in tobacco. We did not anticipate the self-interest of some of the health groups in perpetuating their existence and their fund-raising. Because bashing big tobacco was their fundamental way to raise money.

PIRES: Tobacco-Free Kids?

SCRUGGS: Well, they were on board, strangely enough, but some groups like the American Lung Association saw their fund-raising threatened by a tobacco solution.

PIRES: Because when you take away their core issue, people are not happy. If I said to Dennis, "Is education the problem, is that what we're looking for?" and you say, "Let's educate two generations of black folks," people say, "I'm not happy with that. That doesn't do it."

SWEET: I just want you to realize what Chris Rock is saying. It's more complex than one thing.

HITT: If Congress intervened in this case, Dennis, would you be happy in return for a generous reparations deal to eliminate all minority-preference programs?

SWEET: There's so little left. Sure.

SCRUGGS: What I have envisioned would be a super-affirmative-action program, much more than traditional affirmative-action programs.

HITT: So you agree with Krauthammer?

SCRUGGS: I really have not read this gentleman's work, and I may be doing it a disservice. But if he's just offering money in return for eliminating affirmative action, then no.

HITT: No?

SCRUGGS: I think that's tokenism. Reparations doesn't mean just a bunch of cash payments. The word means to repair. I'm talking about programs. Straight-out payments will create the excuse for future Congresses to say, "We've done it, and what did they do with the money? They went through it; they blew it like other groups have."

SWEET: I agree with you. But you have to be careful about the remedy. It's like the Ayers case, a higher-education case in Mississippi in which the judge said, "The black schools are not being funded properly. The white schools are being funded more properly." The judge says, "Okay, we can show liability. Now let's do the remedy." Hell, the remedy kicked us in the butts. You know what they've started saying? "Okay, we're going to close this black school, and we're going to close that black school because of improper funding." See, the remedy can be worse than the claim. If the outcome of this suit were to give each black person $5,000, that would be a disaster. Then we would have eliminated any moral claim to criticizing the causes that have led to a widespread African-American poverty, and in return for what?

PIRES: In this case, the money's necessary but the money's not enough.

LEGAL STRATEGY II: MULTIPLE TORTS

SWEET: I get the feeling that everyone wants to start by suing the United States government. But I'd hate to see the federal government be the only defendant in the case.

PIRES: What about the states? We haven't talked about the states.

SWEET: I'll tell you a claim that's ready to go. It's an idea for a state case that would at least serve as a beginning. In fact, the state of Mississippi is a sitting duck on this. I'm talking about the Sovereignty Commission.

GARY: What was it?

SCRUGGS: It was like a Gestapo organization.

SWEET: Back in the fifties and sixties, white leaders got concerned that black people might gain power, like the right to vote. So elected public officials of the state of Mississippi funded this spy organization whose sole purpose was to keep black people down. They spied on anybody who was supposed to be a leader. They participated in the Byron de la Beckwith trial, the man who killed Medgar Evers, by helping people identify jurors. I mean, there is a library full of material documenting their activities.

GARY: And the state financed that activity?

SWEET: The state financed it. You have a secret state agency that was formed whose only purpose was to keep black folks in place.

GARY: Plus it will be a great place to start if we're going to move forward with the overall issue, including the larger suits dating all the way back.

SCRUGGS: This is a state action. You've got a statute of limitations even under the civil rights acts.

SWEET: Yeah, but you have a fraudulent concealment. So legally, the statute of limitations doesn't kick in until after the fraudulent concealment has been exposed. Well, they said that none of the documents and information conducted in here shall be open to the public; it was fraudulently concealed from the public. And they're still concealing some of the Sovereignty Commission's work. So the statute would start from the time those documents were first opened to the public, which is right now.

GARY: So you'd pick up thousands and thousands of people just with that lawsuit. Then you could branch out and pick up families in every state in the union.

PIRES: No problem.

GARY: Just recently American General Life and Accident Insurance Company paid out more than $200 million for overcharging black Americans for standard insurance premiums over the last decades. We could file a couple of those types of lawsuits as well.

HITT: Willie, are you suggesting a strategy of filing, say, a web of lawsuits—the Aetnas, Fleet Banks, on the one hand, and then state and federal governments too?

GARY: I think we could get class representatives from each state in the union. If you're going to go after the government, you do all in one.

PIRES: If we filed a pile of lawsuits, you could put a judge in a position where the statute of limitations would be hard to invoke. If there were a national audience watching, then what judge is going to want to be the man who went down a laundry list of several dozen incredibly powerful and legitimate claims and had to dismiss them on a technicality? It might make it difficult.

GARY: If you've got a public outcry, a political movement behind it, while we're in the process of getting ready to file, I think that can affect the way a judge is going to rule. It can make him not want to rule, it can make him hold and then perhaps Congress will step in and you can talk settlement. There are so many things you'd have to do at once. But you definitely need a massive public-relations program. You'd want your Denzel Washingtons and your Danny Glovers on board. You get the black athletes in the NBA to stand with you, you get the NFL to stand with you. And then you might go to someone high in the ministry, because you want top-flight black people like Reverend Jesse Jackson, NAACP president Kweisi Mfume, all these people to stand with you. Then it's a different ball game.

PIRES: Just prominent black people?

GARY: No, black and white people. And the same with the lawyers. It should not just be black lawyers. Look, right now I'm fighting for a white client down in Orlando, a very conservative area. There aren't going to be any black people on the jury. But I've got an old white fellow who's shuffling around with a walking cane! And I'm helping him in and out of the court. The two of us. Let me tell you, it neutralizes a whole lot of shit.

PIRES: So it would be important to have both black and white lawyers up front?

GARY: That's right. If you need a lawyer today, the best thing you could do is have a black lawyer fight for you. And for this case, black and white lawyers fighting together. Look, it was a long, long time before people came on board with Stevie Wonder when he was fighting all those years to make Dr. Martin Luther King's birthday a holiday. But after a while Barbra Streisand and other people came on board, saying it's the right damn thing to do. And all of a sudden the issue changed.

SCRUGGS: You're right, you can't do this case without a public-relations strategy.

GARY: This is the type of case where if you bring certain pressure to bear, if you have the right kind of public support for it, both black and white, then nobody's going to say, "Okay, here's another example of the blacks just trying to get something for nothing."

HITT: When the Jews suing in the slave-labor case were preparing their strategy, that was one of their concerns, that the suit would also promote an old ugly stereotype of Jews and money.

GARY: Same thing here. Blacks trying to get something for nothing. But not if we have a public-relations strategy in place when we begin.

HITT: But correct me if I'm wrong. Overall, are you saying that you file numerous cases at the same time so that one has a fighting chance to change the way people think about this issue? For example, if you file the slavery-era case and maybe Willie's breach of contract and they are dismissed, then does that make, say, the more recent, more economically quantifiable case about mortgage covenants seem that much more possible to win?

SCRUGGS: Exactly. You can make us look downright reasonable by filing some outrageous case over here.

HITT: Yeah. So some of those filings would be to your best case what historians say Malcolm X was for Martin Luther King. Malcolm made King's once-dicey demands look mainstream.

SCRUGGS: That's not a bad strategy. That's something you have to think about. In other words, get Pat Buchanan in your race so that you'll look, no matter how conservative, like you are very reasonable.

GARY: If we file a mess of cases against the states, isn't it also likely that the state would implead the federal government?

HITT: What does that mean?

SWEET: A person charged with a crime can implead other defendants, saying, in effect, Hey, if I did it, this guy did it, too. We should share the punishment.

GARY: The states could bring in the federal government and say, "Hey, wait, we're not going to pick up this tab. We were doing what you all gave us the right to do, all this shit started in Washington, D.C."

SWEET: Neat. The states would try and prove the liability of the federal government for you.

PRO BONO?

PIRES: I have a question for you all, and you should be as honest as you can be. When we put together the black farmers' case, I thought the only way I could get black folks to trust a white lawyer was to give them a retainer agreement that said we would work for free, that they get 100 percent of their recovery. So I got 21,000 retainer agreements with black folks that said they'd get 100 percent of their recovery and we wouldn't get any money from them. And we have to petition the court for legal fees. My thinking was that many black folks, who aren't used to lawyers, would more likely trust us if we didn't take their money. So would you all work for free?

SWEET: What?

SCRUGGS: Um.

GARY: Clients sometimes try to negotiate me down to 10 percent on a case, and I say, "Why would you want me working unhappy for you? I'll get you 100,000 bucks. If you got me happy, I'll get you 2 million."

PIRES: Maybe I'm wrong.

HITT: I guess that issue's resolved.

LEGAL STRATEGY III: DUE PROCESS

SCRUGGS: Before we file a pile of lawsuits, I think there's another way besides a damages lawsuit under traditional theory.

PIRES: Let's hear it.

SCRUGGS: How about a Fourteenth and Fifth Amendment lawsuit against the federal government for either failure to enact sufficient laws to ensure due process or for passing laws that perpetuated the injustice?

SWEET: So a due-process lawsuit?

SCRUGGS: Just like in the sixties when Congress ordered white legislatures in Mississippi and other Southern states to appropriate money for black schools or for school integration. They said no state or local government that discriminated could receive federal aid. They forced the state legislatures to appropriate money.

PIRES: So you're suing for a denial of due process to black people.

SCRUGGS: It would be a case under the laws and the Constitution of the United States, to the effect that under the imprimatur of the United States of America and the protection of the government, black men and women were brought to this country as slaves, against their will and were kept in bondage for a hundred or more years. Remember, slavery existed far longer under the Stars and Stripes than under the Stars and Bars. There were certain half-assed measures taken after the Civil War to try to enfranchise and rectify the injustice that had been done. But they were very ineffectual and incomplete. After the Reconstruction era, when whites regained power in the South, where most blacks lived, they went back to an era of repression, keeping blacks uneducated—

PIRES: And segregated.

SCRUGGS: —and disenfranchised. There were parallel societies, mostly in the South, less so in most Northern cities. Nevertheless, because the federal government failed to enforce the Fifth, Thirteenth, Fourteenth, and Fifteenth Amendments, the state governments were allowed to continue with this disparate treatment of black Americans. And the result is that now blacks are disadvantaged in comparison with whites and most other races in America. The federal government should be compelled to rectify that imbalance by passing legislation that accomplishes certain stated goals. Then there would have to be a federal court order that required the Congress of the United States to accomplish these goals within the satisfaction of the Court pretty much like what the 1964 Civil Rights Act did to Southern legislatures. It required the legislatures in those states to appropriate money for programs that helped rectify the imbalance. If the legislature didn't rectify problems, if it didn't act in good faith, then the states lost federal funding. I think that kind of a lawsuit has a far greater public appeal, and a greater legal foundation, than does simply suing for money for a generation of black people. Because nobody is going to think that will be effective, other than making a few people rich for a short time, not rich, but getting some money in their pockets for a short time. And then the next generation is going to be in the same spot.

SWEET: You know what's nice about this due-process lawsuit? It does away with a lot of the complaints that "we were also done wrong" from the Irish or other minorities precisely because it recognizes the fundamental difference. African Americans were kept down by the force of law, not custom, and then every effort to lift the burden of the law was met with denial of due process. So under this lawsuit, what you're saying is, We're going to give black folks a fair chance to assimilate, just like we gave that opportunity to assimilate to other groups.

SCRUGGS: That's right, that's right.

HITT: We always come back to the technical hurdles. What's your statute of limitations theory?

SCRUGGS: My statute of limitations theory is: continuing constitutional violation. Happens every day. It's like suing your government the same way you sue your doctor for malpractice for not doing his job. The Constitution tells the three branches of government what they're supposed to be doing, what is supposed to be protected. The case law fills that out, records what is supposed to be done. The government is not doing its duty. So, in essence, it's a malpractice case or a mandamus case.

SWEET: Governmental malpractice. Nice.

HITT: A what case? Mandamus? We command, right?

SCRUGGS: Right. In other words, we would argue that what they were supposed to do was not a discretionary function of government. You must do it. The Constitution says, You must. You don't have any discretion, you must do it. That's where I think the remedy lies. Because you force the Congress of the United States to pass laws, whether monetary funding for programs or the creation of programs that pass the courts' scrutiny. Just like Southern states were forced to do thirty years ago.

HITT: Very interesting rendition. So do to the feds what the feds did to the South.

SCRUGGS: That's right. In that way, you're not couching it as reparations; there's judicial precedent for it: it's been done to all the Southern states that were under the Civil Rights Act, okay? I think that's the approach, and we'll have the greatest result in terms of producing, in a few generations, a better society.

HITT: You know, that argument might even have a lot of appeal among Southern whites. It kind of sticks it to the federal government, which, after all, won the war only to set 4 million penniless, property-less, illiterate black men, women, and children adrift in the South. Slavery was evil, but so were the actions of the victors in Washington, D.C., who set in motion the Black Diaspora of 1865 and then walked away from it.

THE META-STRATEGY

GARY: You know, all these theories have something to them. But I don't think we can sit down and figure out the way to legally win this case before we file a lawsuit. I think you've got to put together a concept that can get you there, and it's going to be step by step.

HITT: We have to file a lawsuit without knowing where we're going?

GARY: We need to get some star power on the legal side, make a strong opening case, and get it going.

SWEET: To Congress.

HITT: Is that where you're ultimately headed? Is the idea to file a case that gathers enough momentum that Congress will step in and settle it?

SCRUGGS: I think so.

GARY: You want to get to a settlement.

SCRUGGS: That's how a lot of these cases go, to the legislature.

PIRES: That's what happened with the Japanese internment case. Congress was so embarrassed by the claims that it passed a reparations-settlement bill. Each aggrieved Japanese American received $20,000.

HITT: But that case was easier, legally, to get started.

PIRES: Actually, harder. There was a Supreme Court decision in 1944 declaring the Japanese internment constitutional.

HITT: How did they get around that?

PIRES: They got Congress to open an investigation into the facts that the government supplied the Court in order to make that 1944 decision and found that it was full of deception and lies. So the Court decision suddenly was no longer a roadblock to the case.

SWEET: That's what I was saying at the beginning. There's a way in which educating people about history, through this lawsuit, makes it more possible to file and win such a suit.

PIRES: Congress often gets involved in these cases because there are matters of justice that just can't be litigated fairly within the strictures of our common law and our Constitution.

HITT: What I hear you saying is, ultimately, that many class-action suits are just giant goads to get Congress to deal with politics. You are using the elegance of the law to motivate our legislative branch into doing what, arguably, they should be doing anyway.

SCRUGGS: That's right.

PIRES: It's true. We're getting social change from goddamn lawyers. How the hell did that happen?

SCRUGGS: My view of it is that the guys who wrote the Constitution had just thrown off a dictator, a British king who had exploited them as a colony. They had no rights, no democracy to speak of. They were not about to create a system of government that was going to allow for another dictator. So they created a strong separation of powers so that no one person or one group could gang up on another. More freedom, but at the price of governmental inefficiency. This inefficiency has worsened over time to the point that the political branches of government are capable of solving only the most compelling and broad national problems.

PIRES: Like what?

SCRUGGS: War and peace, things like that. And what's happened is that issues like what we're talking about now, big issues that are very important to people, like abortion, like HMOs, you name it, are—

PIRES: Avoided?

SCRUGGS: No, no. They're exploited, by both political parties. So what's happened is that anything that's going to get solved is punted to the court system.

PIRES: I believe that.

SCRUGGS: The courts have become a safety net. Those in charge of the political branches aren't interested in solutions, only in exploiting the issues for fund-raising purposes.

HITT: Perhaps that explains why the makeup of the Supreme Court is, if you think about it, the only thing our two presidential candidates deeply differ on.

GARY: Getting this to Congress also solves the statute-of-limitations problems.

PIRES: And sovereign immunity.

GARY: Congress can do whatever it wants to do in terms of waiving this and that.

PIRES: It can pass any law it wants.

SCRUGGS: Getting a political solution is the cleanest way to get it done. But today's Congress must be forced to act.

PIRES: Congress won't get there until you get there.

HITT: Very Zen.

SWEET: That's why Congress won't pass the Conyers resolution out of committee.

HITT: What's that?

SWEET: Michigan Representative John Conyers Jr. proposed a bill to apologize for slavery. Congress won't even do that.

HITT: Why not?

SWEET: Probably because it also seeks to authorize a congressional study group to look into slavery.

HITT: A study group?

SWEET: You have to remember: The last congressional study group like this was the one looking into the Korematsu decision. It exposed all the injustices underlying the Supreme Court decision permitting the Japanese internment. By the time the study group finished its work, it was clear that a court case was possible. Congress won't apologize or allow the study group, because it's afraid of precisely this lawsuit—that the lawyers in this *Harper's* forum might reconvene, and not just to chat.

PIRES: And maybe file a complaint before a court. You know, it all gets back to the lack of understanding by the people you mentioned at the beginning, Jack. The people who say, "How could you possibly sue for slavery?" True, you've got all these technical legal problems, the statute of limitations, sovereign immunity, class-certification problems, defining the damages, and the rest. But you have to remember that the judiciary is the only branch of our government that has nothing to do. It sits there, waiting. The legislature writes laws, and the executive carries them out. But our judges sit and wait for us to come with a complaint, which is a kind of prayer. It says, "Judge, I have this story to tell. It's a story of an injustice. It's a new story, a new way of understanding an old injustice. And I ask you today to hear this case, to listen to my story." Sometimes they do. Sometimes, if you play it right, they hear your prayer.

EDITOR'S NOTE: The two sections that follow are a part of the original Harper's *article.*

The following letter was published in *The Freedmen's Book,* a collection of African American writings compiled by the abolitionist Lydia Maria Child in 1865. The letter is a response to a slave owner who has written to his former slave at the war's end, asking him to return to work in Tennessee.

To my old Master, Colonel P. H. Anderson, Big Spring, Tennessee. Sir. I got your letter, and was glad to find that you had not forgotten Jourdon, and that you wanted me to come back and live with you again, promising to do better for me than anybody else can. I have often felt uneasy about you. I thought the Yankees would have hung you long before this, for harboring Rebs they found at your house. I suppose they never heard about your going to Colonel Martin's to kill the Union soldier that was left by his company in their stable. Although you shot at me twice before I left you, I did not want to hear of your being hurt, and am glad you are still living. It would do me good to go back to the dear old home again, and see Miss Mary and Miss Martha and Allen, Esther, Green, and Lee. Give my love to them all, and tell them I hope we will meet in the better world, if not in this. I would have gone back to see you all when I was working in the Nashville Hospital, but one of the neighbors told me that Henry intended to shoot me if he ever got a chance.

I want to know particularly what the good chance is you propose to give me. I am doing tolerably well here. I get twenty-five dollars a month, with victuals and clothing; have a comfortable home for Mandy,—the folks call her Mrs. Anderson,—and the children Milly, Jane, and Grundy go to school and are learning well. . . . We are kindly treated. Sometimes we overhear others saying, Them colored people were slaves down in Tennessee. The children feel hurt when they hear such remarks; but I tell them it was no disgrace in Tennessee to belong to Colonel Anderson. Many darkeys would have been proud, as I used to be, to call you master. Now if you will write and say what wages you will give me, I will be better able to decide whether it would be to my advantage to move back again.

As to my freedom, which you say I can have, there is nothing to be gained on that score, as I got my free papers in 1864 from the Provost-Marshal-General of the Department of Nashville. Mandy says she would be afraid to go back without some proof that you were disposed to treat us justly and kindly; and we have concluded to test your sincerity by asking you to send us our wages for the time we served you. This will make us forget and forgive old scores, and rely on your justice and friendship in the future. I served you faithfully for thirty-two years, and Mandy twenty years. At twenty-five dollars a month for me, and two dollars a week for Mandy, our earnings would amount to eleven thousand six hundred and eighty dollars. Add to this the interest for the time our wages have been kept back, and deduct what you paid for our clothing, and three doctor's visits to me, and pulling a tooth for Mandy, and the balance will show what we are in justice entitled to. Please send the money by Adams's Express, in care of V. Winters, Esq., Dayton, Ohio. If you fail to pay us for faithful labors in the past, we can have little faith in your promises in the future. We trust the good Maker has opened your eyes to the wrongs which you and your fathers have done to me and my fathers, in making us toil for you for generations without recompense.... Surely there will be a day of reckoning for those who defraud the laborer of his hire. In answering this letter, please state if there would be any safety for my Milly and Jane, who are now grown up, and both good-looking girls. You know how it was with poor Matilda and Catherine. I would rather stay here and starve and die, if it come to that, than have my girls brought to shame by the violence and wickedness of their young masters. You will also please state if there has been any schools opened for the colored children in your neighborhood. The great desire of my life now is to give my children an education, and have them form virtuous habits.

Say howdy to George Carter, and thank him for taking the pistol from you when you were shooting at me.

From your old servant, Jourdon Anderson

TIMELINE OF REPARATIONS FOR AMERICAN SLAVERY

1865 General William Tecumseh Sherman issues Special Field Order No. 15, providing forty-acre tracts of captured land along the Atlantic coast, from South Carolina to Florida, for 40,000 former slaves. Congress establishes the Freedmen's Bureau in March to oversee the distribution of land.

President Andrew Johnson reverses the "forty acres and a mule" provision, ordering the Freedmen's Bureau to return the land to the pardoned Confederate landholders. Later, the claim of forty acres and a mule is, oddly, dismissed in many mainstream standard history books as myth. For example, the most recently revised edition of *The Civil War Dictionary* begins its entry with this phrase: "Legend that sprang up among the newly freed slaves ..."

1866 Congress passes the Southern Homestead Act to provide freedmen with land in Southern states at a cost of $5 for eighty acres. Act fails dismally; only 1,000 freedmen receive homesteads.

1867 Republican Representative Thaddeus Stevens proposes H.R. 29, a slave-reparations bill, which promises each freed adult male slave forty acres of land and $100 to build a dwelling. "[The freedmen] must necessarily . . . be the servants and the victims of others unless they are made in some measure independent of their wiser neighbors," Stevens argues.

1915 Treasury Department is sued for $68 million in remuneration for labor performed under slavery. The government dismisses the case on grounds of sovereign immunity.

1955 Activist Queen Mother Audley Moore founds the Reparations Committee of Descendants of United States Slaves after reading "in an old Methodist encyclopedia that a captive people have one hundred years to state their judicial claims against their captors or international law will consider you satisfied with your condition."

1962 Queen Mother Moore's reparations committee files a claim in California.

1969 James Forman, a radical activist and member of SNCC, interrupts Sunday services at Manhattan's Riverside Church and presents his Black Manifesto, demanding that American churches and synagogues pay $500 million in reparations.

1987 National Coalition of Blacks for Reparations in America (N'COBRA) established to seek reparations from the federal government in the form of a domestic Marshall Plan for black Americans.

1989 Representative John Conyers proposes H.R. 3745, the first of several unsuccessful proposals for the formation of a commission to study reparations for American slavery.

1994 Florida agrees to pay $2.1 million in reparations to the survivors of the 1923 Rosewood massacre.

1995 The Ninth Circuit Court of Appeals rules in *Cato v. United States*, holding that the claim for $100 million in reparations and an apology for slavery lacks a legally cognizable basis, and concluding that the legislature, rather than the judiciary, is the appropriate forum for such claims.

1999 Representative Conyers proposes H.R. 40, seeking a formal apology for slavery and providing for a commission to study reparations.

2000 Representative Tony Hall proposes H.R. 356, a formal resolution to acknowledge and apologize for slavery.

A LEGISLATIVE AND JUDICIAL HISTORY OF
AMERICAN SLAVERY AND ITS AFTERMATH

1619 Twenty Africans sold as bond servants in Jamestown, Virginia.

1621 Dutch West India Company given a monopoly of the American slave trade.

1662 Virginia's general assembly determines that "[c]hildren got by an Englishman upon a Negro woman shall be bond or free according to the condition of the mother," effectively sanctioning the breeding of slaves by slaveholders.

1663 Maryland provides that African slaves shall serve for the duration of their lives.

1664 Maryland declares that baptism does not alter slave status.

1672 King Charles II charters the Royal African Company with exclusive rights to provide the colonies with Africans, putting England at the vanguard of the slave trade by century's end.

1688 Quakers in Germantown, Pennsylvania, draft an antislavery resolution.

1705 Virginia confers upon blacks the status of real estate.

1717 Maryland legislates that if any free negro or mulatto marries a white man or woman, he or she becomes a slave along with their children. Whites and mulattoes born of white women who intermarry, however, are consigned to seven years' servitude.

1724 New Orleans establishes the Black Code, with fifty-five articles designed to regulate the behavior of slaves.

1777 Vermont becomes the first American territory to declare slavery illegal.

1778 Virginia outlaws the trafficking of slaves into the commonwealth.

1779 The Virginia Assembly passes Thomas Jefferson's A Bill Concerning Slaves, restricting the movements of slaves and requiring white women who bear mulatto children to leave the commonwealth with their children.

1780 The state constitution of Massachusetts declares colored persons descended of African slaves to be citizens.

1783 Maryland forbids further importation of slaves.

1787 The Constitutional Convention determines that for the purposes of representation and taxation slaves will be counted as three-fifths of a free man.

1790 The first census of the United States records 757,000 black Americans, composing 19 percent of the population. More than 697,000 of them are slaves.

1791 Free Negroes of Charleston, South Carolina, protest severe legal disabilities and request to be treated as citizens.

1792 Construction begins on the White House in Washington, D.C., requiring an influx of slaves to lay the foundation.

1793 Congress passes the Fugitive Slave Act, which allows slave owners to seize runaways in any state or territory and sets fines for the harboring of fugitive slaves at $500. Three slaves are executed in Albany, New York, for antislavery activities.

1797 Congress rejects the North Carolina Slave Petition, the first recorded petition for an end to slavery by freed blacks.

1800 Boston refuses to support black schools.

1804 Underground Railroad begins when a Revolutionary War officer purchases a slave and takes him to Pennsylvania. The slave's mother later escapes and follows her son north. Virginia forbids all evening meetings of slaves.

1808 Congress prohibits further importation of slaves.

1810 Maryland denies free blacks the right to vote.

1817 The American Colonization Society is established to send freed blacks to Africa.

1820 The Missouri Compromise, admitting Missouri as a slave state and Maine as a free state, prohibits slavery in the rest of the Louisiana Purchase north of the 36th parallel.

1822 The American Colonization Society establishes the Liberian colony on the west coast of Africa.

1827 New York enacts gradual emancipation law.

1831 Nat Turner leads a slave rebellion in Southampton County in Virginia, killing fifty-five whites. One hundred twenty blacks are killed in retaliation in less than two days.

   Mississippi law declares that it is "unlawful for any slave, free Negro, or mulatto to preach the Gospel." Violators receive thirty-nine lashes upon their naked back.

1832 Alabama law declares that "any person or persons who shall attempt to teach any free person of color or slave to spell, read or write, shall, upon conviction thereof by indictment, be fined in a sum not less than $250, nor more than $500."

1836 Congress passes a resolution ceding authority over slave laws to the states.

1847 Liberia declares independence.

1850 The Compromise of 1850 results in a new Fugitive Slave Act strengthening slaveholders' ability to capture runaways in the northern free states.

1857 The Supreme Court rules in *Dred Scott v. John F. A. Sanford,* declaring that the Missouri Compromise is unconstitutional, that blacks are not citizens, and that a slave does not become free upon entering a free state.

1862 Congress abolishes slavery in the District of Columbia and the territories.

1863 President Abraham Lincoln issues the Emancipation Proclamation, freeing slaves in the Confederate states.

1865 The Thirteenth Amendment is ratified, abolishing slavery in the United States.

1866 Congress passes the Civil Rights Act on April 9, granting citizenship and equal rights to black Americans.

The Fourteenth Amendment is passed, guaranteeing to all United States citizens due process and equal protection under the law.

1867 Congress grants black citizens the right to vote in the District of Columbia and the territories. The first of several Reconstruction Acts places Confederate states under federal military rule.

1869 The Fifteenth Amendment is passed, guaranteeing black Americans the right to vote.

1875 The Civil Rights Act of 1875 is passed, guaranteeing equal rights to black Americans in public accommodations and in service on a jury. Elected in 1874, black Republican Blanche Kelso Bruce begins serving in the United States Senate from Mississippi.

1877 The Compromise of 1877 ends Reconstruction.

1881 Tennessee segregates railroad cars. Other Southern states follow suit.

1883 The Supreme Court declares the Civil Rights Act of 1875 unconstitutional, holding that the Fourteenth Amendment forbids states, but not citizens, from discriminating against blacks.

1890 The Mississippi Plan requires black voters to pass literacy and understanding tests, leading the effort by Southern states to disenfranchise black citizens.

1896 Supreme Court rules in *Plessy v. Ferguson,* establishing the separate-but-equal doctrine.

1909 The National Association for the Advancement of Colored People (NAACP) is established to advocate for civil rights for black Americans.

1910 Baltimore approves the first city ordinance designating the boundaries of black and white neighborhoods.

1913 President Woodrow Wilson institutes federal segregation of workplaces, restrooms, and lunchrooms.

1917 Marcus Garvey establishes a Universal Negro Improvement Association branch in the United States and launches the Back to Africa movement.

1934 Costigan-Wagner Antilynching Bill defeated in Congress.

1948 Supreme Court rules in *Shelley v. Kraemer,* one of several housing-discrimination cases, that enforcement of restrictive covenants by state courts is unconstitutional. President Harry S Truman integrates the armed forces.

1954 Supreme Court ruling on *Brown v. Board of Education of Topeka, Kansas* strikes down the separate-but-equal doctrine.

1957 Congress passes the Civil Rights Act of 1957—the first since Reconstruction creating a Civil Rights Division in the Justice Department and the Civil Rights Commission to study all aspects of segregation.

1960 The Civil Rights Act of 1960 outlaws interference with desegregation orders and voter rights. The Supreme Court declares segregation in bus and railway terminals unconstitutional in *Boynton v. Virginia.*

1964 The Civil Rights Act of 1964 creates the Equal Employment Opportunity Commission and prohibits discrimination by businesses and employers.

1965 The Voting Rights Act is passed to enforce the Fifteenth Amendment. Two hundred fifty thousand blacks register to vote by the end of the year.

1968 The Civil Rights Act of 1968 prohibits discrimination in housing.

1978 Supreme Court ruling in Regents of the University of *California v. Bakke* strikes down quota system in university admissions.

1995 The Regents of the University of California vote to end affirmative action in university admissions.

1996 California votes in favor of Proposition 209 to ban affirmative action in government employment and college admissions.

1998 Washington citizens vote to ban affirmative action in government employment and college admissions. Similar efforts follow in Florida.

1999 After an investigation reveals that black drivers on the New Jersey Turnpike were five times more likely than white drivers to be stopped by New Jersey State Police, the Justice Department appoints a state monitor.

JACK HITT *is a contributing editor of* Harper's *magazine.*

WILLIE E. GARY *won a $500 million judgment against the Loewen Group Inc., the world's largest funeral-home and cemetery operators, in 1995 and $240 million against the Walt Disney Company last August. He is an attorney with Gary, Williams, Parenti, Finney, Lewis, McManus, Watson & Sperando, in Stuart, Florida.*

ALEXANDER J. PIRES JR. *won a $1 billion settlement for black farmers in their discrimination case against the U.S. Department of Agriculture and is currently working on a multibillion-dollar class-action suit on behalf of Native Americans. He is an attorney with Conlon, Frantz, Phelan & Pires, L.L.P., in Washington, D.C.*

RICHARD F. SCRUGGS *won the historic $368.5 billion settlement for the states in their suit against tobacco companies in 1997 and is currently building a class-action suit against HMOs. He is an attorney with Scruggs, Millette, Bozeman & Dent, P.A., in Pascagoula, Mississippi.*

DENNIS C. SWEET III *won a $400 million settlement in 1999's "fen-phen" diet-drug case against American Home Products Corporation and $145 million against the Ford Motor Company. He is an attorney with Langston Sweet & Freese, in Jackson, Mississippi.*

ROBERT WESTLEY

Many Billions Gone: Is It Time to Reconsider the Case for Black Reparations?

THE ECONOMIC PREDICATE FOR
BLACK REPARATIONS

At the conclusion of his exhaustive examination of statistical indicia of Black socioeconomic disadvantage in relation to whites, the historian and political economist Manning Marable aptly observes that "[s]tatistics cannot relate the human face of economic misery." Buried in the jungle of statistical disparity are the life circumstances, impossible choices, and tedium of deprivation. As a democratic socialist, Manning takes aim in his book *How Capitalism Underdeveloped Black America* at both the legacy of indifference to Black disadvantage fostered by the history of white racism and the exploitive dimensions of capitalist accumulation in which a substantial segment of the Black population is forced to serve as a symbolic index of the distance between working-class whites and the abyss of absolute poverty. Hard-core poverty, poverty resistant to all attempts at amelioration, is thus indexically related to a segment of the Black population (and in some social imaginaries, all Blacks). In the sociological literature, this segment of the Black population is often isolated by the terms "underclass" or "ghettoclass" or "ghetto poor." Although there are substantial reasons to demarcate analytically class or economic distinctions within the Black population, the primary focus of the following analysis is the continuing existence of major disparities in the economic condition and life opportunities of Blacks and whites.

Just as there can be no doubt that such interracial disparities weigh most heavily upon the underclass, there can be no doubt that the persistence of those disparities is due in large measure to legally enforced exploitation of Blacks and socially widespread anti-Black racism. The achievements of Blacks who have prevailed against racist odds to improve their economic condition should not be minimized, but neither should the impact of the history and perdurance of racism on Black economic opportunity be trivialized. Despite well-publicized success cases like Oprah Winfrey, Michael Jackson, Bill Cosby, Michael Jordan, and others, Blacks as a group have not reached anything approaching economic equality or equality of opportunity with whites. Given the glacial and limited nature of economic reform, this is unsurprising. Because racism, in addition to its psychological aspects, is a structural feature of the U.S. political economy, it produces intergenerational effects.

Highlighting the intergenerational effects of structural racism in the United States political economy, [Harvard psychologist] Thomas Pettigrew notes that three useful generalizations can be made about the current situation of Black Americans. First, current statistics on Blacks, when compared to earlier data, show substantial improvement in Black living conditions. However, these same statistics pale when compared to current data on whites. Second, most of the "progress" of the past twenty years reflects the establishment of a solid, sizable, and skilled Black middle class which, crucially, is able to pass on its human capital to its children. Conversely, the most bleak statistics reflect the desperate situation of the unskilled Black poor or underclass. Third, modern forms of racism, to a greater extent than in the past, have become more subtle, indirect, procedural, and ostensibly nonracial. Pettigrew focuses on the analysis of traditional inequality factors, such as income, education, housing, employment patterns, and so forth, and how these factors operate in the context of the new racism. However, the burden of the reparations argument, for which material inequality may serve as a first predicate, is to show that current disparities in material resources are causally linked to unjust and unremedied actions in the past. Rather than merely highlighting intergenerational effects based on traditional inequality factors assumed to be causally linked to past racial discrimination against Blacks, the following discussion seeks to elucidate a key causal element in the maintenance of structural racism: the economic determinant of wealth.

The above observations form a set of concerns for reparations policy and political action this article attempts to address in the two sections below. Under the heading The Underclass Question: General Statistics and the Human

Face of Misery, I will present some of the current data on Black disadvantage that leads me to conclude that equality between Black and white Americans, even those who are considered middle class, has not been achieved. At the same time, I argue that the neoconservative attack on the poor and the instrumentalization of the Black middle class in pursuit of conservative agendas fail to account for the structural and intergenerational dimensions of racial disadvantage and privilege. Under the heading The Racist Restatement, I will sketch the vocabulary and practices of the new racism that set the context in which the reparations struggle must take place.

A. THE UNDERCLASS QUESTION:
GENERAL STATISTICS AND THE HUMAN
FACE OF MISERY

In his highly acclaimed monograph [*Two Nations: Black and White, Separate, Hostile, Unequal*], political science professor Andrew Hacker notes that in the minds of most white Americans, "the mere presence of [B]lack people is associated with a high incidence of crime, residential deterioration, and lower educational attainment." Even though most whites are willing to acknowledge that these characterizations do not apply to all Blacks, most whites prefer not to have to worry about distinguishing Blacks who would make good neighbors from those who would not. Housing segregation and educational disadvantage, therefore, remain dismally high.

Pettigrew, for instance, reports that the modest housing gains of Blacks do not begin to achieve parity with white housing. A "nationwide pattern of residential apartheid" continues to be the rule rather than the exception. Thus, throughout the 1960s and 1970s, urban Blacks were residentially segregated from their fellow Americans far more intensively than any other urban ethnic or racial group. Moreover, the improvements seen in the Black housing stock are primarily attributable to the ability of the expanding Black middle class to buy older houses left behind by suburban-bound whites. Thus the Black middle class, as well as the Black working class, have been victimized by this massive discriminatory pattern in housing.

The white American perception of Blacks as a "bad risk" was openly reflected in federal governmental housing policy until 1948, when the Supreme Court struck down judicial enforcement of one of the most blatant tools of racial discrimination in housing, the restrictive covenant. As Chief Justice Vinson explained, restrictive covenants were private agreements among home owners which have as their purpose the exclusion of persons of designated race or color from the ownership or occupancy of real property. Although the Court only considered judicial participation in the enforcement of such agreements to be illegal, as a consequence of the Court's decision, the Federal Housing Authority discontinued its open policy of subsidizing mortgages on real estate subject to racially restrictive covenants in 1950. But by then, thousands of Black families had already missed out on millions of dollars in wealth through equity accumulation, while whites benefited handsomely from discriminatory federal housing subsidies.

The practice of government-enforced and private "redlining" in the home mortgage industry continued after 1950 through less blatant means than the restrictive covenant, leading to the current urbanization and ghettoization of Blacks, and the suburbanization and relative economic privileging of whites. Based on discrimination in home mortgage approval rates, the projected number of creditworthy Black home buyers, and the median white housing-appreciation rate, it is estimated that the current generation of Blacks will lose about $82 billion in equity due to institutional discrimination. All things being equal, the next generation of Black homeowners will lose $93 billion.

As the cardinal means of middle-class wealth accumulation, this missed opportunity for home equity due to private and governmental racial discrimination is devastating to the Black community. Wealth, although related to income, has a different meaning. Wealth is "the total extent, at a given moment, of an individual's accumulated assets and access to resources, and it refers to the net value of assets (for example, ownership of stocks, money in the bank, real estate, business ownership, etc.) less debt held at one time." Income, on the other hand, refers to the flow of dollars over a set period of time. Just as substantial income, over time, may produce wealth, substantial wealth produces income and all the advantages in life that make up material well-being. Crucially, for the current situation of the Black community, wealth disparities between Blacks and whites are both cumulative and vast. It is a gap that earned income alone cannot close, and a gap that fundamentally supports structural distinctions of status between the white middle class and the Black middle class.

As Oliver and Shapiro argue [in *Black Wealth/White Wealth*], middle-class status "rests on the twin pillars of income and wealth." Without either one or the other, that status can be quickly eroded or simply crumble. On average, Blacks who hold white-collar jobs have $0 net financial assets compared to their white counterparts, who on average hold $11,952 in net financial assets. Black middle-class status, as such figures indicate, is based almost entirely on income, not assets or wealth. Thus, the Black middle class can at best be described as fragile.

Structural advantages accrue to a wealth-based white middle class over an income-based Black middle class. Whether poor or "middle class," Black families live without assets, and compared to white families, Black families are disproportionately dependent on the labor market to maintain status. In real-life terms, this means that Blacks could survive an economic crisis, such as loss of a job, for a relatively short time. Thus one structural advantage that accrues to a wealth-based white middle class over an income-based Black middle class is relative independence within and security from a fluctuating labor market. Another advantage of wealth over income is the possibility to reproduce middle-class status intergenerationally through gift or inheritance. The overall advantage of wealth to income is in the ability both to meet current needs and to plan concurrently for future needs.

Not only are middle-class Black families more fragile, precarious, and marginal than the white middle class due to a lack of wealth, Oliver and Shapiro also demonstrate that poverty among Blacks and whites often means very different things. Poverty-level whites control nearly as many mean net financial assets as the highest-earning Blacks. The importance of this disparity among the Black and white poor would not be revealed by an analysis that focused entirely on income. The importance of this disparity is that it shows that even those at equivalent income levels can have vastly different life prospects, depending on their access to wealth resources. With no assets to rely on, and earning barely enough to survive, an edge of desperation is added to the plight of the Black poor. These disparities are important because they highlight the cumulative effects of societal and government-sponsored racial discrimination.

When we consider the living conditions and life prospects of the Black underclass, we confront a population that is able neither to meet its current needs without public assistance (or private charity) nor to plan effectively for future needs. To many neoconservative critics, the disparity between the Black middle class and the underclass is explicable in terms of the culture-of-poverty thesis. According to the culture-of-poverty thesis, poor Blacks are responsible for their own immiseration due to their cultural pathology and lack of values. Black middle-class success is juxtaposed to Black underclass failure to acquire the skills and discipline necessary to move ahead. And yet, the neoconservative attack on the poor and the instrumentalization of the Black middle class in pursuit of conservative agendas fail to account for the structural and intergenerational dimensions of racial disadvantage and privilege.

Ignoring the structural and intergenerational dimensions of racial advantage and disadvantage, neoconservatives push the idea that racial inequality has little (or nothing) to do with racism, but lots to do with bad individual choices and inappropriate cultural values (or no values at all). Furthermore, neoconservatives assert that government policies aimed at providing subsistence for the poor, such as Aid to Families with Dependent Children, contribute to their demoralization, and for that reason should end. Neoconservatives subscribe to a reform framework that focuses on elimination of poor subsistence support by the government, including the minimum wage, and promotion of self-help.

There are at least three problems with self-help that bear mention in the context of developing solutions to racial inequality. First, there is no assurance that self-help will ever bring about substantive equality between Blacks and whites. Given the scope and extent of current inequality, Blacks generally, and the underclass particularly, may be permanently economically subordinate to and dependent upon whites. Second, even if self-help achieved equality, again, the current disparities are so great that generations would endure unjust deprivations. By contrast, taking account of the structural and intergenerational dimensions of racial advantage and disadvantage implies a reform framework that does not simply blame the victims of societal discrimination and overtly racist government policies. Third, and most importantly, self-help provides no redress for unjust expropriations and denials of equal opportunity. Where the implementation of racist policies has a substantial and continuing impact on the ability of a social group to achieve equality, as they clearly do in the case of Black Americans, reparations is a just remedy.

For Pettigrew, statistics on the state of Black Americans do not augur the "declining significance of race," but the growing significance of the interaction between class and race in American race relations. One feature of this interaction is that because the new Black middle class has typically gained its status through employment in predominantly white institutions, many whites, especially those of higher status, now meet and come to know members of the Black middle class. But Black poverty remains largely out of the intellectual and experiential purview of the vast majority of whites. Pettigrew writes: "The fact that whites know the [B]lack 'success cases' but not the [B]lack poor undoubtedly contributes to the widespread current belief among whites that racial discrimination is now minimal" and ". . . the chances for [B]lacks to get ahead have improved greatly . . ." [citation omitted]. Both at the individual and institutional levels, racism is typically far more subtle, indirect, and ostensibly nonracial now than it was in 1964.

B. THE RACIST RESTATEMENT

In developing a vocabulary to characterize the new racism, Pettigrew isolates the following six features based on his social scientific research: (1) rejection of gross stereotypes and blatant discrimination; (2) normative compliance without internalization of new behavioral norms of racial acceptance; (3) emotional ambivalence toward [B]lack people that stems from early childhood socialization and a sense that [B]lacks are currently violating traditional American values; (4) indirect "micro-aggressions" against [B]lacks which are expressed in avoidance of face-to-face interaction with [B]lacks and opposition to racial change for ostensibly nonracial reasons; (5) a sense of subjective threat from racial change, and (6) individualistic conceptions of how opportunity and social stratification operate in American society.

Pettigrew explains that compliance in the racial context means that whites follow the new norms only when they are under the surveillance of authoritative others who can reward and punish. Internalization means that whites have adopted the new norms as their own personal standard of behavior and will follow them without surveillance. He notes that Black Americans, too, must learn the new norms. This process often entails unlearning past lessons and overcoming suspicions.

Exemplifying the new forms of anti-Black racism, Pettigrew points to the fact that about 90 percent of white Americans believe Black and white children should attend "the same schools," and that 95 percent favor equal job opportunity. However, in 1978 only 24 percent believed the federal government should "see to it that white and [B]lack children go to the same school." Furthermore, this percentage declined from 43 percent in 1966. "Likewise, in 1975 only 34 percent agreed that the federal government should 'see to it that the [B]lacks get fair treatment in jobs,' a percentage that remained constant from 1964." So while an overwhelming majority of whites may currently oppose blatant discrimination, it is likewise the case that they oppose concrete remedies to discrimination. Few would perceive this apparent contradiction as "racist." This perception informs Pettigrew's conclusion that whites experience deep emotional ambivalence toward Black people, while at the same time rejecting gross stereotypes. Whites have a sense of subjective threat from racial change that is inconsistent with the new norms of racial acceptance. Whether, as Pettigrew asserts, the ambivalence of whites toward Blacks is entirely shaped by an individualist conception of opportunity in America, this factor is of notable importance.

Pettigrew's research reveals that (1) spatial discrimination, (2) cumulative discrimination, and (3) situational discrimination are three (often interrelated) ways in which indirect and ostensibly nonracial racial discrimination operates. An example of cumulative discrimination is racially different access to mortgages. Unsurprisingly, spatial segregation results in Black voter dilution through annexations, redistricting, or the like. It produces housing discrimination through decentralization of governmental services or resource distributions as laundered through private preferences in housing and rental markets. Situational discrimination refers to those pervasive and largely unconscious (to the perpetrator, at least) circumstances where white "microaggressions" against Blacks come into play. Pettigrew describes this phenomenon as "triple jeopardy." In face-to-face interracial situations within predominantly white institutional settings, Blacks often encounter three interrelated hardships that make their inclusion difficult. First, Blacks must face the intransigence of racist stereotypes imposed by whites that limit their ability to perform. Second, Blacks experience the stress of occupying solo roles. And finally, Blacks must endure the opprobrium associated with being a token of affirmative action.

The importance of Pettigrew's research consists not merely in development of a framework and a vocabulary by which to examine the modern expression of anti-Black prejudice. Racism in America has frequently been characterized as a "sickness." To the extent that this view of racism is correct, Pettigrew's research pathologizes perspectives which would otherwise be regarded as purely political—for example, the dominance of individualism in American political and social life—or purely personal—for example, the choice of school, profession, or neighborhood. Less frequently in modern discourse, racism is considered to be an intellectual position based on the belief in the inherent superiority of whites. This alternative view, however, is racism's history. Pettigrew reveals that such a view remains racism's practice.

The pervasiveness of white supremacist structures cannot be limited to the social spheres examined by Pettigrew. They inhabit our literature and the canons of literary interpretation; they inhabit our speech; they inhabit popular culture, from films and television, to music, dance, and fashion; they determine classroom curricula throughout the educational system; they influence the friends we make, the restaurants we choose to eat in, the places we shop; they establish national priorities and the means employed to resolve social problems; often, they define what it means to be a problem. White supremacist structures insinuate their presence into the most intimate encounters among people, especially sexual ones; they inform critical standards in art and philosophy, legal standards in politics, educational standards in school and professional standards in employment.

It is difficult, if not impossible, to expose indexically the many blatant and recondite ways racism has entered the lives of Americans. This much is clear: structures of white supremacy have asserted hegemony over numerous aspects of social, political, and personal life in the United States. This is the reality that lies behind the statistics. Racism, as the practice of white supremacy, cannot be circumscribed by the petty injustices that individuals commit against individuals. Racism is a group practice. The theory of that practice is the viability of the race idea, and the anomalous belief that group harms may be legally remedied solely through redress to individuals. To show just how anomalous the belief is that individual redress can adequately remedy group injuries, we should consider three historical moments of group oppression after each of which an attempt was made to compensate serious harms to groups: the Japanese Internment, the Jewish Holocaust, and Black Reconstruction.

1. Japanese Americans

In 1942, under the authority of President Franklin D. Roosevelt, 120,000 people of Japanese ancestry from the West Coast were ordered to be evacuated, relocated, and interned by the U.S. military. Approximately two-thirds of those interned were native-born American citizens. The internment order was issued in direct response to the bombing of Pearl Harbor by the Japanese Empire. The research of Professor Peter Irons revealed that the government fraudulently concealed its actual reasons for internment of Japanese-American citizens from the Supreme Court in initial litigation challenging the internment order. Subsequent litigation efforts have overturned cases upholding the government's authority to enforce the internment order. History has shown that greed, prejudice, and "race" hatred had more to do with the internment of Japanese Americans than concern for national security.

The indignities suffered by Japanese Americans due to their internment were not confined to their loss of freedom. They lost both real and personal property. They lost businesses and employment income. They lost pets and farm animals. They were forced to wear identification tags, and many endured living conditions unfit for animals. They suffered the disruption of familial life and customs. They suffered disease and hardship from exposure to the elements, poor sanitation, and poor diet. They lost all rights to privacy, even to the extent of performing ordinary bodily functions. They suffered shame. They lost educational opportunities. They lost freedom of expression and the ability to communicate freely with others outside the camps. They were denied the right to use the Japanese language or read Japanese literature other than the Bible and the dictionary. They lost control over their own labor. Even the moral conscience of Japanese Americans was invaded by conditioning release on swearing an oath of loyalty to the United States. Many internees, especially the elderly, endured these conditions for as long as four years. Many died. Upon release, hostility toward Japanese Americans continued, though the majority had neither homes nor businesses nor jobs to which to return.

Despite their tremendous collective losses, the government initially provided only minimal assistance to help those who had been interned return to normal life. Most received train fare and $25. In 1948, Congress enacted the American-Japanese Evacuation Claims Act. This piece of legislation remained the only official attempt by Congress to compensate Japanese-American property losses for over forty years. It was flawed primarily for the following reasons. First, it required the attorney general to limit any award to $100,000 upon a showing that damage or loss of property was "a reasonable and natural consequence of the evacuation or exclusion. . . ." Second, it required that compensation be paid only for loss of property that could be proved by records. Finally, once a claim had been paid under the act, the claimant waived his or her right to make any further claims against the United States arising out of the evacuation.

On August 10, 1988, President Reagan signed the Civil Liberties Act of 1988 into law. In doing so, he set in motion the statutory means by which Japanese Americans would begin to receive federal reparations payments. Although deficiencies remain in how the government has implemented this legislation, the importance of the legislation lies in the precedent established for compensation of wronged groups within the American system. Crucially, the Civil Liberties Act pays compensation to the group (surviving internees and their next of kin) on the basis of a group criterion. The act acknowledges that Japanese Americans were harmed as a group; that they should be compensated as a group; and that they should be made whole economically for the injuries they suffered on the basis of group membership. In addition to monetary compensation, the law also authorized institutions by which the injustice done to Japanese Americans may be memorialized.

Memorializing injustices committed in the past is not only an obviously important way of preventing those same injustices from occurring in the future; it also provides public recognition of suffering, a chance for victims and their ancestors to mourn their loss in a social space that symbolizes respect, and a constant reminder to potential aggressors or the destructively indifferent that history will not overlook grievous abuses of human dignity.

Perhaps there are some lessons in the Japanese-American reparations experience for those seeking reparations for Black Americans. In "Racial Reparations: Japanese American Redress and African American Claims," Professor Eric Yamamoto suggests that Japanese-American claims succeeded, as did those of Blacks who were the survivors of the Rosewood massacre, because they, unlike the reparations claims of Black Americans generally, fit tightly within the individual rights paradigm of the law. He proposes that successful claims must fit the traditional individual rights paradigm of the law by satisfying the demand for identifiable victims and perpetrators, direct causation, damages that are limited and certain, and acceptance of payment as final. The demand for identifiable victims and perpetrators and direct causation is difficult (if not impossible) to meet from a class whose reparations claims include acts that occurred hundreds of years ago, and many of whose members were not yet born when the most egregious violations were occurring.

Importantly, however, a tight fit with the individual rights paradigm may be considered a legal prerequisite to success only in the context of judicially imposed redress. A tight fit is not a moral prerequisite, nor is it a legal barrier to legislative redress. It is noteworthy that even Japanese-American claims were denied by courts and ultimately awarded by Congress. Additionally, the survivors of the Rosewood massacre received reparations as a result of the action of the Florida legislature. In the context of legislative action, the demand for a tight fit may be a practical or political, rather than a legal, prerequisite to success. Political realities change. As in the case of reparations for Japanese Americans, political realities changed partly as a function of the passage of time (allowing an abatement of anti-Asian hostility), partly as a function of concerted effort by community activists who challenged the status quo (demanding that American society live up to its professed ideals), and partly as a function of shifts in international relations (at the time that Japanese-American reparations were approved, Japan had become an important U.S. ally and a major economic force). Standing alone, a tight fit with the individual rights paradigm of the law could not persuade American courts to award group reparations even to identifiable victims of racial injustice.

2. European Jews

If, *arguendo*, the example of the Japanese-American internment, followed by legislation enabling Japanese Americans to receive reparations and public recognition of their suffering, can serve as a limiting case of the United States' willingness to redress wrongs committed against a group with group remedies, then the example of *Wiedergutmachung* for the Jewish survivors of the Holocaust should be considered the model from which it is drawn. The Nazi attempt to exterminate European Jewry stands as the centerpiece of the twentieth-century conception of genocide. In this regard, we can say with confidence that all the suffering Japanese Americans endured at the hands of the white American establishment, European Jews certainly suffered under the viciously corrupt government of Nazi Germany.

While the number of Jews who lost their lives as a result of the Nazi campaign of genocide is staggering, the methods employed by the Nazis to accomplish their goals evince an irredeemable degree of hatred and cruelty. But the shocking and gruesome means by which the Nazis slaughtered millions of Jews cannot distract our observation of Jewish material and economic losses. Those losses too were staggering.

Germans plundered Jewish property in a variety of ways. Jews were forced to hand over their jewelry and other valuables, their bank accounts were frozen, they were not allowed to inherit, and they were subjected to collective levies and fines. Jews, fearing their property would be seized, tried transferring it—in toto or in part—to non-Jews by fictitious sales or else sold it at prices far below its real value. Others, deprived of their source of livelihood and in need of wherewithal to go on living, were forced to sell off their belongings. After the greater part of their property had been taken from them in the guise of a "Flight Tax" (*Reichsfluchtsteuer*), those who emigrated could only take a small sum of German money and that too was converted to foreign currency at the lowest possible rate. Fleeing Nazi persecution, tens of thousands of Jews abandoned homes, businesses, and personal property. Germans confiscated Jewish possessions by concentrating the Jews in ghettos and other sealed-off areas. At the point that the Germans began deporting Jews to concentration camps, they often had very little left.

Even before the end of World War II, plans were being formulated by Jewish organizations and personalities outside Germany for compensation to individuals and reparations to the Jewish people as a whole. The eventual claimants who signed the Luxemburg Agreements in September 1952 were the State of Israel, on behalf of the half million victims of the Nazis who had found refuge in its borders, and the Conference on Jewish Material Claims against Germany [hereinafter the Claims Conference], on behalf of the victims of Nazi persecution who had immigrated to countries other than Israel and of the entire Jewish people entitled to global indemnification for property that had been left heirless. The Luxemburg Agreements became the basis of an unprecedented piece of legislation known as *Wiedergutmachung*.

Wiedergutmachung was unprecedented in several respects. First, international law did not require Germany to make reparations payments to victims of the Holocaust. Nor did the Allied Powers exert pressure on Germany to accede to the Luxemburg Agreements. The treaty obligation by which Israel was to receive the equivalent of $1 billion in reparations from West Germany for crimes committed by the Third Reich against the Jewish people reflected

Chancellor Konrad Adenauer's view that the German people had a moral duty to compensate the Jewish people for their material losses and suffering. Secondly, the sums paid not only to Israel, but also to the Claims Conference, showed a genuine desire on the part of the Germans to make Jewish victims of Nazi persecution whole. Under Protocol No. 1 of the Luxemburg Agreements, national legislation was passed in Germany that sought to compensate Jews individually for deprivation of liberty, compulsory labor and involuntary abandonment of their homes, loss of income and professional or educational opportunities, loss of (World War I) pensions, damage to health, loss of property through discriminatory levies such as the Flight Tax, damage to economic prospects, and loss of citizenship. The elderly, the needy, and the disabled were to receive priority in payment. Near heirs were eligible to assert the claim of a persecutee who died without receipt of payment. Real property was to be restored, with extremely limited protection of "good faith" purchasers, and identifiable personal property was also to be restored or compensated. In matters of proof of possession, equitable consideration was given to persecutees whose files and documents had been lost or destroyed.

Finally, *Wiedergutmachung* was remarkable and unprecedented for the principle it established. As David Ben Gurion was to say after signing of the agreements: "There is great moral and political significance to be found in the Agreement itself. For the first time in the history of relations between people, a precedent has been created by which a great State, as a result of moral pressure alone, takes it upon itself to pay compensation to the victims of the government that preceded it. For the first time in the history of a people that has been persecuted, oppressed, plundered and despoiled for hundreds of years in the countries of Europe a persecutor and despoiler has been obliged to return part of his spoils and has even undertaken to make collective reparation as partial compensation for the material losses."

The principle, then, was that when a state or government has through its official organs—its laws and customs—despoiled and victimized and murdered a group of its own inhabitants and citizens on the basis of group membership, that state or its successor in interest has an unquestionable moral obligation to compensate that group materially on the same basis. Jews were persecuted and oppressed in Germany as a group. Germany sought to compensate them both individually and as a group. Much of the impetus behind the Jewish demand for group compensation was the realization that, because so many of the Nazis' Jewish victims had perished, the new German state would reap the material benefits of Nazi crimes. Like abandoned Japanese property on the West Coast which escheated to the state and was auctioned off, heirless Jewish property in Germany provided yet another classic example of unjust enrichment. *Wiedergutmachung* in the form of reparations to the entire Jewish people significantly diminished the extent of this injustice.

It is unlikely that David Ben Gurion, in stating that the Luxemburg Agreements represented a "first" in the history of human society, was unaware of the situation of Black people in the United States. Blacks have never received any group compensation for the crime of slavery imposed upon them by the people and government of the United States. As in the case of the Japanese, Jews received not only material compensation for their losses, but their victimization was also publicly memorialized in Germany, Israel, and in the United States (even though there was no legitimate claim of oppression or genocide that Jewish survivors of the Holocaust might assert against the United States). The only "memorial" dedicated to the suffering of Black slaves and the survivors of slavery in the United States is contained in a series of legislative enactments passed after the Civil War. The history of Black Reconstruction shows how these enactments were successively perverted by the courts, and by Congress itself.

3. Black Americans

After the hostilities of the Civil War ended, Congress pursued a legislative program calculated to secure the social and political equality of the freedmen. In pursuance of its enforcement power under the Fourteenth Amendment, Congress passed the Ku Klux Klan Act of 1871. Congress also passed the Civil Rights Act of 1875 under the Fourteenth Amendment. Its preamble stated: "[W]e recognize the equality of all men before the law, and hold that it is the duty of government in all its dealings with the people to mete out equal and exact justice to all, of whatever nativity, race, color, or persuasion, religious or political . . . [and that it was] the appropriate object of legislation to enact great fundamental principles into law."

In pursuance of its enforcement power under the Fifteenth Amendment, Congress also passed the Civil Rights Act of 1870. This act essentially reiterated the provisions of the 1866 act, adding criminal penalties for violation of the law and a conspiracy section, and seeking to effectuate the right of free suffrage.

At the same time, Congress sought to ensure the future economic independence of Black people. Of the Freedmen's Bureau Acts passed for the economic independence of Black people, the most important aspects were the land and education provisions. Under the first act, Congress made no appropriation for the duties assigned to the bureau. The bureau's income was derived from abandoned lands rented to freedmen and refugees. As President Johnson pursued his policy of pardoning ex-Confederates and restoring their land to them, however, the bureau was gutted of its only source of funding. More importantly for the freedmen, their hope of buying this land from the federal government evaporated.

Congress acted again in the summer of 1866, this time not through Freedmen's Bureau legislation, but by extending the hope of land to the freedmen through the Southern Homestead Act. Under the act, lands in Alabama, Arkansas, Florida, Louisiana, and Mississippi were opened for settlement in eighty-acre plots. Ex-Confederates could not apply for homesteads before January 1, 1867. This gave the freedmen roughly six months to purchase land at reasonably low rates without competition from white Southerners and Northern investors.

Because of their destitution and depressed economic conditions in the South, most freedmen were unable to take advantage of the homesteading program. The majority of the homesteads were taken up by Blacks in Florida, but even there the total number was only a little over three thousand. The lands provided by the Homestead Act were generally inferior for farming purposes. Often the lands were distant not only from transportation lines but also from employment centers where freedmen needed to work until they could become self-supporting. Most homesteaders lacked both the means for a few months' subsistence and the most elementary farming equipment. The homesteading program was thus a miserable failure.

The work of educating the freedmen was first taken up during the war by the benevolent societies of the North, such as the Edward L. Pierce group, the American Tract Society, and the American Missionary Association. By January 1865, 75,000 Black children in the Union-occupied South were being taught by approximately 750 teachers. Nearly all those who received compensation for teaching Black pupils in the South during this time were supported by private charities.

Under the Freedmen's Bureau Act of 1866, Congress provided $500,000 for rent and repair of school and asylum buildings, and decided that the bureau might "seize, hold, lease or sell for school purposes" any property of the ex–Confederate States. To meet the need for permanent schools, the bureau in most states paid for completion of buildings that the freedmen themselves began constructing. Often these structures were located on land that the freedmen had purchased for themselves. Additionally, in order to obtain financial assistance from the bureau, school organizations were required to ensure that the buildings would always be used for educational purposes and that no pupil would ever be excluded because of race, color, or previous condition of servitude. By March 1869, the bureau had either built or had helped to build 630 schoolhouses. It had spent $1,771,132.25. In the next three years, its appropriation for educational expenses amounted to another $2,000,000.

From 1867 to 1870, the bureau furnished $407,752.21 to twenty institutions of higher learning for freedmen and $3,000 to a school for white refugees. Of this amount, $25,000 went to Howard University in the nation's capital. By 1871, there were eleven colleges and universities and sixty-one normal schools in the nation which were especially intended for Blacks.

For the safekeeping of the freedmen's savings and the investment of their wartime bounties, Congress also chartered the Freedman's Bank under the Freedman's Saving and Trust Company Acts. The bank was a miserable failure, which, in the end, deprived many of its trusting depositors of their savings.

Although no federal plans for reparations to the former slaves were ever considered, even by the most "radical" members of Congress, the lands provision of the first Freedmen's Bureau Act was intended to make good on a promise that had first been planted in the minds and hearts of Black people by General Sherman. While the Freedmen's Bureau Act of 1865 had promised to purchasers of the lands only "such title thereto as the United States can convey," once the government assigned plots and collected rents and gave options, the radical politicians would be able to argue that it was morally bound to pay reparations to the freedmen. The government could hardly take back for the sake of slave masters and traitors, they would say, what it had given to freedmen and loyalists.

The purpose of the land redistribution plan, as with many of the programs instituted during Reconstruction, was not only to punish the Confederates, but to create among the freedmen a landowning yeomanry, to indebt the freedmen politically to the Republicans, and to ensure the future economic independence of the freedmen. The purpose of land redistribution, however, was not by any means to pay reparations to Blacks for their loss of freedom and uncompensated labor. Ironically, during its first year of operation, the freedmen financed the efforts of the bureau with the rents they paid and they were expected to buy the lands that the Union had confiscated. Even more tragically, President Lincoln had supported, both before and during the war, a plan to pay slave owners for their lost "property" as a means of ending slavery.

Opponents of land redistribution, rejecting the radical analogy of Blacks to the Indians, stated: "There are many reasons why Congress may legislate in respect to the Indians which do not apply. . . . The Indians occupy toward this Government a very peculiar position. They were in possession of the public domain; they had what the Government recognized as a possessory right. . . ."

Congressional critics of Freedmen's Bureau legislation also objected that the position of the freedmen within the American polity was not *sui generis*, and therefore "class legislation" on their behalf was neither justified nor in the spirit of the American constitutional system.

The desire for landownership was both natural and strong among the freedmen. They had cultivated the land on Southern plantations for generations. They had fought in the war to gain their own freedom. Despite the abuses they endured from white Southerners, they thought of the South as their home. In fact, the desire for land was so strong, the belief that the government would deliver so great, and the freedmen's knowledge of government protocol so poor, that carpetbaggers were able to sell fake land deeds to the former slaves. The freedmen were sometimes sold painted sticks which supposedly had been distributed by the government for the purpose of staking out the negroes' forty acres. One spurious land deed proclaimed: "Know all men by these presents, that a naught is a naught, and a figure is a figure; all for the white man, and none for the nigure. And whereas Moses lifted up the serpent in the wilderness, so also have I lifted this [damned] old nigger out of four dollars and six bits. Amen. Selah! Given under my hand and seal at the Corner Grocery in Granby, some time between the birth of Christ and the death of the Devil."

There is no need to recount here the horrors of slavery. Suffice to say that, if the land redistribution program pursued by Congress during Reconstruction had not been undermined by President Johnson, if Congress's enactments on behalf of political and social equality for Blacks had not been undermined by the courts, if the Republicans had not sacrificed the goal of social justice on the altar of political compromise, and Southern whites had not drowned Black hope in a sea of desire for racial superiority, then talk of reparations—or genocide—at this point in history might be obtuse, if not perverse.

As things stand, however, the South pursued a policy of racial separation with the sanction of the Supreme Court and the silent consent of Congress for a century after the official abolition of slavery. The expedient of the lynch mob secured for white supremacists the twin goals of control and exploitation of Blacks on the one hand, and extermination of Blacks on the other. Since Blacks (or "disloyal" whites) could be lynched, beaten, castrated, or burned to death with basic impunity, usually on the pretext of rape of a white woman, the twin goals were met. Total annihilation was never forced to an issue. Even during Reconstruction, Blacks had very little to say about what was owed to them as a group that the white man was bound to respect. That situation has changed remarkably little.

The material bases of the claim for group reparations to Blacks are (1) the value of the uncompensated labor of generations of slaves and (2) the century-long violation of Black civil rights through state-enforced segregation. As Boris Bittker [in *The Case for Black Reparations*] argued succinctly in 1973, the claim for reparations cannot be limited to the outrageous exploitation of Blacks perpetrated during slavery. The ugly facts of the recent past and contemporary life also require redress and compensation. The legacy of Jim Crow is still with us, as the statistics from Pettigrew quoted earlier demonstrate. The psychological inheritance of slavery still exercises the image of the Black in the white mind.

Though slavery officially ended, the attitudes toward intrinsic Black character, based on ideologies of race, persisted. One of the best contemporary articulations of this persistent belief in the duality of Black character occurs in James Baldwin's *Notes of a Native Son*, in the essay "Many Thousands Gone." There Baldwin writes: "In our image of the Negro breathes the past we deny, not dead but living yet and powerful, the beast in our jungle of statistics. It is this which defeats us, which lends to interracial cocktail parties their rattling, genteel, nervously smiling air: in any drawing room at such a gathering the beast may spring, filling the air with flying things and an unenlightened wailing. . . . Wherever the Negro face appears a tension is created, the tension of silence filled with things unutterable."

Blacks deserve reparations not only because the oppression they face is "systematic, unrelenting, authorized at the highest governmental levels, and practiced by large segments of the population," but also because they face this oppression as a group, they have never been adequately compensated for their material losses due to white racism, and the only possibility of an adequate remedy is group redress.

This final part of the argument for Black reparations addresses the nettlesome objection to reparations based in concerns about distributive justice. Doctrinal objections to reparations rooted in the complex question of the identification of victims and perpetrators often serve as a proxy for concerns about redistributive fairness. Distributive justice will not uphold the status quo in which the privileged benefit from past wrongs committed by others. When, moreover, those wrongs were committed with the assistance, support, or acquiescence of government, a claim for redress is appropriately directed to the government. However, any redress awarded by government to victims of group oppression will inevitably be to some extent overinclusive and underinclusive. In this respect, I contend, Black reparations resemble affirmative action, but the arguments in favor of reparations are more compelling than those in favor of affirmative action as a form of redress. In the course of my argument for a plan of group reparations, I consider the ways in which Black reparations avoid some of the pitfalls and drawbacks of affirmative action. Finally, I conclude that Black reparations should be considered a prerequisite to civil equality.

C. REDISTRIBUTIVE FAIRNESS AND BLACK REPARATIONS

In arguing for reparations to Blacks on the model of *Wiedergutmachung,* and drawing upon the precedent of the Civil Liberties Act of 1988, several issues of redistributive fairness must now be faced squarely. Both the Jewish and the Japanese-American experiences contain features that diverge from the reality of Blacks. From the Japanese and Jewish experiences it is clear that the courts are an inappropriate body before which to submit a claim for reparations. Moreover, even though reparations were paid to Japanese Americans on the basis of a group criterion, each eligible claimant received an individual payment. For their part, Jews received both individual and group compensation from the West German government.

The problem of who legitimately represented the material claims of Japanese Americans and Jews was settled in two different ways. In the case of the Japanese Americans, it was settled by structuring the legislation so that individual claimants were compensated. In the case of the Jews, it was settled by structuring the agreements so that individuals were compensated, and a recognized Jewish state, the government of Israel, was compensated on behalf of the group. Because Israel existed as a state, Jews were prepared to accept nonmonetary compensation in the form of goods and services.

The questions raised by the claim of reparations for Blacks from the standpoint of redistributive fairness are what form should reparations take, and what amount of overinclusiveness and underinclusiveness should a plan for Black reparations permit.

1. A Plan for Group Reparations

Because it is my belief that Blacks have been and are harmed as a group, that racism is a group practice, I am opposed to individual reparations as a primary policy objective. Obviously, the payment of group reparations would create the need and the opportunity for institution building that individual compensation would not. Additionally, beyond any perceived or real need for Blacks to participate more fully in the consumer market—which is the inevitable outcome of reparations to individuals—there is a more exigent need for Blacks to exercise greater control over their productive labor—which is the possibility created by group reparations.

Most of the earlier catalogued disabilities that Black people face in contemporary America are traceable to the economic question. Blacks are unemployed or underemployed because they have insufficient Black industries to turn to for jobs when white-controlled industries discriminate against them. Black business is undercapitalized and dependent on government because Blacks have no strong financial institutions willing and able to invest in their development. Blacks are uneducated and undereducated because they cannot, as a group, afford the cost of quality education. For the same reason, inability to pay, Blacks suffer from poor quality health care or no health care at all. The Black image in the white mind cannot be changed in a direction that Blacks would prefer so long as Blacks do not exercise significant control over the media that produce, package, and market representations of Blacks. Each disability, from failure to exercise fully the franchise, to homelessness, poverty, disease, and occupational disadvantage, has an economic component and admits (at least partially) of economic solutions. But the security of these solutions depends on group reparations.

It is one of the aims of this inquiry to demonstrate that Blacks are a cognizable group for purposes of recognition of their rights as a group and group redress. Thus, the question of group status is one which cannot be answered simply. The irony posed by the very question of Black national group status is that in ordinary social and political discourse, Blacks are treated as a group for every purpose other than rights recognition. Even as we profess the values of color blindness, it is common and accepted usage to maintain a catalogue of "racial" firsts, failures, accomplishments, and defects. The contradiction is neither accidental nor a remnant from an earlier period of "race" consciousness.

None of this is to say that the question of Black national group status is an easy matter to resolve. Ideally, however, one could settle the problem of legitimate representation for purposes of obtaining group reparations through, first of all, seeking the endorsement and support of established Black organizations, and secondly, through a plebiscite of intended beneficiaries. Given the current conditions under which the majority of Black people live, a plebiscite would be effective only if the work of educating the masses were carried out with meticulous care. Blacks, and whites, desperately need to understand the basis of the claim for group reparations, the historical precedents, and the future potential of a successful campaign.

On the issue of accepting nonmonetary compensation, it must first be pointed out that, in a sense, that is what affirmative action has been about.

Affirmative action, by providing Blacks with educational and employment opportunities that they would not otherwise have due to white racism, "compensated" Blacks for the injustices they suffered. This sort of nonmonetary compensation is unacceptable for the following reasons: (1) many whites and some Blacks believe that affirmative action is a "handout" and not compensation, thus perpetuating discrimination against all Blacks, and not just those who benefit personally from affirmative action programs; (2) very few Blacks actually benefit personally from affirmative action, thus all Blacks are not compensated; (3) affirmative action, by its nature, must ultimately be administered by a judiciary which increasingly believes that affirmative action is "reverse discrimination," unconstitutional if not narrowly tailored to redress specific acts of discrimination by identified violators, and furthermore, that rights are (or should be) individual, and that "race" is the wrong basis on which to assign benefits and burdens under the law; (4) affirmative action subjects Blacks participating in it to Pettigrew's "triple jeopardy" threat; (5) the fact that affirmative action is out of line with mainstream white American values means that politically it could not be maintained for a long enough period to compensate Blacks adequately; (6) affirmative action is, or is intended to be, meritocratic; (7) affirmative action, in most cases, is discretionary, situational and sporadic, not uniform and systematic.

Compensation to Blacks for the injustices suffered by them must first and foremost be monetary. It must be sufficient to indicate that the United States truly wishes to make Blacks whole for the losses they have endured. Sufficient, in other words, to reflect not only the extent of unjust Black suffering, but also the need for Black economic independence from societal discrimination. No less than with the freedmen, freedom for Black people today means economic freedom and security. A basis for that freedom and security can be assured through group reparations in the form of monetary compensation, along with free provision of goods and services to Black communities across the nation. The guiding principle of reparations must be self-determination in every sphere of life in which Blacks are currently dependent.

To this end, a private trust should be established for the benefit of all Black Americans. The trust should be administered by trustees popularly elected by the intended beneficiaries of the trust. The trust should be financed by funds drawn annually from the general revenue of the United States for a period not to exceed ten years. The trust funds should be expendable on any project or pursuit aimed at the educational and economic empowerment of the trust beneficiaries to be determined on the basis of need. Any trust beneficiary should have the right to submit proposals to the trustees for the expenditure of trust funds.

The above is only a suggestion about how to use group reparations for the benefit of Blacks as a whole. In the end, determining a method by which all Black people can participate in their own empowerment will require a much more refined instrument than it would be appropriate for me to attempt to describe here. My own beliefs about what institutions Black people need most certainly will not reflect the views of all Black people, just as my belief that individual compensation is not the best way to proceed probably does not place me in the majority. Everybody who could just get a check has many reasons to believe that it would be best to get a check. On this point, I must subscribe to the wisdom that holds, if you give a man a loaf, you feed him for a day. It is for those Blacks who survive on a "breadconcern level" that the demand for reparations assumes its greatest importance.

2. The Overinclusiveness and Underinclusiveness of Black Reparations

Just as affirmative action has been criticized by some for rewarding undeserving middle-class Blacks at the expense of underclass or poor whites, a plan for Black reparations could be attacked on the ground that Blacks who enjoy relatively privileged and discrimination-free lives would benefit at the expense of underprivileged whites and non-Black nonwhites. This criticism raises the issues of overinclusion and underinclusion to show that a valid concern of redistributive fairness is not satisfied in a remedial scheme based on the equality principle that excludes any segment of the poor or underprivileged and includes any segment of the privileged.

This criticism substitutes class status rather than racial group status as the proper basis for remediation. One difficulty with this approach is racially identifiable class stratification, with Blacks disproportionately absent from and whites overly represented among the privileged classes. Racial group status, therefore, cannot be simply shoved aside as irrelevant to concerns about equality among economic classes. Moreover, the class-over-race approach to redistributive fairness ignores that the central claim of Black reparations is redress for exploitation through government-sanctioned white supremacy. The claim is one of entitlement, not need.

Nonetheless, it would be undeniably troubling if a relatively privileged group insisted on pressing its entitlement claims in a context in which the underprivileged and truly disadvantaged would have to pay. We can imagine a scenario in which a more powerful social group unjustly exploits a less powerful group, and then later finds itself in a less advantageous economic position than those who had been wronged in the past. Perhaps those who had been wronged, through their own industry and efforts or by windfall, managed to surpass the achievements of those who had been their oppressors. A valid claim for redress might exist among the newly prosperous group against their former exploiters, and yet it might seem unjust to pursue the claim. Because of the class differences, redress may impair the achievement of social equality between the two groups in a society where equality was a normative value.

The problem of overinclusion within the beneficiary group in a plan for Black reparations hardly approaches the level of a threat to the values of social equality among groups. Nor would a plan for Black reparations that was marginally overinclusive in the sense that some are compensated who suffered no harm seriously impair the ability of society to achieve social equality. On the contrary, the goal of social equality is enhanced when the beneficiary group is also a group such as Blacks that suffers continuing economic subordination despite advances made by some individuals.

Overinclusion within the group who must pay reparations also presents problems of the troubling but not irresolvable variety. Some might object that their ancestors had nothing to do with enslavement of Blacks or actually opposed racism and discrimination. Moreover, if reparations are drawn from general revenue, beneficiaries who are also taxpayers will pay a part of their own redress. From the standpoint of redistributive fairness, this may seem unjust. On the other hand, given the near impossibility (because of administrative costs) of obtaining redress only from those who perpetrated racism and exploitation, the focus of reparations doctrine needs to be on the role of government. This was the case with respect to both Japanese and Jewish reparations.

Reparations to Blacks is an obligation of the American government for its role in slavery and the violation of Black rights. Government obligations are paid with taxpayer funds. Taxpayers do not typically have a right to pick and choose among specific governmental expenditures they wish to support; nor should they. In order to preserve government at all, policy makers must be allowed to make allocation decisions in the best interest of their constituents and society as a whole. In my view, the alternative inexorably leads to freerider dilemmas and social fragmentation or immobilization. Thus, Black reparations, as with other government obligations, may justly be paid out of general revenue consistently with redistributive fairness.

The class-over-race approach to redistributive fairness also raises the issue of underinclusiveness. If Blacks receive reparations for wrongs done to them and their ancestors, shouldn't other poor and underprivileged groups, including some white ethnic groups, also receive reparations? This question becomes an objection to Black reparations when it suggests either that America has too many victims to compensate them all, and therefore should not compensate any, or that Black reparations could be paid only at the expense of harming the non-Black poor. In other words, the former assumes zero balances and the latter zero sum.

The basis of the claim for Black reparations is not need, but entitlement. Need is not irrelevant, but it is by no means central to the claim. Reparations as a norm seeks to redress government-sanctioned persecution and oppression of a group. In that regard, it has been a workable norm for groups other than Blacks. Compensation of such groups need not be a zero-balance endeavor given the variety of compensatory possibilities and circumstances to which groups seeking redress respond. Sovereignty, land, money transfers, tax breaks, educational scholarships, and medical and housing subsidies all lie within the compensatory arsenal of government. Their extensive use outside the context of reparations belies the assertion that their implementation within the context of reparations would overburden national revenues.

Moreover, reparations to Blacks would not inevitably harm the non-Black poor. Racist exploitation has contributed to the persistence of poverty among Blacks and the unjust privilege of whites. Redressing these harms through Black reparations would help to alleviate part of the problem of persistent poverty. To the extent that poverty remains a problem among non-Blacks and Blacks alike, it is both just and consistent with the equality principle to demand adequate social welfare, equal educational opportunity and access to jobs. Other national goals, like space exploration or defense, may need to be downsized in order to fulfill the moral obligation of social justice.

Recognition of reparations as an enforceable legal norm, available to any similarly subordinated group, upholds justice without placing an undue burden on the valid concerns of redistributive fairness. In addition to its positive deterrent effect, payment of reparations contributes inestimably to the norm of social equality because of how it might change the lives and perspectives of the subordinated. Equal treatment is not a shibboleth but a real possibility. Injustice against groups is not tolerated or rewarded. Opportunity free from stigma and disadvantage is the norm in my country. Discrimination and racism may end. In my view, these effects alone, if realized, make implementation of Black reparations as an enforceable legal norm long overdue.

D. BLACK REPARATIONS AS PRECONDITION TO CIVIL EQUALITY

One final argument may be advanced in defense of Black reparations. The genocidal conditions under which Black people have been forced to live during their tenure in the United States have been ameliorated but not ended by the demand for civil rights. The miserly development of that sorry and treacherous history was highlighted at the beginning of this article. A crucial but seldom considered defect of all civil rights legislation is the fact that it needs to be administered and enforced. Many Blacks (and whites, too) appear to be under some delusion that once Congress passes civil rights legislation, Blacks are protected from discrimination and white racism. Nothing could be further from the truth, as the history of Black Reconstruction clearly shows. Every measure passed by Congress during Reconstruction for the social and political equality of Blacks—with the possible exception of the Thirteenth Amendment—was subverted or made null and void before the turn of the century.

During the heyday of the 1960s Civil Rights movement, Blacks again received legislation from Congress which, in turn, is being methodically nullified by invidious court opinions. Since President Bush vetoed the Civil Rights Act of 1990, and Congress failed to override that veto, Black people will not begin to get back the civil rights lost during the Reagan administration for some time. A pattern of gain and then loss has unquestionably revealed itself. To attribute this pattern to the fact of administration is not to overlook or discount other ideological components of legislative failure. It is merely to acknowledge the perverse operation of that timeworn saw: the price of freedom is eternal vigilance.

Civil rights legislating is an open-ended enterprise, with no end in sight so long as such laws must be administered by those whose commitment or resources are seldom great, and enforced against those who are determined to discriminate. I fear that I am correct in supposing that Blacks will always need civil rights legislation, and there will always be "new patterns of racism," that is, until Blacks as a group obtain something approaching economic parity with whites. This, it appears to me, is the precondition of achieving autonomy and respect as a group in the United States. Also, until white people make peace with their racist and exploitative past, they will never accept the responsibility for racism and exploitation, both present and portended.

Each year the government fails to pass Black reparations legislation the debt increases rather than diminishes and the obligation to redress wrongs inflicted on the Black community becomes more difficult to satisfy. Congress's failure even to hold hearings on the need for such legislation may be attributed to selective indifference to Black social justice claims on the one hand and racial antipathy to Blacks on the other. As with the Supreme Court, a ruling majority seems to believe what Justice Bradley articulated over one hundred years ago in the Civil Rights Cases, that: "When a man has emerged from slavery, and by the aid of beneficent legislation has shaken off the inseparable concomitants of that state, there must be some stage in the progress of his elevation when he takes the rank of a mere citizen, and ceases to be the special favorite of the laws, and when his rights as a citizen, or a man, are to be protected in the ordinary modes by which other men's rights are protected."

As Justice Harlan expressed then in dissent, and we today may acknowledge, "[i]t is scarcely just to say that the colored race has been the special favorite of the laws"; and less than twenty years out from slavery, in any event, did not mark the point in Black progress at which equality with whites no longer should have been a national concern. The maintenance of a system in which "any class of human beings" is kept "in practical subjection to another class, with power in the latter to dole out to the former just such privileges as they may choose to grant[,]" marks Justice Bradley's statement as not only unreasonable, but also unjust. It is no less unreasonable and unjust today.

However, for those who long for the millennium in which Black equality with whites ceases to be the American dilemma and becomes the American reality, reparations contain within them at least the promise of closure. The closure afforded by reparations means that no more will be owed to Blacks than is owed to any citizen under the law. This is the effect of any final judgment on the merits. Once reparations are paid, Blacks will be able to function within American society on a footing of absolute equality. Their chance for public happiness, as opposed to private happiness, will be the same as that of any white citizen who currently takes this concept for granted because the public so utterly "belongs" to him, so utterly affirms his value, his humanity, his dignity and his presence.

ROBERT WESTLEY *is an associate professor of law at Tulane University. He has lectured widely on how reparations to African Americans might be systematically distributed in the form of trust funds. This article was first published in the* Boston College Law Review, *December 1998, Volume XL, Number 1. The original article has over 170 footnotes, which the editor encourages the reader to examine for details of Westley's arguments.*

KEVIN OUTTERSON

Slave Taxes

> *Taxes are what we pay for civilized society.*
> —OLIVER WENDELL HOLMES[1]

D uring the eighteenth and nineteenth centuries, state, local, and national governments taxed slaves and slave commerce. Slave taxes[2] proved to be very valuable: from colonial times to the Civil War, American governments derived more revenue from slave taxes than from any other source. Given the uncivilized treatment of slaves in America, perhaps a refund is due.[3]

If this essay is successful, it will begin a debate and illustrate the type of research that could be pursued under a federal reparations study commission.[4]

I. SLAVE TAXES AND BLACK REPARATIONS

Reparations have been paid to many groups over the past sixty years. A partial list would include: Japanese Americans interned during World War II; the State of Israel by the Federal Republic of Germany; Swiss banks to Holocaust-era depositors; German and Italian insurance companies for Holocaust-era policyholders; German companies to industrial war slaves; Maori restitution by New Zealand; multiple indigenous peoples suits in the United States settlements and restitutions; and Korean sex slaves in Japanese-occupied territory.[5]

To this list I would add affirmative action, which was originally defended as redress for slavery and discrimination.

Many of the arguments currently offered to oppose black reparations have been successfully resolved in the reparations listed above. These arguments include sovereign immunity (waived by Germany, New Zealand, the United States in Japanese American redress and in Native American settlements, and to a very limited degree by Japan); the fairness of imposing the burden of reparations on a later generation (all cases); the methodology for documenting claims without strict legal proof (all cases); and tailoring the reparation to the crime and the needs of the population (all cases). Even the most ardent opponent of black reparations should admit that slavery was a crime against humanity. The Holocaust was sufficient to overcome these barriers to reparation; slavery and its aftermath deserve the same evaluation.

And yet the black reparations movement faces daunting prospects. Congress has been reluctant to even study the issue. What is different about black reparations that makes the nation shrug rather than acknowledge responsibility? Why can't the nation take the bare minimum step to study the question in a federal reparations study commission?

Three arguments emerge as likely explanations. First and foremost, the argument goes, slavery was a long time ago, ending with the Civil War. All of the people involved are dead and it wouldn't be fair to visit the sins of the fathers on the great-grandchildren, or to reward people whose ancestors were slaves.

The second argument takes a different tack: it acknowledges that slavery and discrimination were evil, but argues that the debt has been paid, namely by the Civil War, emancipation, the New Deal, the Great Society, and by affirmative action. Another version of this argument says it would cost too much to pay adequate black reparations.

The third argument is a holdover from the Middle Ages: sovereign immunity, the king can do no wrong. In slavery, the governments of Europe, America, Asia, and Africa committed grievous wrongs. To say that sovereign immunity prevents discussion of black reparations is disingenuous. Waiver of sovereign immunity is a political act, as it was in many of the reparation claims mentioned above, and building political consensus for a waiver requires factual analysis and discourse. These are the goals of a federal reparations study commission. In any event, sovereign immunity may already have been waived in this case, particularly since slavery and the slave trade was recognized as a crime against humanity. The U.S. government has previously invoked this principle from a position of righteousness. Now the shoe is on the other foot.

This essay is designed to respond primarily to the first argument. While the individuals involved may all be dead, some legal persons remain, particularly governments and corporations that benefited from slavery. Despite the passage of time, these beneficiaries of slavery are still among us, and could face legal claims for their complicity.[6] If a government benefited from a crime against humanity, it should be held responsible.

II. SLAVE TAXES IN AMERICAN
GOVERNMENTAL FINANCE

A. The Beginning of Atlantic Slave Taxes

The first English Atlantic slave voyage was Sir John Hawkins's expedition in 1562–1563, capturing three hundred Africans.[7] The English Crown chartered companies in the seventeenth century to bring slaves to English colonies in the New World.[8] The English were latecomers to the slave trade. The Portuguese pioneered the Atlantic slave trade in the fifteenth century, establishing their first post on the Gold Coast in 1482, followed by the Spanish. These governments profited from the granting of trade company charters as well as direct investment in slave trading. Spain also imposed taxes at the rate of 20 percent on some slaver voyages in the late fifteenth century.[9]

In the late sixteenth century, other foreign powers, notably Dutch private traders, joined the Portuguese and Spanish in the Atlantic slave trade. The first Dutch West India Company, chartered in 1621, paid a dividend to each of the town councils comprising the Netherlands as a form of tax. By the 1660s, the English, Danes, Swedes, French, and Germans had joined the Dutch, Portuguese, and Spanish in the struggle for control over the African trade,[10] with the participation of some African nations.[11] None of the European nations were able to establish a monopoly on the African trade, although they were able to capture significant tax revenues.[12] At that time, the slave trade to North America was just beginning. As the Atlantic slave trade approached its 150th anniversary, the vast majority of the slaves of African origin in the New World were in Brazil, Mexico, Peru, and the Caribbean Islands, producing sugar, silver, and gold for export to Europe.[13]

Spain used its New World colonies to raise revenue. In 1713, Spain sold the monopoly on the Spanish colonial slave trade to the Royal African Company for 200,000 crowns plus a duty of 33 crowns per slave imported. The plan required the importation of at least 144,000 African slaves over a thirty-year period, with profits to be split between the Royal African Company and the Crowns of England and Spain.[14]

Once African[15] slaves began to arrive in Virginia and Maryland, they joined indentured Britons in the cultivation of tobacco.[16] In the 1660s, duties on the export of tobacco amounted to 25 percent of worldwide English customs revenues and 5 percent of the Crown's entire income. Colonial tobacco duties raised £100,000 per year by the mid-1670s for Charles II. By 1699, the tobacco duties reached £400,000 annually.[17] Maryland laid an import duty on white servants and Negro slaves as early as 1695.[18] The duty on Negro slaves was four times greater than the duty on white servants.

As the Atlantic trade in slaves and slave commodities boomed, the sponsoring governments and their colonies reaped huge financial rewards.

B. Slave Taxes in the Colonial States

In 1775, of the 331,000 blacks living in British North America, 310,000 lived in the Carolinas, Georgia, Virginia, and Maryland.[19] Not only did slaves produce taxes for the states, but in an era of monetary shortage, slaves provided an important form of taxable wealth that could be bought and sold.[20] Slaves also produced the crops—first tobacco and then cotton—which generated colonial wealth.

1. North Carolina
In North Carolina in 1763, 75 percent of the colony's revenue was derived from a poll tax on every white male sixteen or older and on all blacks, slave or free, male or female, over the age of twelve. Taxes on property (including slaves) continued during the Revolutionary War, when an *ad valorem* tax was imposed in 1777.[21]

2. South Carolina
Reliance on slave taxes was a consistent feature of South Carolina's government finance. In the 1730s, the slave duty was £50[22] per imported slave, sufficient to fund two-thirds of the needs of government.[23] In the years prior to the Revolutionary War, the slave duty still raised from a quarter to one-half of the financial needs of the government.[24]

Fear of slave revolts motivated the legislature to occasionally impose prohibitively high tariffs, which affected slave duty revenue. South Carolina alternatively raised and lowered the tariff many times, sometimes offering lower duties for African slaves, who presumably knew nothing of the slave insurrections in the West Indies.[25] From 1695 to 1705, six colonies established control policies favoring newly captured Africans over revolt-prone West Indian slaves, a pattern that was repeated throughout the slave period.[26]

In addition to slave import taxes, South Carolina raised funds through a property tax on land and a poll tax on slaves in 1777.[27] After the Constitution banned export taxes, South Carolina emphasized property taxes on slaves. In the decades prior to the Civil War, direct slave taxes raised approximately 60 percent of the revenue of South Carolina's state, county, and local governments.[28]

3. Georgia

In the colonial era, Georgia's financial needs were quite modest, averaging only £2,215 per year from 1763 to 1773. The tax on each slave was equalized to the tax borne by one hundred acres of land.[29] In 1849, 49 percent of Georgia's state property tax came from the slave property tax. Eight years later, it was still 42.3 percent of the total.[30] This allocation of the taxing burden generally reflected the fact that slaves were nearly half of the wealth of the state.[31]

In addition to the property tax, the state imposed a poll tax of $0.39 per slave.[32] Poll taxes in this period had little to do with voting, but rather operated in the case of slaves as a property tax paid by their owners.[33]

4. Virginia

The notable feature of Virginia's colonial finance was the success of the tobacco export tax, reaping £7,000 in 1770.[34] Since the late seventeenth century, slaves were producing the bulk of Virginia's tobacco. The Revolutionary War disrupted tobacco commerce, leaving Virginia scrambling for revenue to meet war requirements. Virginia responded with taxes on land, silver and gold plate, slaves, horses, and mules.[35]

In the midst of a very difficult period for the Continental Army, Virginia devised a plan to recruit additional soldiers by promising a personal slave as a bounty for enlistment. Virginia planned to obtain these slaves by taking 5 percent of the slaves from any Virginian who owned more than twenty slaves. Large slaveholders opposed the plan, and the slave bounty did not pass.[36]

5. Maryland

Tobacco also figured prominently in Maryland's tax system. In 1775, Virginia and Maryland exported 220 million pounds of tobacco.[37] Maryland imposed a poll tax of 30 pounds of tobacco for the benefit of Anglican clergy, raising £8,000 sterling a year in 1766. Maryland also imposed an *ad valorem* tax on slaves, servants, and merchants' stock-in-trade.[38]

6. Other Southern States and Southern Cities and Counties

In Florida, Louisiana, and Mississippi, although the slave populations were smaller, the story was similar: direct slave taxes raised from 30 to 40 percent of the revenue.[39] Alabama levied a 0.2 percent property tax on the value of slaves, land, and other property, with slave values ranging from $100 to $550 each. At those rates, slightly less than half of Alabama's state property tax came from slaves, while less than a quarter was on agricultural land.[40]

Most state governments also allowed the local units of government to tax slaves as well, often at a percentage of the state rates. For example, the City of Charleston in 1838 received over 22 percent of its total revenue from direct slave taxes.[41]

7. New York

As early as 1709, New York taxed both slaves and chimneys under the same law.[42] New York placed duties in 1734 on imported cider, beef, and pork and ordered a census and poll tax of one shilling on slaves aged fourteen to fifty. If the owner did not pay the slave tax, the collector could levy and destrain the slave, and if not redeemed within four days, sell the slave at a public auction to the highest bidder to pay the tax. The tax was to be used to build fortifications to defend the colony.[43] New York collected impost taxes averaging about £5,000 a year in the period 1760–1774 from duties upon imported rum and other liquors, wine, cocoa, slaves, and dry goods from Europe.[44]

8. New Jersey

Slavery existed in New Jersey from the beginning and continued in limited form until the ratification of the Thirteenth Amendment, making New Jersey the last Northern state to abolish slavery.[45] New Jersey taxed slave imports, beginning in 1714.[46]

9. Other Northern States

In 1796, slaves were taxed in New Jersey, Pennsylvania, Maryland, Virginia, Kentucky, North Carolina, South Carolina, and Georgia, but not in Vermont, New Hampshire, Massachusetts, Rhode Island, Connecticut, New York, or Delaware.[47] Pennsylvania and Massachusetts taxed imported slaves prior to the Constitution,[48] with a view to discourage the traffic.[49] Rhode Island's slave import duty was used to pave the streets of Newport, to build bridges, and other municipal improvements.[50] One early Massachusetts poll tax repaired "their majesties' castle, upon Castle Island, near Boston."[51] While Massachusetts, Rhode Island, and Connecticut disfavored slavery at home, their vessels plied the Atlantic slave trade for commercial gain.[52]

10. Other Governments

The American colonies were not unique in their reliance on slave taxes for government finance. The British West Indies colonies used slaves to produce sugar for export to Europe. High duties were placed on competing East Indian (non-slave) sugar, resulting in an artificial government subsidy of slave-produced sugar to the amount of £4 million per year in the 1820s.[53] The French and the Danish also operated sugar colonies with slave labor and tax incentives,[54] while the Spanish and Portuguese possessions in the Western Hemisphere were highly dependent upon slave labor. African governments also derived revenue from the slave trade.[55] The sultan of Zanzibar levied a tax on each exported slave. This tax was such a significant portion of his revenue that Zanzibar resisted British efforts in the nineteenth century to abolish the slave trade. Zanzibar eventually signed an abolition treaty in 1873, but only after Britain agreed to pay Zanzibar £8,000 per year in compensation, a payment that continued until 1968.[56]

C. Slave Taxes and Racist Social Policy

1. Tax Preferences for Importation of Docile Slaves for Breeding

In addition to raising revenue for the government, taxes on slaves served racist social purposes as well. Throughout the South, import duties were modified to restrict the importation of troublesome slaves from Caribbean islands that had experienced slave revolts. Tax policy preferred slaves fresh from Africa.[57] In the Danish slave colonies, talk of abolition prompted fears as to whether the slave population could keep up with the death rate if importation became illegal. The 1792 Danish Royal Decree abolished the Danish slave trade and slave imports into the Danish West Indies, but delayed the legal effect until 1802 to allow for enough slaves to be imported in the ensuing decade to keep the sugar plantations staffed. Exports of slaves were forbidden, and tax incentives were granted to encourage the importation of female slaves, particularly female field hands, so that slave breeding would be more prolific.[58]

2. Tax Discrimination Against Free Blacks

It seems that in the early eighteenth century, the taxation of free blacks differed little from whites and slaves. Immediately before the Revolution, the use of special taxes on free blacks for discriminatory purposes took root and grew.[59] In all the states of the Lower South save Louisiana, free blacks were subjected to a higher poll tax under a deliberate social policy to reduce the free black population.[60] In Georgia, much higher taxes were placed on free blacks in order to encourage them to leave the state.[61] Slaves were taxed at $0.39 each while free blacks were required to pay $5, a very significant sum in the 1850s.[62] Free blacks who failed to pay had their property seized and sold at auction; those with taxes remaining were themselves sold at auction into involuntary servitude to pay the tax.[63]

The poll tax was pressed into service to restrict voting in the 1890s and early 1900s, complementing literacy tests and residency requirements.[64] Disenfranchisement was not focused solely on blacks: it was also an anti-Populist measure against poor white farmers as well.[65] The states with the most significant poll tax for disenfranchisement were Virginia, Mississippi, and Alabama (in that order), followed by Arkansas and Texas.[66]

In addition to poll taxes to discriminate against free blacks, many Southern states and cities enacted occupational taxes on slaves and free blacks that favored white labor in the skilled professions.[67]

3. Class Conflict Over Slave Taxes

The level of taxes on slaves represented a political compromise within particular Southern states between rich planters and poor white farmers.[68] This tension was often expressed in debates over the assessed value for tax purposes of slaves and land.[69] In this manner, poor white farmers who were not slave owners themselves directly gained from slave taxes to the extent that the tax burden fell someplace else. Every white person had a stake in the slave tax system.[70]

For example, the Tennessee Constitution of 1796 established that no slave could be taxed at a rate higher than two hundred acres, double the rate of a white man.[71] The level of slave taxes throughout the South allowed taxes on white men, land, and other property to remain at quite low levels.[72] Benefits were not limited to the South; the federal direct taxes on slaves also directly reduced taxes on land, houses, and other property throughout the country, including the North. Northern farmers and homeowners received a financial benefit from direct federal slave taxes from 1798 to 1802 and again during the War of 1812.

4. Black Land Ownership

One effect of slavery and the prohibition on free black land ownership in many states[73] was that most blacks were locked out of the most significant wealth-building opportunity of the eighteenth and nineteenth centuries: the availability of cheap public lands for sale in the United States. It was a historic opportunity cost.

For example, a major source of revenue in Georgia was the sale of lands "ceded" by Native Americans to the state. These lands were sold to white citizens in lotteries (free blacks were excluded).[74] The lucky farmer selected in the lottery could purchase a parcel of 202.5 acres at an attractive price. From the early 1800s to 1832, Georgia collected $1.7 million on these land sales. Given the small size of the state budget, $1.7 million was a huge sum: between the War of 1812 and the Civil War, taxes paid less than half of the state revenue of

Georgia.[75] In his comprehensive study of public policy in nineteenth-century Georgia, Peter Wallenstein noted:

> No phenomenon better captures the fundamental race bias of public policy in Georgia than the most important undertaking of the state government during its first half-century under the U.S. Constitution: the transfer of land from Indians to whites, land that was in turn, much of it, worked by slaves.[76]

Slavery and discrimination prevented the vast majority of blacks from purchasing state and federal lands during the eighteenth and nineteenth centuries. Although slave taxes had contributed to the funding of the Louisiana Purchase,[77] slaves had no opportunity to settle a homestead themselves.

D. Federal Slave Taxes

> The blessings in which you, this day, rejoice are not enjoyed in common.
> The rich inheritance of justice, liberty, prosperity and independence bequeathed by your fathers, is shared by you, not by me.
> The sunlight that brought life and healing to you, has brought stripes and death to me.
> This Fourth [of] July is yours, not mine. You may rejoice, I must mourn.
>
> —FREDERICK DOUGLASS, "WHAT TO THE SLAVE IS THE FOURTH OF JULY?" JULY 5, 1852[78]

1. The Continental Congress

The continued existence of the Continental Congress was fully dependent upon acceptance of slavery. Embracing slavery was the price of liberty from the British.

On July 30, 1776, Congress was discussing raising taxes to fund the war. The discussion had centered on using population as a method to apportion the tax burden amongst the states. A question arose whether to count slaves as people for tax purposes. James Wilson feared that not counting slaves would "be the greatest Encouragement to continue Slave keeping."[79] Thomas Lynch of South Carolina then delivered what came to be known as the Lynch Ultimatum: "If it is debated, whether their Slaves are their Property, there is an End of the Confederation."[80]

The Continental Congress did not have the power to tax directly, but instead assessed amounts against the various states, who then voluntarily paid requisitions out of their own funds.[81] Southern states were heavily dependent on direct slave taxes, and all of the states derived significant revenue from slave taxes.[82] These funds were then passed on to the Continental Congress to fight the Revolutionary War. In 1783, for example, South Carolina levied a property tax *in specie* on slaves, other property, and imported goods. South Carolina was to pay 52 percent of its budget to Congress that year to meet its requisition for Revolutionary War debts.[83] Collection was difficult in this postwar period, and many taxes were collected by execution and public auction.[84] If the Revolutionary War was "the price of liberty,"[85] a significant portion of that price was paid with slave taxes, including slaves sold on the auction block.

2. Slave Taxes Under the Constitution

> *[T]he power to tax involves the power to destroy.*
> —SUPREME COURT CHIEF JUSTICE JOHN MARSHALL[86]

The present Constitution was not ratified until 1789, replacing the Articles of Confederation. One question faced by the drafters was whether slaves would count as persons for apportioning direct taxes among the states. Counting slaves would increase the direct taxes paid by the South.[87] Southerners, anticipating Marshall's dictum, worried about slavery being taxed to extinction,[88] but they wanted the additional representatives in the House that would result from counting slaves as people. The eventual compromise counted three-fifths of the slaves for both purposes.[89]

Since most federal taxes from 1789 to the Civil War were not direct taxes subject to apportionment, the Southern concession on taxes cost them little, while the South was guaranteed additional representation in Congress.

The Constitution grants the federal government exclusive control over import duties,[90] but the South won another protection against "the power to destroy" in that the federal import duty on slaves could not exceed $10.[91] In the first Congress, Josiah Parker of Virginia introduced a motion to impose the $10 federal slave duty as a means to raise revenue. South Carolina and Georgia reacted strongly, as the burden of the tax would fall on slave importers in their states. The ensuing arguments threatened to derail the first U.S. revenue act, until it was referred to a committee chaired by Parker and never heard from again.[92] The United States never imposed the $10 federal slave duty.

After the slave trade was abolished in 1808, the slave trade continued under lackluster enforcement. If a slave ship was captured, the ship was taken to the nearest port. If that port was in a slave state, the slaves could then be sold on the auction block rather than freed, with the proceeds paid to the federal treasury.[93]

3. Direct Federal Slave Taxes

> [T]hese debates demonstrate the centrality of slavery to the de-
> sign of a national tax system. From the 1776 attempt to draft
> Articles of Confederation to the end of the Federalist era, slav-
> ery intruded into every effort to create a tax system at the na-
> tional level.
>
> —ROBIN L. EINHORN, UNIVERSITY OF CALIFORNIA, BERKELEY[94]

After the adoption of the Constitution, Congress on two occasions imposed a direct federal tax on slaves. In 1798, Congress was preparing for a potential naval war with France in the wake of the XYZ Affair.[95] Since customs duties were not raising enough funds to support military preparedness, Congress passed a direct tax on land, houses, and slaves.[96] The Act of July 14, 1798, imposed a federal tax on slaves between the ages of twelve and fifty at $0.50 each.

In the debate, New Englanders opposed the slave tax, while Southerners insisted upon it because "[a]n apportioned tax on land and slaves would mute class conflict in the South but exacerbate class conflict in the North. This could only help Republicans and hurt Federalists."[97] The compromise was to add "dwelling houses" to the formula. In the final vote, the slave tax was supported by the congressional delegations from Vermont, Delaware, New York, Pennsylvania, Kentucky, Maryland, North Carolina, Tennessee, Virginia, Georgia, and South Carolina. It passed 67 to 23, gathering a majority of both Federalists and Republicans.[98]

This law required the first national tax return, listing "in alphabetical order, the names of all persons owning, possessing, or having the care of any slaves, with the number of slaves, as aforesaid, owned by, or under the care of each person: And the forms of the said lists shall be devised and prescribed by the department of the treasury."[99] The direct tax on houses, land, and slaves was expected to raise $2 million, which would have been about 26 percent of the federal government's total expenditures in 1798 and about 18.5 percent of federal expenditures in 1800.[100] About 11 percent of the direct tax was expected to come from the tax on slaves.[101] While collections proved harder than assessment, direct slave taxes remained a small but significant part of the federal budget from 1798 until 1802 and beyond. Within three years, four-fifths of the tax had been collected.[102]

In the election of 1800, Thomas Jefferson unseated John Adams as president, and Congress repealed the measure early in Jefferson's first term, on April 6, 1802.[103] Despite repeal, collections from the tax continued to come in for a number of years and helped to shore up the financial health of the American government while Jefferson negotiated the Louisiana Purchase. Gallatin financed the Louisiana Purchase in 1803 through $2 million in cash surplus, $11.25 million in new 6 percent stock redeemable after fifteen years, and $1.75 million in temporary borrowing.[104] U.S. customs revenue blossomed during the Napoleonic Wars, but the revenue from the direct federal tax was still a significant factor in U.S. federal finance during the purchase of the Louisiana Territory.[105]

In anticipation of the War of 1812, Secretary of the Treasury Gallatin proposed reenacting various versions of the direct tax in order to support the war effort. On July 22, and August 2, 1813, with the war already under way, Congress passed a $3 million direct tax, again on houses, land, and slaves, with some modifications to improve collections.[106] One modification was the abandonment of the per-head slave tax and the progressive house tax in favor of an *ad valorem* approach for both.[107] The annual revenue amount was doubled to $6 million in the act of January 9, 1815,[108] and an additional $3 million in the act of March 5, 1816.[109] The direct tax collected by December 1817 totaled $10,469,992[110] and financed 40 percent of the War of 1812.[111] Although the War of 1812 soon ended, the taxes remained for a while to pay down the war debt. The direct tax on land, houses, and slaves was repealed in the act of December 23, 1817,[112] but collections continued in arrears until 1848.[113]

Congress attempted to impose a direct tax to finance the Mexican War, but could not agree on the proper level of slave taxes.[114]

E. Indirect Slave Taxes

This essay up to this point has focused on direct slave taxes such as property and poll taxes. These taxes provided significant governmental revenue, but in most states they comprised a minority of the taxes collected. Direct slave taxes are not the full measure of the impact of slavery on governmental revenue, however. Without slavery, much of the other wealth of the states would not have been available for the coffers of government. Indirect slave taxes played a major economic role in government finance.

1. Slave Improvements
Property taxes were levied throughout North America on land, houses, capital, luxuries, and other personal property in addition to slaves. Much of this

property was either constructed by slaves or slave profits, or had its value enhanced by slaves. Slave owners benefited, as did the governments that taxed property.

Slaves and slavery were crucial to public revenue. Their labor gave value to the land, as public domain was converted to private farms and public funds. Moreover, taxes on slaves constituted nearly half of all the tax revenue of antebellum Georgia's state and local governments, and much of the rest of tax revenue came from land recently acquired from Indians.[115]

For plantation owners, the houses had been built with slave labor, or from the profits of slave labor. Plantation land itself was valuable only when worked by slaves, generally in the production of cotton, tobacco, or rice. Even the luxuries taxed by the Southern states—such as carriages—were either built by slave labor or purchased with income from slave labor. Once these indirect taxes are accounted for, slaves and slave labor supported the bulk of the revenue needs of Southern governments.

2. Shipping and Commerce

Shipping and financial interests benefited from the slave trade and slave commerce. Yankee and British ships carried many slaves across the Atlantic or participated financially in the voyage, contributing to the fortunes of Newport, Boston, Philadelphia, Liverpool, and London. Most of the exports of slave commodities—cotton, tobacco, and rice—were carried in American shipping. By the 1750s, 95 percent of the commerce between the West Indies and the colonies was carried by American ships. The American merchant fleet carried 75 percent of the manufactured goods imported from London and Bristol. The economic benefit from slave-related commerce was important to the governments of New England and the rest of the states. In 1787, Samuel Hopkins wrote:

> The inhabitants of Rhode Island, especially those of Newport, have had by far the greater share in this [slave] traffic, of all these United States. This trade in human species has been the first wheel of commerce in Newport, on which every other movement in business has chiefly depended. That town has been built up, and flourished in times past, at the expense of the blood, the liberty, and happiness of the poor Africans; and the inhabitants have lived on this, and by it have gotten most of their wealth and riches.[116]

3. Customs Duties and Imposts

The primary export crops of the South were slave commodities—tobacco, rice, and indigo—with cotton growing to preeminence after the development of the cotton gin near Savannah, Georgia, in 1793.[117] Slave commodities were the key exports from North America in the decades prior to the Civil War, earning the foreign exchange necessary to purchase European-manufactured goods. In the fiscal year ending September 30, 1827, the United States exported $59.9 million in goods to foreign countries. Of that amount, raw cotton accounted for $29.4 million (49 percent); tobacco $6.6 million (11 percent); and rice $2.3 million (3.8 percent).[118] These three Southern crops represented nearly two-thirds of all U.S. exports in that year.[119] By 1829, the slave states produced 29 percent of the world's supply of cotton, over 160 million pounds. By 1860, American slave cotton accounted for 66 percent of the world's supply, some 2.3 billion pounds. Exports of raw unmanufactured cotton alone were 40 percent of all U.S. exports in 1816, rising to 57.5 percent of all U.S. exports in 1860.[120]

While exports of slave commodities were crucial to the economy of the Republic, the benefit to the government was not as simple as a tax on the export of slave products. The South feared discriminatory tariffs and had demanded that the Constitution prohibit taxes or duties on exports.[121] But these crops were the major source of U.S. exports and foreign exchange earnings, and the proceeds were used to purchase goods from abroad. In the absence of exported slave cotton, the United States could not have imported so freely from Europe. From 1789 to 1815, the federal government derived 90 percent of its revenues from duties on imported goods.[122] These duties were the primary source of federal revenue throughout the slavery period.

From the birth of the Republic to the eve of the Civil War, the majority of federal revenues were customs duties on goods imported with funds earned with the export of slave commodities. As General Pinkney said to the Constitutional Convention: "[South Carolina] has in one year exported to the amount of £600,000 Sterling all which was the fruit of the labor of her blacks."[123] The fruit of slave labor purchased goods abroad, which, when imported into the United States, funded the needs of the federal government until the Civil War.

III. CONCLUSION

Slave taxes played a major role in American economic history, financing government during the colonial, Revolutionary, and pre–Civil War periods. Slave taxes were not a minor or incidental component, but a crucial source of revenue for improvements, wars, and government finance such as the Louisiana

Purchase. Slave taxes were neutral or technocratic, but were designed and applied to further racist social policies, including oppression of free blacks.

The stated goal of this essay was to begin the process of answering the first argument against black reparations: Slavery was a long time ago and everyone involved is now dead.[124] It can now be seen that some legal persons that benefited are very much alive, namely governments.[125] This research is very superficial and preliminary; much more work needs to be done to flesh out the role of slave taxes. Let the debate begin!

More fundamentally, American democracy is generally unwilling to view the government as merely a legal person, but looks *behind* the government to the true source of legitimacy, the people.[126] Surely, governments were not the only beneficiaries of slave taxes. Free white citizens and immigrants used the improvements such as the Erie Canal; worked in factories using imported British machinery; drank tea with slave-produced sugar; dressed in slave-produced cotton; settled in lands taken from Britain and Native Americans in wars or in the Louisiana Territory purchased from France; and otherwise enjoyed the benefits of the system. Slave taxes benefited white American society in fundamental and enduring ways. If one claims the heritage of freedom, one might also accept responsibility for its history.

KEVIN OUTTERSON *is an associate professor of law at West Virginia University. He received his LL.M. at the University of Cambridge and a J.D. at Northwestern University School of Law.*

OMARI L. WINBUSH

Reflections on Homer Plessy and Reparations

You want to know if the dog has teeth, pull his tail.—JAMAICAN PROVERB

EDITOR'S NOTE: *Using storytelling to explain legal theory is a technique pioneered by legal scholar Derrick Bell and is firmly rooted in the African American oral tradition. Bell uses it to simplify legal theory, which is inherently enmeshed in convoluted language. The following essay was born when the editor sought legal historical advice about the impact of* Plessy v. Ferguson *on the history of reparations. Considered one of the three most important Supreme Court decisions involving the civil rights of Africans in America (the other two being* Dred Scott v. John F. A. Sandford, 1857, *and* Brown v. Board of Education, 1954), *the 1896* Plessy v. Ferguson *continued the economic, social, educational, and health assault of Africans in America by consigning them to a "separate but equal" status that accelerated economic, educational, and health gaps between white and Black Americans and laid the basis for reparations lawsuits.*

This essay arose from a dialogue between the editor and his attorney son after I consulted him about the role of the decision in securing reparations for African Americans and uses Bell's question-and-answer strategy to explain this legal history. The editor wishes to thank Annelee J. Winbush for clarifying several legal arguments here.

FATHER: *So what is at the heart of the world's great legal systems?*
SON: The key to achieving reconciliation for past and present damage is compensatory measures, that is, reparations for the damage. The principles of Mosaic, Islamic, and even English common law are all rooted in *compensation for injury* caused by another individual. The common law allows for compensation of injuries sustained by either the intentional injurious act of an individual or the negligent act of an individual that results in injury. Any successful reparation campaign must allege *all* of the elements to

either type of tort. For intentional torts, the elements are unlawful intent to injure and injury. For a cause to arise out of negligence there must be a duty, a breach of duty, causation, and damage.

FATHER: *Isn't the United States hypocritical in its condemnation of human rights toward other nations, and why doesn't it just apologize for its past wrongs against Africans in America?*

SON: American history is replete with acts of oppression and denial of basic human rights to Black people. The Middle Passage, destruction of family units, beatings, lynchings, stigmatization, and other heinous acts of oppression against Blacks in America constitute perhaps the longest unpunished genocide in world history, longer than the Holocaust, longer than the extermination of indigenous people, longer than and more bloody than the Tutsi-Hutu clashes of the early to mid-1990s, and the Balkan conflict, which continues today. However, in all of the previously mentioned genocidal occurrences, one continuing theme rings true: the United States has consistently maintained a condemning spirit toward those who perpetuated the genocide and in many instances supported reparations for victims of such crimes.

Meanwhile, America racially is not willing to apologize for its continuing participation in denial of basic human rights to a large segment of its own population. Why? Because with apologizing comes financial accountability, and that would require a massive transfer of funds that approaches the incalculable. For example, the African World Reparations and Repatriation Truth Commission meeting in Accra, Ghana, in 1993 consisting of attorneys, economists, and social scientists issued the Accra Declaration, which calculated that enslavement, colonization, the appropriation of land, extraction of raw materials and minerals, for example, diamonds, gold, oil, mahogany, cocoa, et cetera, out of Africa by Europe and the United States cost the African continent *$777 trillion*. Whatever the figure it is astronomical, and for a country predicated upon continued economic growth the prospect of paying such an amount to Black people is staggering. This is the strongest reason against a national apology for slavery, for the simple apology inevitably leads the wrongdoer toward reconciliation and repayment. Apologizing in and of itself is not difficult, fulfillment of the requirements associated with it, is.

Is the apology necessary? Yes, as W. E. B. Du Bois stated many years ago, the problem of the twentieth century is the problem of the color line. That color line cannot be erased or eradicated, however, unless the problems associated with the color line are reconciled by compensatory measures toward those who were its victims. I think those measures will take two basic forms: (1) *Reparations Trust Funds* that will be available for small businesses, tuition, and health care. This approach is favored by legal scholar Robert Westley and championed by activist Randall Robinson. It allows for set-aside monies for African Americans who are members of a class of persons in specific geographical areas who were disenfranchised because of slavery, violence, or fraud on a massive scale. The Tulsa, Oklahoma, riot of 1921 is a good example of this, as is the Rosewood, Florida, massacre occurring two years later. (2) *Reparations to Individuals* where direct payments to persons who can trace land theft, a house burning, to a specific individual or institution responsible for the crime.

FATHER: *What about torts and negligence? How are these related to reparations?*
SON: In the legal universe, compensatory measures come only if it is necessary to right a wrong. Wrongful acts are born from either an individual's or institution's affirmative action to cause injury or negligent behavior leading to intentional and negligent torts respectively. For an action in tort to be deemed intentional there must first exist intent to harm and then injury caused by the actor. For an action to lie in negligence there must exist a duty, a breach of duty, causation, and injury. In either case if an injury occurs the injured party can seek and may be entitled to compensatory damages intended to restore the injured party to his or her status prior to the injury occurring. The injured party may also be entitled to punitive damages intended to punish the tortfeasor for the heinous nature of the tort, as well as discourage future behavior, which may result in injury to others. This is the social aspect of tort law that so many large businesses would like to see eradicated with tort reform.

Armed with the basics of tort law, it is now necessary to identify the injurious act that can be linked to an intent to harm Black people (intentional tort) or a duty to prevent injury (negligence) to Black people living in America. Without the injurious act there can be no action in tort. Identification of the act must include an examination of the tangible effects of the act upon Black Americans, for any injury suffered must be real and not hypothetical. Hypothetical damage, or intellectual discussions as to what should have happened in the past, will lead to nominal damage awards due to the fact that the injured party cannot quantify the extent of his or her injury.

FATHER: *So how do these damages exist in the real world today?*
SON: The real damage is made manifest every day in the inner cities of the south and north. The results of this damage show themselves through a casual ex-

amination of the prison population of the United States. Also in new initiatives designed to address, among other things, racial profiling and years of discrimination. It's evident even in my *reasonable* fear of all police. The truth is that equality in America has not been achieved despite the many laws, court decisions, and executive orders designed to bring about the equality that eludes so many Black Americans. Inequality is the damage but what was the injury? When and how did it occur? Who is the injured party? These are the questions that make up the stuff of America's Biggest Tort Action.

FATHER: *Are you referring to* Plessy v. Ferguson?

SON: Yes. Many constitutional scholars, Black and white, liberal and conservative, have long berated the 1896 *Plessy v. Ferguson* decision as the single most devastating Supreme Court decision relating to Black people living in America. The decision was the ill-begotten child of the post–Civil War/Jim Crow era. Homer Plessy was, as the court referred to him, a citizen of the United States with seven-eighths white blood and one-eighth Black blood. The Supreme Court of the United States made sure to point out that the presence of the one-eighth African blood was not discernible in Mr. Plessy. Plessy bought a ticket for passage on the East Louisiana Railway. His trip was to take him from New Orleans to Covington, Louisiana. As he boarded the train, Plessy took a seat reserved for white passengers. This resulted in Plessy's ejection from the train, trial for violating an act of the general assembly, and a guilty verdict as to the charges. The basis of the trial court's decision was that the Constitution permitted the many states to make laws that could treat their citizens differently based on race.

The *Plessy* case should have allowed the Supreme Court to give full faith and credit, so to speak, to the Civil War amendments. Instead, the Court, when given the opportunity to examine these amendments and find a duty to provide equality to Black people contemplated by the Thirteenth, Fourteenth, and Fifteenth Amendments, held that *separate but equal* was constitutionally acceptable. The holding ruling had the following three major components related to the Thirteenth, Fourteenth, and Fifteenth Amendments:

Amendment 13. That it (Louisiana law) did not conflict with the Thirteenth Amendment, which abolished slavery and involuntary servitude, except as a punishment for crime. A statute that implies merely a legal distinction between the white and colored races has no tendency to destroy the legal equality of the two races or reestablish a state of involuntary servitude.

Amendment 14. The object of the amendment was undoubtedly to enforce the absolute equality of the two races before the law, but, in the nature of things, it could not have been intended to abolish the distinctions based upon color, or to enforce social, as distinguished from political, equality, or a commingling of the two races upon terms unsatisfactory to either. Laws permitting, and even requiring their separation, in places where they are liable to be brought into contact, do not necessarily imply the inferiority of either race to the other, and have been generally, if not universally, recognized as within the competency of the state legislatures in the exercises of their police power. Perhaps the most telling statement contained in the *Plessy* decision is the following:

> It is claimed by the plaintiff in error that, in any mixed community, the reputation of belonging to the dominant race, this instance the white race, is property, in the same sense that a right of action or of inheritance is property. Conceding this to be so, for the purposes of this case, we are unable to see how this statute deprives him of, or in any way affects his right to, such property. If he be a white man, and assigned to a colored coach, he may have his action for damages against the company for being deprived of his so-called property. Upon the other hand, if he be a colored man, and be so assigned, he has been deprived of no property, since he is not lawfully entitled to the reputation of being a white man.

FATHER: *So why is this paragraph so critical in understanding the impact of* Plessy *on us?*

SON: With this single paragraph Africans in America have been permanently stymied in their attempts to capture the myth of equality. *Plessy* leads to but one conclusion: states are free to prohibit persons of one race from occupying the place assigned for occupancy of another race. Simply stated, there is a place for Black people and there is a place for white people in the United States. Neither Black nor white people should be out of place. According to *Plessy*, this approach to separation of the races was constitutionally acceptable as long as the respective places occupied by Blacks and whites were equal. In so holding the Court began what I call the *Jones Approach to Equality*. The Jones approach to equality is a simple theory: *make everyone think that equality can be attained by having the same stuff as others.* Put another way, equality of the races can never be attained as long as what one race *has* is the measure of equality. Simply put, white people have white skin, therefore individuals possessing Black skin can never be equal, in the sense of its possession. *Plessy*, therefore, reduces equality to a mere goal of Black people achieving the same position of white people with the caveat of this achievement occurring in a separate place.

Constitutional scholars viewed the decision as a vote for discrimination, and continued subjugation and oppression of Black people in America. This view is reasonable when examined from the side of what Black Americans have always wanted to have—the same opportunity as white folks. Unfortunately that view—the "keeping up with the Joneses" view—continues to this very day. Black people have always viewed equality from the standpoint of having what white people have. This perpetuates the slave-instilled mentality that somehow what white people have is better than what Black people have.

FATHER: *This is an intriguing argument. How then do you view equality?*

SON: Equality does not require the seeker of it to keep up with the Joneses. The seeker in effect *becomes* the Joneses, and does not have to work any harder than Jones to merely have what Jones has. *Plessy* codified the "keep up with the Joneses" mentality and made it possible to sweep the real issue of *inequality* under the rug. Think about it. If grade schools for Black children in the South were equally funded and equipped prior to the *Brown* decision, then Black children would have gained access to the same educational opportunities as white children. While a young William Jefferson Clinton grew up in Arkansas on the "white" side of the railroad tracks, there would have been another Black boy who grew up on the other side of the tracks with the same opportunities as Bill. *Plessy* made Black Americans more concerned with sitting where they sat on a train than where that train was taking them.

So the *Plessy* decision began what should be rightfully deemed an impossible goal. The Jones approach to equality was later affirmed by the *Brown* decision in 1954, despite the Warren's court overruling of *Plessy*. *There really has been no fundamental change in America's approach to the pursuit of equality since the* Brown *decision.* And Black Americans continue the futile attempt of keeping up with the Joneses, which focuses on the material aspects of equality. Brown *and all attempts thereafter have failed because America and its highest court have really never intended to address the problems of Black people in America.* The *Brown* decision effectively served as a red herring and Black people found themselves more concerned with fictional moves to equality than with realizing their *inherent* right to exist freely as human beings. Critical race theorist Derrick Bell has outlined this eloquently in his important book *Faces at the Bottom of the*

Well. In all honesty, the *Plessy* decision only works when the decision is taken very literally, that is, that separation of the races is acceptable as long as respective separate place, condition, or status is equal.

The Plessy *court held that the Thirteenth, Fourteenth, and Fifteenth Amendments abolished the slavery of Blacks, naturalized Blacks, and ensured equal protection for all individuals under the law, including Blacks. It did not mean that the same law worked to require commingling of the races.* Instead, it stated that the individual state legislatures were free to determine the social and political place of black people living in America. In effect, the court found that the sole purpose of the Thirteenth and Fourteenth Amendments was to grant citizenship to Black people.

FATHER: *So what is* Plessy *really about?*

SON: While many may see the *Plessy* decision as counter to the African American struggle for freedom and independence, real examination of the decision will yield a different result. The Court's ruling that separate but equal is constitutionally acceptable serves as the first affirmative duty of the United States government to make Black skin acceptable to American society. The concept of separate but equal implies that it is acceptable to keep the races separate as long as both races have the same access to *opportunity,* and by implication, *outcome.* It was not until later that the Supreme Court in *Brown* found that separate but equal was flawed. I believe that *if* separate but equal were to actually have happened in America, Blacks in America would have without question attained some of the social, economic, and perhaps even political clout associated with true equality. What needed to be bussed in the early 1970s were dollars, not Black students. This is why reparations—the way of compensating for historic patterns of economic discrimination—have become a bona fide movement, since they force the United States to expose the lies of its racist past. *Brown* was a two-edged sword in that it took from us the power to control our schools and our curriculum, to an unequal status when measured against white schools.

There are perhaps two duties owed to Black Americans in the wake of the *Plessy* decision; I think they form the basis for the emerging prosecution for reparations for Africans in America. These duties arise out of the *Plessy* Court's recognition that the Thirteenth and Fourteenth Amendments did in fact provide Black people with all the rights and privileges associated with being American citizens. The *Plessy* Court stated the following in reference to the Thirteenth Amendment:

[T]his amendment was regarded by the statesmen of that day as insufficient to protect the colored race from certain laws which had been enacted in the Southern states, imposing upon the colored race onerous disabilities and burdens, and curtailing their rights in the pursuit of life, liberty and property to such an extent that their freedom was of little value; and that the fourteenth amendment was devised to meet this exigency.

This statement provides rounds of ammunition needed to prove negligence. First, it contains an admission. In its majority opinion, the Court admitted that the Thirteenth Amendment in and of itself was powerless to change the position of Black people in American society. The Court admitted a prevailing sense of pointless freedom for Blacks, especially Blacks in the South, due to rules that existed to maintain notions of the white supremacy status quo. This admission is well after the passage of the Thirteenth and Fourteenth Amendments and is important evidence of knowledge that the United States was complicit in its allowance of discrimination toward its Black citizens, which violated the equal protection mandate of the Fourteenth Amendment. Knowledge that extends itself from the Court into the halls of Congress, where the drafters of the Thirteenth Amendment were forced to draft a subsequent amendment in order to deal with the hapless and impotent law granting Blacks a meaningless freedom. The Court implied that Congress, with full knowledge of the impotency of the Thirteenth Amendment, set about to unburden Black people now American by the Thirteenth Amendment by drafting the Fourteenth Amendment.

Through this statement we gain insight into the intent of the drafters of the Fourteenth Amendment. The amendment was an exigency plan, a catchall approach to the problems associated with freeing a race of people. However well prepared Congress hoped it would be with its new exigency plan Fourteenth Amendment, nothing could prepare America for the change of mind that was necessary to make the Fourteenth have real impact. The fact is Congress, the Court, the Executive, governors, and their state legislatures were not interested in implementing the words of the Fourteenth Amendment. In fact, the Court, almost in the same breath that it admits the impotence of the Thirteenth Amendment and the Fourteenth's necessary passage, states the following with regard to statutes that segregate:

> *A statute which implies merely a legal distinction between the white and colored races, a distinction which is founded in the color of the two races, and which must always exist so long as white men are distinguished from the other race by color has no tendency to destroy the legal equality of the two races, or re-establish a state of involuntary servitude.*

Essentially, the Court renders the Fourteenth Amendment powerless with this statement. The statement is made after the Court's contemplation of the purpose of the Fourteenth Amendment. It is made with the reader's thoughts still fresh as to the purpose of the Fourteenth Amendment. The Court then states plainly that the amendment designed to deal with the burdensome and onerous disabilities that plagued the Black race has no impact on statutes that distinguish between the white and colored races, the very type of statutes that had the tendency to curtail the right of Blacks to pursue life, liberty, and property. These statutes were the ones that made living as a Black person in America onerous. Homer Plessy was incarcerated because he pursued a seat on a train that was designated for a white passenger. That is denial of his liberty. He paid for a seat with his money or his *property*. He was not free to sit where he wanted. This was a denial of *liberty* and it occurred despite the lack of some compelling state interest necessary to accomplish a larger legitimate societal goal. Separation of the races in and of itself was no longer a legitimate societal goal since the Fourteenth Amendment had the effect of ending the distinction of race. The goal of the statute was simply to prohibit the commingling of the races.

FATHER: *It sounds as though you're saying that there was never any sincere attempt by the U.S. government to ever grant equality to Black people.*

SON: Correct. *And given this thesis, Black people are entitled to reparations for the denial of the opportunities associated with the United States' never actually implementing the separate but equal doctrine.* The lack of implementation, whether intentional or negligent, constitutes the damage requiring compensation. A demonstration of the disparity is so obvious that it was taken for granted. One-room schools versus many-room schools. New textbooks versus old textbooks. Cutting-edge technology versus the complete lack of technology. Access to information versus information deprivation. The disparities were not caused by a lack of intellect, as racist scholars such as turn-of-the-century Stanford psychologist Lewis Terman, Berkeley psychologist Arthur Jensen, and author of *The Bell Curve* Richard Herrnstein conveniently argued. The problem stemmed from the wholesale denial of rights under the Constitution of the United States and the failure of the Executive and Congress to pass and enforce law to end the disparity. The country's highest court readily recognized the problem as an inability to undo the societal imposition upon the colored race of all the stereotypical burdens heaved upon its shoulders.

FATHER: *But the damage has been done. Many Black people unconsciously view themselves as having and being "less than" white folks.*

SON: You're the psychologist and you know more about that than I do, about how Africans in America respond to white supremacy. I believe that separation of the races does not necessarily require that either race must look to the other as better than or less than. The inequality begins when that separation occurs and the two races are not equally equipped, which is what was made law after the 1896 *Plessy* decision. The statement in reverse would go something like this: integration does not lead to automatic unqualified equality among the races. This would imply that the stereotypes and the long history could be undone with the passage of laws. Fanciful though it may seem, if this were the case, this nation arguably would have never enslaved Black people, since its forefathers would have had the innate sense that Black people were in fact men created equal and endowed with certain inalienable rights.

FATHER: *So you are admitting that separate but equal did have an adverse impact on our people, because the equality called for in* Plessy *never actually took place.*

SON: Exactly. The fact is that Black people have suffered due to Black skin and the passage of legislation has never remedied the situation. This is not to say that Black people have not been the benefactors of programs designed to undo the impact of racism. Affirmative action has helped to provide some with opportunity, but with minimal impact on the masses. As you might expect, some white people erroneously refer to this as reverse discrimination. Affirmative action, although necessary, has deepened the resentment of Blacks by white society, who believe in the rumors of inferiority discussed by Jeff Howard and Ray Hammond in their classic 1985 *New Republic* article. It has played off the already deep divide of racism, and rubbed salt into an already gaping wound. White people are given the excuse to complain, and the Courts listen. It is ironic that less than forty years after passage of the Civil Rights Act, it is more difficult for a Black person to *allege* discrimination based on race and win than it is for a white person to allege reverse discrimination and lose. This nation has placed equality out of the reach of many Black individuals by the utilization of clichés and catchphrases like "color-blind society" and "compassionate conservatism." In fact the *Plessy* decision, although mired in the inequities of the late nineteenth century, went a lot further to ensure equality of the races than the *Brown* decision did in the mid twentieth century. *Plessy* stated separate but equal was constitutional. *Brown* made the blanket decision that separate races could never be equal. The decision in theory is the correct decision; however, in practice it is limited in its ability to effect real and necessary change. Because *Brown* focused more on *separation* than *equality,* the United States was led to believe that the race problem would be cured with integration, when truthfully, the separation issue was only one-third of the challenge. The other two-thirds can be simply referred to as *caste.* Yes, caste. The nation relegated Black people to a certain place in society. With that place came the boundaries associated with caste. Caste is inescapable. It does not allow for self-determination or the vaunted notion of rugged individualism or the *Kujichagulia* spoken of in the *Nguzo Saba* without a Herculean struggle. It inhibits self-improvement. There are no bootstraps long or tall enough to lift oneself up from a predetermined destiny. The grip of caste is a death grip. And caste is not eliminated with integration. Indeed, it is exacerbated by it. The victim of caste is made to see how little his or her life matters, while the individual whom caste benefits begins to see the great differences between himself and the lower caste person. If left to one's place, caste seems little more than the inability to elevate oneself in the social hierarchy. It is not until the person from the lower caste begins to see what is missing from his life that caste becomes painful. Integration unclouded the vision of Black people. It made it plain to see just how much white skin made a difference. We always knew there was a difference but *Brown* allowed many of us to see how much of a difference it really made. The inequality was and still is painful to experience when all the rules say you're the same. Integration did not fix the inequities, it exposed them, and in so doing, the great divide between Blacks and whites grew even bigger. It grew bigger not because of *hardened* white hearts but because of *enlightened* Black and white hearts on their perceived status in a white supremacist society. . . .

FATHER: *That's true. Caste is both social and psychological.*

SON: Yes. The *Brown* decision did not provide equality but it did deepen caste. Implementation of *Plessy* would have undoubtedly ended caste in America, and we would not have needed affirmative action or the Civil Rights Act in order to accomplish it. After *Brown,* this country undertook a number of calculated attempts designed to root out and stem the tide of racism and bigotry in American society. These efforts by and large have failed because they approach racism from the integration side rather than the side of caste and the white supremacist ideology embedded in the soul of the nation. The inequities of caste are as ever-present today as they were pre-emancipation, when most individuals would admit America's caste system thrived.

It was not until *Plessy* that Black Americans could begin to look at the caste system as an implementation of the vestiges of slavery. The decision stated that although men are created equal, sometimes they need to be separated on the basis of race. This also marks the first legal decision whereby empirical measures of inequality could occur if separate proved *not* to be equal. *Plessy* argued forcefully that separate could be equal and that it was the duty of the government to enforce this. That duty should have ensured *equal* infant mortality, *equal* medical care, *equal* accommodation, *equal* acceptance to *equal* schools, *equal* education, *equal* graduation, *equal* hiring, *equal* salary, *equal* firing, *equal* unemployment per capita, *equal* incarceration per capita, *equal* death sentences per capita.

Black people are entitled to reparations from the United States government for the wholesale failure to implement the separate but equal doctrine outlined in Plessy. The lack of implementation, whether intentional or negligent, constitutes the damage requiring compensation. Separate but equal was never inherently evil, however, but the blind eye that America turned toward racism and bigotry was and remains so today. This is the breach of the duty owed to Black people.

For too long Black people in America have been made to shoulder the racism and bigotry and prejudice associated with having Black skin. America must begin to undo the long-lasting legacy of racism, bigotry, and stereotypes that have become all too commonplace in this society. As psychiatrist Frances Cress Welsing argues, racism (white supremacy) is reflected in all of our institutions. Education, entertainment, religion, law, politics, our views about sexuality, who we wage war against, business and who works at what in the United States are all embedded in racism. Reparations will begin the long-overdue task of correcting these institutions. The great divide that is race must be bridged through recognition of the problem, repentance of the evildoing, and conciliatory measures toward the sufferer. Reparations will do that. Everything that has happened in the past is only another stone in the foundation for this bridge. The divide is great, but the consequences of failing to bridge it are even greater. Continuing to sweep racism under the rug, to play it safe by not confronting white supremacy, and to avoid this difficult conversation means that the past will continue to haunt every American. The true land of equal opportunity and equality is achieved only when neither side views the other as somehow gaining the benefit of some tool the other has not been given.

OMARI L. WINBUSH *is an attorney and consultant in Tennessee. A graduate of the University of Alabama School of Law, he has an interest in historic civil rights decisions and their present consequences.*

PART III

Voices for and against Reparations

ARMSTRONG WILLIAMS

Presumed Victims

Firewood will not collect itself.
—NAMIBIAN PROVERB

The notion that blacks are owed reparations for the awful crimes that slavery visited upon them two centuries before was barely a blip on the screen during the Clinton administration. Now, with the ascension of a white, Republican president, who happened to be born into a wealthy political clan, black intellectuals are pushing the issue into the mainstream across the country. Plainly, the civil rights leaders know that they can capitalize on free-floating white guilt, and the black voting public's distrust of the Republican Party to rally the troops around the reparations issue. Along the way, they can empower themselves by making the current ruling structure seem as racially insensitive as possible.

The implications are truly frightening: Despite the marked socioeconomic progress black Americans have made in this country over the past half century, the reparations movement, at bottom, encourages minorities to believe that they are really lost souls. The leaders of this movement do not talk about how such a distant crime has led to specific damages in the present lives of most minorities. For them, feelings of victimization in general, not damages in the specific, are the point. So they fervently maintain that all full-grown, capable minorities ought to blame the missed opportunities of their lives on the slavery that transpired centuries ago as though their pains were interchangeable with those endured by slaves.

Otherwise stated, the reparations movement encourages minorities to regard themselves, collectively, as helpless victims. By extension, advocates argue that the current ruling structure owes all minorities financial reparations as though the current government is interchangeable with the English monarchy that first brought slaves to America in the seventeenth century.

Sadly, this romantic notion of the white descendants of slave owners paying reparations ignores the fact that various African tribes directly participated in slave raids so as to solidify their own local rule. Even today, rival African tribes in Rwanda continue to commit genocide upon one another, while a brutal civil war in the Sudan has fueled a new slave trade in that area. Furthermore, thousands of black Americans owned slaves during the antebellum period. Bottom line: Several groups, including no small amount of Africans and black Americans, share responsibility for the mid-Atlantic slave trade.

Nonetheless, the preachers of racial despair continue to insist that the white ruling structure owes them money. As an American who also happens to be black, I find this equation between being a minority and being a victim to be not only self-limiting, but also inherently racist. Plainly I do not wish to be defined by a stereotype of failure, nor do I wish to shout from the rooftops that I am, by virtue of the hue of my skin, damaged goods.

Sadly, the notion of lining one's coffers with reparation money has sent minorities across the country scurrying to adopt the mantle of victim. This rallying cry has resonated in Chicago and Detroit, where legislators are debating a bill that would give blacks a $330,000 tax credit to make up for the government's false promise of forty acres and a mule, offered to newly freed slaves during the Reconstruction era.

One would think that if this reparations movement is to have any meaning, it must be limited to instances of specific damages (as has been the case with reparations paid to Holocaust survivors and Japanese Americans interned during World War II). The black reparations movement sees things differently. What matters is not personal injury, but rather the feeling of victimhood that you arrive at. The major implication: The Civil Rights movement has veered from individual rights to social retribution.

Supporters argue that such reparations payments would publicly acknowledge the atrocities committed by the English government (America was a colony at the time) when they descended upon Africa in the seventeenth century, tearing apart the continent's cultural configurations and forcibly enslaving their people. And indeed, it is true that the systematic destruction of another culture needs to be raised to consciousness for examination. It is equally true that there remains quite a bit of hangover in this country from the cultural division that slavery wrought. I firmly believe that the racism of today isn't as much about skin color as it is about the racial hierarchies that a shared history of slavery and discrimination ingrained into our national identity.

This rousing point was not lost on those civil rights activists of the '60s, who sought to redistribute those social hierarchies through social activism. The landmark legislation that they secured during the '60s was animated by a single idea: equality amongst white America and its former inferiors. Men like Rev. Martin Luther King advocated not for special treatment, but merely for those basic human rights that we associate with happiness: equality and individual freedom. He dreamt not of revenge, but of a more perfect union where the sons of former slaves and the sons of former slave owners would be able to sit down together at the table of brotherhood.

Thirty years later, the Civil Rights movement has shifted from Dr. King's quest for individual equality to a cry for collective retribution. Consider that the reparations movement seeks to penalize our current government for what white slaveholders did centuries ago. By failing to draw a distinction between past and present, the reparations bill encourages the view that all blacks are victims, and that all whites are collectively responsible. Rather than staking its claim upon the proof of specific damages, the reparations bill simply takes for granted that all minorities are victims.

Again, as an American black, I find this stereotype of failure not only to be personally insulting, but also to be radically destabilizing. Simply to regard all members of a group as victims neatly removes such terms as "character" and "personal responsibility" from the cultural dialogue. After all, what need is there for individual striving when it is plainly understood that all the difficulties that blacks suffer are the direct indisputable result of incidents that occurred centuries ago. The real danger with reparations, then, is that it presumes that being black is synonymous with being inferior. By extension, all blacks are encouraged to identify themselves as victims, even those who are plainly among society's privileged.

This became clear during my participation in a recent reparations debate sponsored by the Harvard Black Law Students Association. At one point, an attendee barked out, "Don't you think African Americans have been victimized by the white man and his racist system?"

The question caused me to wonder aloud how much these "victims" pay a year for their Harvard education.

"Thirty-five thousand dollars a year," a student responded.

I shook my head incredulously. "What precisely about your thirty-five-thousand-dollar-a-year education has taught you to believe that you are a victim? I mean, why even go to college if you are already defeated?"

The questions went unanswered. Instead, a panelist participating in the event asserted that many black women suffer from breast cancer and a lack of health benefits—twin facts that she attributed to the lack of reparation payments.

"So let's get this right," I responded. "If reparations had been paid, no black women would suffer from breast cancer or find themselves in need of health care? The disease would somehow leap over black women if reparations were paid?"

Ripples formed across the student's forehead.

"Many of our civil rights leaders," I continued, "have created the myth of retribution, the idea that seeking payback will somehow create equality. That's how they stay in business—by nurturing other people's anger and ignorance. But this is dangerous because it encourages society to regard all members of a fixed group as victims."

If that's the case, then it occurs to me that our post–civil rights culture has become terribly soft. Just name your problem, sit back and blame centuries-old crime. Personal responsibility will seem to fall by the wayside when one is obsessed, often to the exclusion of all else, with his victim status.

I quote my mentor, Justice Clarence Thomas, who once observed that the [civil rights] revolution missed a larger point by merely changing the status [of minorities] from invisible to victimized.

Plainly, the Civil Rights movement did not teach blacks to identify themselves as inferior. Nor did it adopt as a rallying cry, "Victims." Herein lies the rub: the Civil Rights movement has been hijacked lately by racial hustlers like Randall Robinson, Jesse Jackson, Kweisi Mfume, and John Conyers, who empower themselves by dispensing a warm drug, a surrender of the will to the feelings of victimization.

The leading voice for the reparations movement, Randall Robinson, gave the game away when he titled his manifesto *The Debt: What America Owes to Blacks*, in which he asserts that "America's socioeconomic gaps between the races remain, like the aged redwoods rooted in a forest floor, going nowhere, seen but not disturbed, simulating infinity, normalcy. Static." For Robinson, being black is forever code for being poor, for being a victim.

In reality, minorities in this country are by no means living in a state of static socioeconomic despair. As Stephan and Abigail Thernstrom observed in *America in Black and White*, "No ethnic group in American history has ever improved its position so dramatically in so short a period, though it must be said in the same breath that no other group had so far to go."

A few examples: Since 1940, the poverty rate for blacks dropped from 87 percent to 26 percent. Six decades ago, 60 percent of all employed black women were household servants; presently, that same percentage is employed in white-collar positions. The income gap between black and white families has closed to within 12 percent over that same period. In 1940 there were only a handful of black politicians. Today, there are over 8,000 black elected officials representing diverse districts throughout the country. Six decades ago, only a handful of blacks lived in the suburbs. Today, nearly 33 percent of blacks live in the suburbs and over 40 percent own their own homes.

Bottom line: the socioeconomic trajectory of black Americans since 1940 has been anything but static (the word Robinson employs to justify his reparations claim).

This is not to diminish the importance of race-related problems in this country. As Stephan and Abigail Thernstrom also note, there is no shortage of genuine problems, ranging from poverty to disintegrating family structures to the education gap. But a great deal of inequity has been overcome. Particularly illuminating is that much of this progress predates the race-based quota systems of the '60s, implying that the very practical merits of self-help supercede even legislation. Given this progress, it seems a peculiar time to eschew the self-help narrative in favor of the bleak view that "black" and "inferior" are synonymous.

Yet, that's precisely what the reparations advocates are doing when they proclaim that all blacks should be publicly singled out for their collective burden and given some cash. As such, reparations make no distinction between the American black who has been successful and the American black who has foundered. It holds as a basic tenet that every man, woman, and child with black skin should be outed by the government as a victim whether they know it or not. It is upon this stereotype of failure that the reparations movement chooses to dedicate their resources and efforts.

According to Robinson, pursuing "reparations will be a cathartic act of self-discovery that will do wonders for the spirit of African-Americans." Yet one wonders how embracing victim status will imbue the collective black cultural identity with newfound dignity and self-esteem, let alone bring us together. In reality, the reparations movement only recapitulates the separatists' message that regards minorities as colors first, humans second.

Never mind that to blame white domination for much of one's problems implies that whites have all the answers. That in itself empowers a person to believe that race controls your destiny. Of course, one would be foolish to argue that racism and sexism don't exist. But to make people believe that they are forever the victims of a centuries-old crime is worse than stupid, it is harmful, degrading, and a passive form of enslavement; it passively reinforces those old racial hierarchies that deem whites the masters.

Never mind that, as a matter of sheer practicality, reparations raises more concerns than it assuages. One wonders, for example, what percentage of black blood would entitle a citizen to reparations? What reparations, if any, would Africa be required to pay for selling their own citizens into slavery? An estimated 3,000 slave owners in this country were black. Would their ancestors still be entitled to reparations? Would American Indians be able to stake a similar claim? How about the various religious groups that the Puritan settlers persecuted? Would modern-day members of the occult be entitled to reparations, to make up for the fact that our founding fathers regularly burned their relatives at the stake? Should ancestors of those who immigrated to this country after the antebellum period still have their tax dollars funneled into reparations? Precisely how much money is required to heal the shared history of slavery? Have reparations already been paid in the form of affirmative action and various racial quotas?

Bottom line: If it literally paid to be a victim, countless people would rush forward to adopt the mantle. Plainly, forcing this government to pay reparations to the biological, cultural, or religious offshoots of every group that they wronged over the past two hundred years would bankrupt this country. For this reason, reparations have no chance of becoming a reality.

It does not matter. The idea of receiving money from the government appeals to most. Consequently, those who stir racial tensions for a living are paying lip service to the idea.

One wonders, when are these cultural prophets going to start focusing on what it takes to move forward in this country? Because it occurs to me that until a black man begins defining himself by his own unique experience, rather than as a victim of a centuries-old crime, he remains but half free.

As a child growing up on a farm, I was taught that personal responsibility was the lever that moved the world. I understood that a neutral human being—one without will and responsibility—simply could not move forward and grow. That is why it pains me to watch the Civil Rights movement go from attempting to overcome victimization to regarding it as an imminent function of history.

It should perhaps be noted that those seventeenth-century slave traders who first descended upon Africa held similar views. That is to say, history had conditioned the European invaders to regard blacks as inferior savages. By extension, the imperialists felt they were merely upholding the natural order of things by subduing the black inferiors and uniting the world beneath the manifest glory of one dominant, civilized tribe. To embrace reparations is to embrace the notion that history has indeed conspired to make blacks inferior.

The Civil Rights movement taught us an entirely different lesson. It demonstrated, in no uncertain terms, that social activism, will, and personal responsibility could be the engines that propelled American blacks toward equality. Nearly half a century later, and the country has moved imminently closer to achieving Dr. King's vision of a more perfectly integrated union. Still, serious issues remain: the disparity in prison sentencing between blacks and whites, the disintegration of the family unit amongst American black households, the disparity in political representation, etc. I fear that these issues will never be resolved, if we simply rest our heads on the warm pillow of victim status. That is why it pains me to see the Civil Rights movement shift its emphasis from conventional social activism to retribution. Rather than focusing on what it takes to move forward in this country, our cultural prophets now seem content to revel in the tragedies of the past.

As we haul ourselves through a new millennium, surely there are some American blacks who do not view themselves as inferior, right? I vote NO to victim status, and YES to activism and growth. The alternative is to embrace inferior status, by creating a culture of victimization that never moves beyond the initial steps of the '60s civil rights legislation.

Called "one of the most recognized conservative voices in America" by the Washington Post, **ARMSTRONG WILLIAMS** *is a nationally syndicated columnist (Chicago Tribune Media) and talk show host (www.armstrongwilliams.com).*

CHRISTOPHER HITCHENS

Debt of Honor

This article first appeared in the June 2001 issue of Vanity Fair *magazine. Hitchens wrote a response to "After the Civil War, freed slaves were supposed to get '40 acres and a mule.' " Now the price tag on reparations is estimated at $1.4 trillion. With the debate raging, the author takes on the loudest opponent of compensating slavery's descendants to explain why this debt is not easily canceled.*

A few years ago, I was engaged in writing an introduction to the Modern Library edition of *American Notes*, the most controversial book written by Charles Dickens, and the only boring one. I was having a hard time with it until I came to a heart-freezing passage on slavery. Dickens in 1842 had the inspired idea of quoting directly from the "lost and found" classified ads of the press in the Old South, where those whose slaves had run away could advertise the fact, and those who had found stray blacks could announce likewise:

- "Ran away, the negro Manuel. Much marked with irons."
- "Detained at the police jail, the negro wench, Myra. Has several marks of lashing, and has irons on her feet."
- "Ran away, a negro woman and two children. A few days before she went off, I burnt her with a hot iron, on the left side of her face. I tried to make the letter M."

- "Ran away, my man Fountain. Has holes in his ears, a scar on the right side of his forehead, has been shot in the hind parts of his legs, and is marked on the back with the whip."

I now know the source from which Dickens plagiarized this chapter. It is a tract from the American Anti-Slavery Society entitled *American Slavery as It Is: Testimony of a Thousand Witnesses.* Published in New York in 1839, it is one of the most potent and lucid polemics on any subject ever printed; it's all about the moral and physical consequences of people owning other people. That same year the good ship *Amistad,* with its crew of runaway slaves, put in at New London, Connecticut. In the atmosphere of interest that was created by Steven Spielberg's movie of that drama, *The Hartford Courant,* this country's oldest newspaper, early last year published an exposé of its hometown's leading industry, which is the insurance racket. It turned out that Aetna had been writing policies for slaveholders on the lives of their slaves, and doing so well into the nineteenth century. Score one for the *Courant*—journalism worthy of the tradition of Mark Twain, who worked on *Huckleberry Finn* in Hartford. But then there were conniptions at the newspaper office. It also turned out that until at least 1823 the venerable sheet had carried lucrative ads not just for the sale of slaves but also for the re-capture of runaways. Last July, it ran a 1,500-word apology and explanation on the front page. "The stories about Connecticut's slave profiteers had a glaring omission," said the editors. "The *Courant* itself." As the story went on to say, "It was accepted practice. Slavery was so woven into the nation's economy and social fabric that such ads were probably less controversial than gun or tobacco marketing would be today."

And that, as Randall Robinson would say, is more or less the point. Sitting in his office at TransAfrica, a hybrid somewhere between a think tank and a lobby, this humorous but highly determined man is launching his latest campaign. He initiated the countrywide boycott of South Africa in the 1980s, and also the moves to isolate the military dictatorships in Nigeria and Haiti. Now his new book *The Debt: What America Owes to Blacks* has made the best-seller lists in Los Angeles and Washington, D.C., and in *Essence* magazine—the list to watch for African-American readership. Resolutions to reopen the debate on reparations for slavery have been passed by the city councils of Washington, Detroit, Chicago, and a growing list of other centers of population. Legal scholars such as Professor Charles Ogletree of Harvard, and legal there-must-be-a-word-for-it types like Johnnie Cochran, are getting involved.

"Show Me the Money" may be the slogan of Mr. Robert L. Brock, a veteran reparations activist who tours black audiences, tells them that each of them is owed about $500,000 for what happened to their folks, and then asks a $50 fee for his "claim form." The idea of black Americans getting a windfall from the enslavement and torture of their forebears apparently doesn't strike him as odd, though so far he and his emulators are the only ones to have made any dough from the scheme.

Robinson patiently explains that his own proposal is for a trust, set up by the federal government, from which nobody would get an individual check but by which education and housing and job training would be financed by accumulated back pay. With interest? "With interest," says Robinson. This is, after all, America.

The original proposal to compensate slaves was direct and simple. The victorious General William Tecumseh Sherman, on January 16, 1865, issued Special Field Order No. 15. It resulted in the donation to liberated slaves of "40 acres and a mule." General Rufus Saxton, who believed in the creation of a black freeholding class, supervised the distribution of land in South Carolina, while Congress ratified the order by establishing the Freedmen's Bureau to implement it and other aid to former slaves. It is still, in the collective memory of black America, a lost golden age. It didn't last long. President Andrew Johnson was one of those who thought that it was the slave*holders* who should receive compensation. (That's partly why he was later impeached.) He took the land back, and various forms of disenfranchised sharecropping and peonage became the successor system to slavery until—don't forget—the 1960s. Only two years ago, a class-action suit against the U.S. Department of Agriculture was settled on behalf of 23,000 black farmers for the systematic denial of loans. The federal government had colluded in the writing of "restrictive" racist mortgage rules well into the 1960s.

As a result, the very means that had allowed poor immigrants from Europe to become people of property and inheritance in a generation or two—access to ownership of a plot of land or a house—were forbidden to blacks in recent living memory. And blacks are not the children of immigrants, or at least not of voluntary ones. Ever wonder how to make black Americans become incensed? Tell them to get over it because this was all a long time ago. (The great-grandchildren of the compensated Confederates, on the other hand, get all upset if you tell them that the battle flag of Robert E. Lee doesn't look so hot on the statehouse roof. Bygones be damned.)

Jefferson Davis, first and only president of the Confederacy, was in no doubt about the main issue of the war. Even the limitation of slavery to the South, he replied to Lincoln in 1861 in justifying secession, would render "property in slaves so insecure as to be comparatively worthless . . . thereby annihilating in effect property worth thousands of millions of dollars."

Lincoln himself in his second inaugural was also very conscious of the property aspect. In a brief and telling speech, he spoke of "all the wealth piled up by the bondsman's two hundred and fifty years of unrequited toil," and of the "peculiar and powerful interest" represented by that one-eighth of the population held as chattel.

My friend Adolph Reed, the most caustic and witty of the black voices in print today, has ridiculed the notion of reparations in a recent essay in *The Progressive:*

> On the one hand, this could promote public education about the real history of the United States, although that is a project that does not require the rhetoric of reparations. On the other hand, it fits the Clintonoid tenor of sappy public apologies and maudlin psychobabble about collective pain and healing. . . . Among some strains of cultural nationalists, this view unabashedly reproduces the old "damage thesis" . . . according to which slavery and its aftermath left black Americans without cultural moorings and therefore especially vulnerable to various social pathologies.

Reed's contempt for this self-pity view of history is rivaled by his scorn for the impracticalities. "Who qualifies as a recipient? Would descendants of people who had been enslaved elsewhere (for instance, Brazil or the Caribbean) be eligible? And what of those no longer legally black people with slave ancestors?"

"How much" some people ask, nervously. Jack White of *Time* magazine, who favors reviving some version of the old Freedmen's Bureau, reckons you have to calculate unpaid wages for 10 million slaves, doubles for pain and suffering, and with interest. That would be in the trillions. The economist Larry Neal, making an inflation-adjusted guess about unpaid net wages before emancipation, arrives at a figure of about $1.4 trillion. In other words, it's not just the principle. It's the money of the thing.

Well, my health insurance coverage (such as it is) is with Aetna. And I have recently contributed a couple of book reviews to the distinguished literary pages of *The Hartford Courant.* I came to these shores as an immigrant and have no slaveholding ancestors, but I've still benefited from the many facilities in the nation's capital that were built by unpaid labor. If I knew where to send it, or to whom, I would happily kick in a percentage of what the *Courant* paid me, in order to be quit of that debt. And I'd more than gladly change my health-insurance "carrier." However, gestures are futile. I propose the following way of making up your mind about this.

In February, my old friend and enemy David Horowitz began placing ads in campus newspapers on behalf of his very conservative Center for the Study of Popular Culture. In various places—Berkeley, Brown, and Harvard among them—the ads were either rejected or apologized for by the editors who'd printed them. In some university towns, the papers that carried the ad were stolen or destroyed. As a rebuke to this nastiness, I'll give the ten points of his ad for free and then my own devil's-advocate response.

1. "There is no single group responsible for the crime of slavery."

As is well known, slaves were originally rounded up and sold by Africans and Arabs. A few thousand southern blacks become slave owners, and many poor whites were indentured. However, the Confederacy openly stated both that it was based on the principle of white supremacy and that "African slavery" was biblically warranted. (Thomas Day, who bought *The Hartford Courant* in 1855, wrote in an editorial, "We believe the Caucasian variety of the human species superior to the Negro variety; and we would breed the best stock." Moreover, Mathieu Kerekou, president of Benin, has recently made a public apology for the part played by West Africans in enslavement, and there are dynastic fortunes in West Africa that were founded on trade. Yes, these elements, too, should be included in the bill, if it is ever to be drawn up.

2. "There is no single group that benefited exclusively from slavery."

It is true that black Americans benefit from the overall prosperity of the United States. But nobody is arguing that only white people pay reparations. The federal government, which helped administer slavery and hunt down its fugitives, also took in much of the revenue. But it would act, if it set up a reparations trust, in a color-blind manner.

3. "Only a minority of white Americans owned slaves, while others gave their lives to free them."

Horowitz says that only one white person in five owned slaves in the antebellum South. Actually, J. D. B. DeBow, the superintendent of the census, took care in 1860 to emphasize that the proportion of slaveholders was closer to one-third overall, and more like one-half in rural South Carolina, Mississippi, and Louisiana. Furthermore, David Christy, author of the famous 1855 book *Cotton Is King*, made the decisive argument that the wealth of the nonslave states also derived largely from slavery. "As new grazing and grain-growing States are developed, and teem with their surplus productions, the mechanic is benefited, and the planter, relieved from food-raising, can employ his slaves more extensively on cotton. It is thus that our exports are increased; our foreign commerce advanced; the home markets of the mechanic and farmer extended, and the wealth of the nation promoted. It is thus, also, that the Free labor of the country finds remunerating markets for its products—though at the expense of serving as an efficient auxiliary in the extension of slavery!"

4. "Most living Americans have no connection (direct or indirect) to slavery."

Waves of immigrants, Horowitz points out, arrived after 1880 and 1960. But even the most impoverished Irish or Hungarians were able straightaway to join the building trades or the police departments, from which American-born blacks were excluded. And legally enforced discrimination against the descendants of slaves persisted into the 1960s. In any case, all citizens of the country have benefited from the unrewarded heavy lifting done by kidnapped non-immigrants. Antebellum northerners, too, used to be fond of saying that they were untainted by slavery—even as they quietly reaped indirect dividends from it.

5. "The historical precedents used to justify the reparations claim do not apply, and the claim itself is based on race not injury."

This point has not yet been convincingly answered by the supporters of reparations, except to say that slavery was also based on race. As Adolph Reed points out, it's difficult to establish who are the "descendants" of slaves. But why is that? Black Americans are different "shades" because their maternal ancestors were raped and their paternal ones were sold down the river, and the children forcibly dispersed. Maybe we'd raise more federal and "faith-based" money if this were called reparations for violated family values.

6. "The reparations argument is based on the unsubstantiated claim that all African Americans suffer from the economic consequences of slavery and discrimination."

Horowitz here points out the West Indians also suffered from slavery but, in America, achieve average incomes equal to those of whites. "How is it that slavery affected one large group of descendants but not the other?" Slavery was abolished almost a generation earlier in the British Empire, and West Indians had voting rights and other liberties, with no Jim Crow system, well before black Americans. One might also note that, for a Republican who presumably resents the estate tax, Horowitz is strangely indifferent to the relative inability of American blacks to acquire mortgages or properties that they are able to bequeath. Derrick Jackson in *The Boston Globe* calculated that the average white baby-boomer and the average black baby-boomer can now expect to inherit, respectively, $65,000 and $8,000.

Justice Clarence Thomas, explaining recently to a high-school audience his almost complete silence in Supreme Court deliberations, said that he had been disabled by his native tongue, Gullah, in an English-speaking classroom. Gullah is a compound of West African tongues originally brought to South Carolina and Georgia by people in chains. Either this excuse was true and relevant or it was not. Horowitz might know.

7. "The reparations claim is one more attempt to turn African Americans into victims. It sends a damaging message to the African-American community and to others."

Undecidable.

8. "Reparations to African Americans have already been paid."

Welfare payments, Great Society programs, minority set-asides, and affirmative action are cited here. It's clear that Horowitz doesn't approve of them either. Nor does he approve of the War on Poverty in general, even though the majority of low-income Americans are white. Rural blacks in the South were excluded by law from most of the affirmative action for poor whites that was enacted during the New Deal. They also largely missed out, because of discrimination, on the greatest affirmative-action law ever passed, namely the G.I. Bill.

9. "What about the debt blacks owe to America?"

Smile when you say that, David. "In the thousand years of slavery's existence," he adds, "there never was an anti-slavery movement until white Anglo-Saxon Christians created one. . . . If not for the sacrifices of white soldiers and a white American president who gave his life to sign the Emancipation Proclamation, blacks in America would *still* be slaves."

It would be just as true to say that Christians didn't turn against slavery for almost two millennia: the first anti-slavery petition in America or anywhere else was drawn up by the Quakers of Germantown, Pennsylvania, in 1688. Then there were Thomas Paine (white and Anglo-Saxon, but not Christian) and Frederick Douglass (black, probably fathered by his mother's owner, highly critical of Christian hypocrisy). The WASP abolitionists in general believed that slavery was a curse and a sin and that it would take (note this) many generations to erase. This was because of the rape and degradation and deliberate family breakup that it involved. Mr. Lincoln (see above) outlived the Emancipation Proclamation by fifteen months and signed it only as a limited war measure.

Many blacks, it goes without saying, are tenaciously proud Americans and fought for the Union and the country even when (as from 1863 to 1949) they were allowed to do so only in segregated units.

10. "The reparations claim is a separatist idea that sets African-Americans against the nation that gave them freedom."

Maybe this would have been better as a nine-point statement. But see above.

What is clear is that the argument has now begun. If it is well constructed, it could be an excellent way of revisiting the past and seeing what may be learned from it, as well as what can be salvaged or repaired. This is a generous society as well as a litigious one. The proportions of these qualities will count; the dispute could easily become boring and rancorous, or enlightening and clarifying. Most people, even in liberal Hartford, had until recently no idea how close to the surface the bones lay. Here is what Thomas Jefferson wrote in his *Notes on Virginia,* coolly observing the slaves he owned but never emancipated, and from whom he produced off-the-record and deniable presidential children:

> They secrete less by the kidneys, and more by the glands of the skin, which gives them a very strong and disagreeable odor. . . . They are at least as brave, and more adventuresome. . . . Their griefs are transient.

In Mr. Jefferson's beloved Virginia, until 1967, marriage between blacks and whites was prohibited by law, and sterilization of "inferior" types was practiced by medical men under state warrant until 1979. There is hardly a black American whose grandmother couldn't tell him or her a personal story that would harrow the soul. Some of these griefs are beyond repair, but it would be rash and indeed impolite ever to refer to them as "transient."

CHRISTOPHER HITCHENS, *longtime contributor to* The Nation, *has written his wide-ranging, biweekly column for the magazine since 1982. Born in 1949 in Portsmouth, England, Hitchens received a degree in philosophy, politics, and economics from Balliol College, Oxford, in 1970.*

JOHN McWHORTER

Against Reparations

> *I can buy a big house in an exclusive neighborhood.*
> *I can buy a fancy car or two.*
> *I can send my kids to private school.*
> *I can work hard and empower myself.*
> *Oprah Winfrey pulled herself up by the bootstraps. So if I work hard,*
> *someday I, too, can achieve the American Dream.*

The fundamental problem with this rugged individualist dogma is that I would still be black.

Those sentences were written by a young black woman in an undergraduate newspaper at the University of California at Berkeley, as part of a response to David Horowitz's notorious antireparations advertisement that appeared in several campus newspapers a few months ago. They are an eloquent distillation of what it is that has rendered the American debate about race an eternal stalemate. Our debate is going nowhere, and the latest development in this holding pattern that is masquerading as a "dialogue" is the reparations movement, which in recent years has gathered alarming momentum in the African American community.

For this reason, Randall Robinson's best-selling book is genuinely useful. A close examination of his book can help us to understand why so many African Americans, many of them neither poor nor close to poor, feel strongly that they are owed money denied to ancestors whom they never knew; and why they feel that the payment of this guilty cash will somehow represent a moral triumph for them. In a broader sense, *The Debt* is an important document of the belief, which is more and more widespread among blacks, that a true "dialogue" on race has yet to happen. You might say that Robinson's *cri de coeur* has much to teach by negative example.

The idea of reparations has been kicking around black intellectual and political life since the beginning of the twentieth century, but it has acquired a certain cultural influence in the years since the Black Power era. The first extended treatment of the idea appeared in 1973, in *The Case for Black Reparations* by Boris I. Bittker, a law professor and a white man. Since then, there have been some books on the subject by blacks, most of them not widely distributed and hence not widely influential, or, in the case of Sam E. Anderson's documentary comic book *The Black Holocaust for Beginners* (1995), lacking the gravity necessary to spark a movement. But Robinson's book has overcome both those obstacles, and so it has become the manifesto for a movement recently revivified by Representative John Conyers and pored over by black readers and reading groups across the country.

Robinson's title faithfully conveys the tone of the new reparations movement. Bittker ended his book by saying that "I have sought to open the question, not to close it," but Robinson, while he initially claims "to pose the question, to invite the debate," clearly considers the moral urgency of reparations a closed issue. Bittker made a case for reparations, but Robinson's theme is "the debt," the definite article dogmatically implying the existence of the bill that is owed us. In the face of such righteous certainty, those who question whether there is merit to the idea of reparations are certainly not welcome to join the discussion. What is being described as an exploration is in fact a call to arms. Robinson presents his position as representative of the race, and he sets things up so that the failure of America to heed his call can be explained only by the eternal hostility of white people toward black people.

Yet to say that the foundations of Robinson's argument are questionable is to put the problem lightly. In truth, to embrace Robinson's assumptions about race in America would have the consequence only of perpetuating the very alienation that his book was written to dispel. *The Nation* not long ago promoted Robinson as "a worthy heir to W. E. B. Du Bois," but his book is just another fevered expression of the misguided ideology that the radical left foisted upon black America in the 1960s, a cluster of beliefs that continues to hobble our progress today.

The first of Robinson's assumptions is the denial that there has been any real progress at all. "America's socioeconomic gaps between the races remain, like the aged redwoods rooted in a forest floor, going nowhere, seen but not disturbed, simulating infinity, normalcy. Static." He penned those pessimistic lines six years ago, at a time when almost 50 percent of black families were middle class (defined by Stephan and Abigail Thernstrom in *America in Black and White* as twice the poverty line), in contrast to only 1 percent in 1940 and 39 percent in 1970. In 1990, one in five blacks were managers or professionals. In the three decades prior to 1990, the number of black doctors doubled and the number of black college graduates between the ages of twenty-five and twenty-nine tripled. Static? Hardly. And there are many other indicators of real social, economic, political, and legal advancement.

But Robinson is not really interested in social-scientific indicators. He prefers to deliver his arguments in the form of allegorical "stories," in the vein of his fellow bard of data-shy pessimism, Derrick Bell. A "story" that Robinson uses as a leitmotif in his book involves a certain black boy named Billy. He is being shown around the Mall in Washington, D.C., by a mentor, who is increasingly embarrassed to find that there is no monument on the Mall "about" black people with which he can inspire Billy. Needless to say, Billy is from Southeast, the inner city: the middle-class half (yes, half) of black America is just a lifeless statistic, but poverty-stricken blacks, who in fact represent less than one-quarter of African American families, are what's really goin' down. Nowhere is Robinson's misconception of the economic condition of his race more poignantly clear than in his assertion that, since there are proportionately more poor blacks than poor whites, poverty defines black America. Most black writers decry such a "racialization of poverty" as stereotyping; but after reading Robinson's passage three times, I concluded that he really does accept the terrible equation between "black" and "poor."

In his fable about Billy on the Mall, Robinson revealingly describes "a black woman wearing thick owlish glasses, strolling hand-in-hand with a bookish-looking white man, and two black men with white women." In his tale, all the blacks on the Mall except Billy are "attached to white people." The implication is that all black people who did not grow up like Billy—except, we presume, Randall Robinson—are sellouts who live and marry outside their race, and are probably homely besides. So the problem for Robinson's picture of the world is not merely that black is indistinguishable from poor, but that blacks who are not poor are disloyal and inauthentic. In Robinson's account, poor blacks are admirable victims, while middle-class blacks are suspect.

He has his peculiar reasons for this sentiment; but he is hardly alone in his basic insistence that the growth of the black middle class is somehow "beside the point," leaving the poor minority as the "essence" of black America. *The Debt* is symptomatic of a general implication in most arguments for reparations that even in 2001 "black" is essentially a shorthand for "poor," when this has not been true for decades. Of course, many of the people who are most fervently in favor of reparations are quick to condemn the tendency for whites to think that all black people are poor. And so we are brought up against a savage irony: the reparations movement is founded in large part upon a racist stereotype.

Robinson has another justification for reparations. It is that racism remains "unbowed" in modern America. He makes this claim about a country in which, as of 1993, more than one in ten blacks were married to non-blacks—an important fact even if the black women in question favor "owlish" glasses. Is racism "unbowed" when housing segregation among blacks is now documented to be largely voluntary, and when antidiscrimination cases are regularly and successfully filed on behalf of black plaintiffs by white officials? Of course we have some distance to go: nobody can deny that racism still exists in America. But when the backcountry whites of Jasper, Texas, turn out in droves for the funeral of James Byrd Jr., we must also question the notion that whites are poised to turn the hoses on us again at any moment.

To be sure, phenomena such as the Jasper mourners and assorted statistics and personnel lists may seem more symbolic than substantial. But the signs that racism is abating in America are everywhere. "It's the little things," as Lena Williams instructs in the title of her black victimology primer. Starbucks now includes "Strange Fruit," Billie Holiday's wrenching ballad about lynching, in one of its music mixes, in the understanding that its latte-drinking white customers would consider this heartbreaking song a worthy interruption of their musical routine. Movies for teenagers increasingly depict a world in which, with no particular attention called to it, blacks and whites coexist in easy harmony despite the black students' remaining identifiably "black." (*She's All That* is a recent example.) Black-white romances are becoming downright ordinary on television and in film, and not used as sensational ploys. (In *Save the Last Dance*, a willowy blonde teen falls in love with a black boy as he teaches her how to dance hip-hop style: the Astaire-Rogers trope for a new America.) As I say, we are making progress.

Robinson most likely does not catch these movies, and does not seem to cock his ear to the background music; but the fact remains that this is the America upon which he is fatalistically pronouncing. Unfamiliarity with what's *really* goin' down is what makes Robinson scoff at the efficacy of African American initiative. Thus he imputes all black poverty to the per-durability of racism:

> Modern observers now look at the canvas as if its subjects were to be forever fixed in a foreordained inequality. Of the many reasons for this inequality, chief of course is the seemingly incurable virus of de facto discrimination that continues to poison relations between the races at all levels.

Note that "of course," and its assumption that no reasonable person could allow that today a significant portion of today's black poverty is owed to self-defeating handout policies in the 1960s that denied people the initiative to strive upward. I do not mean to deny that black despair is real; I mean only to challenge the historical analysis that keeps it alive, and to contest the insistence that social inequity is a sentence of doom rather than an impediment that can be overcome.

Robinson's argument is also predicated upon a fervent Africanism. In his view, I am at heart an "African" person, more intellectually and spiritually akin to Nigerian immigrants than to anyone born in the only country that has ever been home to me. Never mind that I grew up comfortably middle-class in Philadelphia speaking nothing but English: I am to consider it a denigration of "myself" when *The New York Times* downplays a story about a lethal pipeline explosion in Nigeria, because in such cases "we don't know what happened to us and no one will tell us."

Robinson the proud Africanist weirdly overlooks the fact that "Africa" is not a single culture. This sense that being dark-skinned and speaking a language unlike English somehow renders all the groups in Africa the same is rather similar to the view that "All Coons Look Alike to Me," as the troublesome old song went. (The song happens to have been written by a black composer.) If a newspaper headline reads "Asians Found Adrift on Raft," most of us spontaneously recoil at the notion that Chinese, Japanese, Vietnamese, Korean, and Cambodian peoples have been grossly lumped together. Yet throughout *The Debt* we are taught—by a black man—that the residents of four dozen countries speaking over one thousand languages are all simply "Africans."

There is an ideological reason for this lumping tendency. If we treat "Africa" as a single culture, then we can claim the literate and technologically advanced societies of ancient Egypt and Mali as "our ancestors." Thus Robinson devotes another one of the allegories in his book to a hypothetical forebear of ours from the civilization that built libraries in Timbuktu. But what about the societies from which the ancestors of black Americans actually came? It is safe to say that not a single African American is descended from an ancient Egyptian, and only a very small proportion of slaves were brought to America from as far north as present-day Mali. The English and American slavers drew the vast majority of their slaves from Senegal down through present-day Sierra Leone, Liberia, Ghana, Benin, Nigeria, Congo, and Angola: preliterate societies with little technology, in no sense comparable in material or intellectual advancement to Europe or even to the Mayan cultures of Central America. As the founder and president of Trans-Africa, Robinson is surely aware that there is a profound difference between the history of Ghana and the history of Egypt.

One cannot avoid the sense that Robinson considers the actual cultures from which most American slaves were taken as insufficiently "advanced" to serve as a basis for a case of aggrieved deracination. But current work in anthropology demonstrates that the reason most West Africans (and many other of the world's peoples) had not created the kinds of "civilizations" that Europeans and some other groups had created was largely an accident of geography. The plants and the animals that thrived at a particular temperate latitude were uniquely amenable to cultivation, and thus yielded a surplus that swelled populations, which in turn facilitated the emergence of densely hierarchical societies in which certain classes had the leisure to create technology as less-fortunate others worked the land. Findings such as these—they are masterfully presented in Jared Diamond's *Guns, Germs, and Steel*—leave me with no sense of "shame" that my West African ancestors most likely lacked libraries, pyramids, and muskets. To marginalize our actual ancestors in favor of Alexandria and Timbuktu is to base the case for reparations upon a false conception of our history, and to abase the people whose lives were ruined to create us.

But the most glaring omission in Robinson's uplifting depiction of my alleged African homeland is the fact that Africans themselves were avid and uncomplaining agents in the selling of other Africans to whites. Instead Robinson depicts the slave trade as having been based primarily on "catching" individual slaves unawares: in his Africa allegory, an aging African dismayed at the decay of his society at the hands of white predators bemoans that "our young people cannot sit still and listen to tales of glory from a dying old man while they fear being caught."

Robinson is hardly alone in this misconception: *Roots* and *Amistad* and other mythmaking artifacts of popular culture have intimated that whites acquired most slaves by lassoing people while they were out walking. The sad reality is that this method would hardly have netted Europeans enough slaves to furnish dozens of colonies of plantations, with each plantation often requiring as many as several hundred workers. (Wouldn't Africans have stopped going for walks?) The primary sources on the slave trade demonstrate with painful clarity that slaves—not some slaves, but most slaves—were obtained by African kings in intertribal wars, and were sold in masses to European merchants in exchange for material goods. This is an incontrovertible—and not exactly unknown—truth of history. Not once in his book does Robinson so much as mention it, since it would get in the way of his portrait of Africans as a preternaturally perfect people.

The historical truth about the origins of slavery also undercuts Robinson's notion of "African" as a single cultural identity from which we were brutally wrested. Traditionally, Africans were just like other humans: they, too, regarded people who spoke other languages and practiced different customs as alien. Indeed, the monolithic notion is a construction of the very essentialist worldview that Robinson considers to have gutted black America's soul.

Essentialism, after all, is a form of dehumanization, as Robinson is well aware; and for him few things better indicate the extent to which African Americans have been stripped of their humanity than the alleged suppression by America of their history as an African people. For example:

> Since this nation's inception, taxpayers—white, black, brown—have spent billions on museums, monuments, memorials, parks, centers for the performing arts, festivals, and commemorative occasions. Billions have been spent on the publication of history texts, arts texts, magazines, newspapers, and history journals. Formulaic television and large-screen historical fiction treatments virtually defy count. Almost none of this spending, building, unveiling, and publishing has been addressed to the needs of Americans who are not white.

Such melodrama requires an almost staggering indifference to reality. Robinson's grievance here is empirically outrageous. The National Endowment for the Arts and the National Endowment for the Humanities have long had an outright bias toward funding projects oriented toward the African heritage of black Americans. Moreover, this portrait of an America in which blacks' origins in Africa is invisible is made possible only by deftly restricting one's purview to projects funded by federal taxes. The America in which I have spent my life is a country in which museums in large cities frequently have exhibits of African art and performances by African dance troupes. Just a few years ago the media was abuzz with reviews of Hugh Thomas's *The Slave Trade,* a dense tome whose publication was feted as a national event; and Basil Davidson's briefer and more readable *Black Mother: The Years of the African Slave Trade* has been in print since its appearance in 1961. (Robinson includes both of these books in his list of sources.)

Again, consider "the little things." *Scientific American* has a page on which it prints excerpts from past issues. Most of these citations are naturally about science, but in a recent issue the magazine chose to feature a quotation from March 1851, giving it the headline "Open Sore": "The population of the United States amounts to 20,067,720 free persons, and 2,077,034 slaves." This suggests a spontaneous consciousness of our country's racial history in the editors of a journal neither dedicated to social and political issues nor aimed at a black audience. Examples could easily be multiplied. No literate American can help regularly stumbling across small but real signs that mainstream America is quite aware that a portion of its population was brought here in chains.

Robinson draws a certain conclusion from this presumed concealment of our African roots. He believes that the poverty and the despair in which many black Americans are mired is a direct result of our sense of rootlessness, of having been plucked from our African homeland. Robinson notes often that a people must have a sense of belonging to a particular "culture" in order to thrive. One might then investigate how blacks will embrace the culture of their homeland, but Robinson is making a different point. He claims that belonging to American culture is impossible, because "the armaments of culture and history that have protected the tender interiors of peoples from the dawn of time have been premeditatively stripped from the black victims of American slavery." Shepherding Billy around the Mall, Robinson's "mentor" sees the monuments as a statement from whites: *"This is who we are. This is who we are."* But there are no statues of African kings on the Mall, and so Billy is bedeviled by the question: *"Who am I? Just who the hell am I?"*

The italics are Robinson's: Billy always speaks a little hysterically. And this argument, central to Robinson's presentation, is the most dangerous one in his book. Never mind that it is not exactly obvious that most whites interpret the Washington Monument as a lesson in who they are. (Americans are not known for being a historically minded people.) More problematically, Robinson's position comes close to an advocacy of ignorance for blacks. In his anger and his sarcasm, he panders to the very sense of separation from "learning for learning's sake" that is the prime source of the gap between black performance and white performance in American schools.

More than once Robinson takes potshots at what he mockingly describes as "Punic this, Pyrrhic that." In his book he repeatedly dismisses Hegel on the basis of a racist statement that was, after all, typical of a man of Hegel's day. His readers may be forgiven for believing that racism was the entire substance of the philosopher's work. But turning his readers away from Hegel (and I am not suggesting that Hegel has any particular light to shed on the racial question in America) is a fine way of turning them away from other "dead white male" thinkers whose ideas are central to the philosophical heritage of the only society that most black Americans will ever consider home. It is a short step from this to the observation of the Berkeley undergraduate who bemoaned that so much of what she is expected to learn on Berkeley's "racist" campus is "white."

Robinson predictably falls in with those who attribute all of the problems of black students in school to social inequalities. He pauses to note that black students are lagging severely in school performance in Prince George's County, Maryland, owing to "grinding, disabling poverty," when in fact Prince George's County is a notoriously well funded district in which the persistence of low grades and low scores even among black students of comfortable backgrounds has been well covered by the local media for twenty years. It is widely documented, moreover, that much of this problem may be traced to the feeling of many black students that school is a realm fundamentally separate from the essence of being "black." A study by Clifton Casteel notes that while white adolescents tend to say that they do their schoolwork to please their parents, black students tend to say that they do their schoolwork to please their teachers.

This awful situation is a product of a race-wide pull away from the integrationist ideal. In the 1960s it may have made sense to define ourselves against "whitey"; but in those militant days few could foresee the awkward results that this ideology would have as time went by, and one of those results is an ingrained sense in black peer culture that school is something that "whitey" does. "Punic this, Pyrrhic that": this sentiment has a lot more to do with black students' problems in Prince George's County schools than the "poverty" of their middle-class suburban existences.

Finally Robinson's argument is crippled by the fallacy that it would make a whit of difference in Billy's psychological well-being to be taught that his essence is that of an Igbo boatman in the seventeenth century. I have rarely read a book by a black writer that demonstrates so little pride in the heritage of black people right here on these shores. "Far too many Americans of African descent," Robinson observes, "believe their history starts in America with bondage and struggles forward from there toward today's second-class citizenship." Hear, hear—but Robinson's assumption that redressing this means harking back to African villages is mistaken. For the purpose of black uplift, nothing could be less promising than the view that we will lack inner pride until we studiously equate ourselves with people who do not talk, eat, move, dress, or see the world the way we do.

Ultimately the Swahili lessons and the rest are merely a kind of theater, self-affirming in some ways, but largely in a gestural sense. Most black Americans see themselves neither as "African" nor (obviously) as "white." The truth is that black Americans think of themselves as a new race altogether. For Robinson, of course, this amounts to an obliteration of the self, the working-class black man in Cincinnati denying his primal urge to get back to Lagos. But this only demonstrates that the old "one drop" rule is now more fiercely wielded by blacks than by whites, which is not exactly "progressive."

"*What about us?*" Billy moans as he is trotted around the Mall, as if he were a Ghanaian village boy. One might make this reply to the poor lad: it was "us" who worked the American system against great odds and survived, who appealed to its philosophical foundations in sparking a civil rights revolution that few blacks could have imagined even a decade before. But this is not what Robinson wants the Billys of America to hear. He would be even less enthusiastic to have it said within earshot of Billy that our ultimate ideal is for Americans of all colors to consider the monuments on the Mall to represent the history of "us." There is still a long way to go before we reach that blessed moment, obviously; but it is the only direction in which we can set out if we seek true interracial harmony.

And what about recasting our vision of what happened *after* we were brought here? What "far too many Americans of African descent believe," in no small measure because of books such as *The Debt*, is that blacks have never been able to accomplish much of anything in America, except for the likes of Frederick Douglass and Colin Powell, who are portrayed less as ideals of African American history than as freaks of African American history. Robinson's analysis cannot account for the thriving black business districts in certain cities just two generations after emancipation, or for the revolution of American popular music that African descendants sparked, or for the fact that in the late 1800s black university students were well known for taking top prizes over white students not in athletics or music, but in oratory. Indeed, in classical oratory: Pyrrhic this and Punic that!

All of this is marginal for Robinson, because he is operating according to a studiously defeatist paradigm that restricts his view to a limited body of data. He enthusiastically subscribes to the orthodoxy of so many black writers today: that individual initiative is a matter of luck or a matter of extraordinary ability, and that it will remain so until American society presents no obstacles whatsoever to advancement—until there is no longer any racism in any white person's heart. This notion would have perplexed most of the civil rights leaders of the past; but its appeal is that it offers a balm for the insecure by providing an everlasting explanation for adversity that will distract one from the examination of one's own inadequacies.

We are hollow chocolate bunnies, beached in an alien culture: blacks have embraced this view out of pain and doubt, and Robinson teaches us not to conquer it, but to cherish it. In the end *The Debt* is founded on a paradigm of black existence rooted in shame. Robinson's view is that imperfect conditions render black success meaningless, and never mind the pride and the resilience that are common features of African American experience. Robinson actually offers a rare example of this destructive philosophy made explicit:

> There are always those special few who achieve (or fail) against all odds. There are those, like me, whose families successfully defy mainstream society's low expectation of us. The exceptions, however, would

not be numerous enough to allow the closing of the income gap, even if the coarse and tangible old brand of discrimination were to go tomorrow into some period of long-term miracle remission. This is so because a static, unarticulated, insidious racial conditioning, to which all Americans are subject, lifts the high-expectation meritless . . . and, more often than not, locks down in a permanent class hell the natively talented but low-expectation black.

This sense of racism as rendering all black success "accidental" is ultimately the *primum mobile* of the reparations movement. Thus the black scholars Robert Chrisman and Ernest Allen Jr., arguing the case for reparations, proclaim that "racism continues as an ideology and a material force within the U.S., providing blacks with no ladder that reaches the top."

This belief that there is no path for blacks to the top accounts for Robinson's sour attitude toward blacks making progress. He can only see self-advancement as a kind of self-abnegation: to gain passage into the world of the gatekeepers one must become one of them. One of the most appalling passages in *The Debt* describes Robinson at a commencement ceremony at Howard University. He records that he was appalled when a black undergraduate speaker said "thank you" in French, German, and Italian, rather than in Swahili, Chichewa, and Wolof. "She was not a European American of any variety: She was an American of *African* descent. Why on earth was she iffing herself European?" No, sir: this woman was "iffing" herself a new race entirely, one with a heritage as richly Western as African. Indeed, since no slaves were brought to America who spoke Swahili or Chichewa, the study of Swahili or Chichewa would no more return her to her roots than the study of European languages. Our new Du Bois might recall that the old Du Bois was fluent in German, and would have had choice words for anyone who told him that such cultivation was not a "black" thing.

In his contribution to the "Who's got the bigger Holocaust?" competition, Robinson has it that slavery "has hulled empty a whole race of people with intergenerational efficiency. Every artifact of the victim's past, every custom, every ritual, every god, every language, every trace element of a people's whole inherited identity, was wrenched from them and ground into a sharp choking dust." As often in *The Debt*, the music has a certain pull—but this is a grievous insult to four centuries of black Americans. Could Robinson look Denmark Vesey, Sojourner Truth, Frederick Douglass, James Weldon Johnson, Louis Armstrong, Mary McLeod Bethune, Duke Ellington, Paul Robeson, Thurgood Marshall, Rosa Parks, Ralph Ellison, Jacob Lawrence, and Martin Luther King Jr. in the eye and tell them that they were hulled empty? Could he even say this to the middle-aged black woman of a certain age working at the post office, or to the black middle manager at Pacific Bell with a house and family, or to Condoleezza Rice, or even—looking in the mirror honestly, for once—to himself?

Which brings us to the money. There are many obvious retorts to the notion of reparations as a practical matter. They include: that many whites in America today arrived after emancipation; that many whites owned no slaves; that racial mixture would render the very question of who qualifies as a "black" person tricky at best and arbitrary at worst. I feel uncomfortable with the idea of taking money meant for someone I never knew. I feel "black American," but I do not feel African, and I certainly do not feel that I am just a few steps past being a white person's property. My connection to my ancestors six generations back, about whom I know nothing, feels more academic than spiritual; and I would feel the same way if my ancestors were wealthy white barons. So I for one could not take the money.

But Robinson does not dwell long on the "back pay" angle. He and the reparations crowd have their responses to these objections, but they recognize how unlikely it is that we will reach a consensus that will extract huge sums of money from the national government. For this reason, the discussion of reparations has moved toward appealing less to slavery than to the effects of slavery, specifically to segregation and disenfranchisement. In this regard, the movement has returned to the position advocated in 1973 by Bittker, who emphasized the effects of *Plessy v. Ferguson* as grounds for reparations.

But here we run up against another objection to the argument for reparations: that for almost forty years America has been granting blacks what any outside observer would rightly call reparations. When Robinson grouses that "once and for all, America must face its past," one wonders what he thinks of the War on Poverty that Lyndon Johnson launched, with Adam Clayton Powell Jr. dedicatedly steering sixty bills through Congress in five years as chairman of the Education and Labor Committee. For surely one result of that new climate of the 1960s—of the official recognition that America owed its black citizens some sort of restitution—was a huge and historic expansion of welfare.

As begun in the 1930s, welfare policies were primarily intended for widows. Chrisman and Allen get this right, adding that at the time more whites than blacks received welfare. (They could have added that through the 1950s institutional racism ensured that black widows often got lower payments than white widows.) Yet they sail over the fact that in the mid-1960s welfare programs were deliberately expanded for the "benefit" of black people, in large part due to claims by progressive whites that the requirements of the new automation economy made it unfair to expect blacks to make their way up the economic ladder as other groups had. Federal and state governments have since poured billions of dollars into welfare payments. None of this was termed "reparations," in the technical sense, but it certainly provided unearned cash for underclass blacks for decades (as well as jobs for the many others who staffed the bureaucracy that the policy created).

It is now plain that this policy was not successful in pulling significant numbers of blacks out of poverty. But still America has not given up on the effort: today welfare programs are being recast as temporary stopgaps, with welfare mothers being trained for work. The governor whose version of this program was the most successful is now the head of the Department of Health and Human Services. Time will tell how successful this revision of welfare will be, but the signs are good as I write, and the advocates of reparations have yet to propose any more realistic solution. The funds and the efforts devoted to welfare-to-work, then, represent a concrete acknowledgment of the effects of "structural" poverty. A society with no commitment to addressing the injustices of the past would restrict welfare payments to the temporarily unfortunate, 1930s style. It would have no welfare-to-work programs aimed at poor blacks.

And there is also the policy of affirmative action—a reparative policy if ever there was one, designed to address the injustices of the past. Chrisman and Allen snap that "so-called 'racial preferences' come not from benevolence but from lawsuits by blacks against white businesses, government agencies, and municipalities, which practice discrimination." This is nonsense. Lyndon Johnson was a white man the last time I checked, and it was he who entered into the history books the famous remark that "you do not take a person who, for years, has been hobbled by chains and liberate him, bring him up to the starting line in a race and then say, 'you are free to compete with all the others,' and still justly believe that you have been completely fair." Affirmative action was initiated as a call to recruit and to hire and to admit qualified blacks. It quickly transmogrified into quota systems, with lesser-qualified blacks often given positions and slots over better-qualified whites—but then we cannot help but suspect that many reparations advocates would laud these preferments as their just deserts.

"Once and for all, America must face its past": has Robinson noticed that whites are often as horrified as blacks at the prospect of any contraction or alteration in welfare or affirmative action? It was Peter Edelman and Mary Jo Bane who resigned from the Clinton administration in protest over welfare reform; but no black members of that administration are on record as having considered taking down their own shingles. Many whites in elected office are fiercely clinging to the policies of affirmative action—so fiercely, indeed, that Republican pols have backed away from their opposition to affirmative action for fear of offending white women voters. It was two white men, William Bowen and Derek Bok, who produced the most serious defense of the policy in their book *The Shape of the River*, and it was University of California president Richard Atkinson who recently suggested working around the outlawing of racial preferences in California by eliminating the SAT. The most strident student organization seeking to reverse the ban on racial preferences in California, By Any Means Necessary (BAMN), has barely a black person in it. Surely all of this demonstrates that the analysis of black poverty in terms of its "root causes" is now a central element of white thinking about race. A significant and powerful contingent among whites now considers it a moral imperative to compensate blacks through set-asides.

If I were assigned to develop a plan for black reparations, here is what I would do. I would institute a program supporting poor black people for a few years while stewarding them into jobs—which is currently in operation. I would have the government and private organizations channel funds into inner-city communities to help residents buy their homes—which is what Community Development Corporations have been doing for years, working underpublicized miracles in ghettos across the country. I would give banks incentives to make loans to inner-city residents to start small businesses—which is what the Community Reinvestment Act has been doing since 1977. I would make sure that there are scholarships to help black people go to school—which are hardly unknown in this country. I would propose that affirmative-action policies—of the thumb-on-the-scale variety designed to choose between equally qualified candidates—be imposed in businesses, where subtle racism can still slow promotions. Most importantly, I would ensure that black children had access to as good an education as possible. I do not believe that blacks should be left simply to pull ourselves up by the proverbial bootstraps. Our grim history is real. Yet so, too, are the reparations that we have already secured in the form of all these government programs and government monies.

Robinson and the reparations crowd do not regard all these momentous changes as worthy of address, because their true interest is less in helping blacks than in assuaging the sense of inferiority that gnaws at the black soul. For them, all the payments, all the grants, and all the set-asides that have been directed to their people do not count as "reparations," because they did not come explicitly labeled as an apology for four hundred years of black suffering, and as an acknowledgment that whites are responsible for anything that ails anyone black in America. Robinson ardently hopes that we can "wear the call as a breastplate, a coat of arms." But he prefers the call for change to change itself. There is no other way to explain the most stunning aspect of

Robinson's book, namely, that he devotes less than three pages to a discussion of the actual form that reparations might take.

This is what Robinson proposes, concretely: a trust fund dedicated to education; the recovery of funds from companies that benefited from slave labor; continuing support for current civil rights advocacy (onward and upward . . . !); and the making of financial amends to Africa and the Caribbean. That's it. More than two hundred and fifty preceding pages are devoted to Robinson's fantastical portrait of an America not a millimeter past *Plessy v. Ferguson,* plus desultory recountings of Robinson's trips to Cuba and Africa. Bittker devoted several chapters of his book to careful legalistic argument exploring how Section 1983 of Title 42 of the United States Code might be applied to obtaining reparations for blacks; but Robinson—a graduate of Harvard Law School—announces only that "my intent is to stimulate, not to sate," having, "by necessity, painted basic themes with a broad brush."

Once again, the tragedy of what passes for "civil rights" in our moment stands out in sharp relief. Hooked on the satisfactions of victimhood, too many black "leaders" today have forgotten that the protests of the late 1950s and the early 1960s were driven by a commitment to forge a new paradigm, to build new programs and new institutions, to work toward interracial harmony. For thirty years now, it has been considered "authentically black" in many circles to indulge in year after year of ceremonial agitprop while whites developed almost all of the policies—successful or not—that have attempted to improve the lot of the race. Enterprise Zones or Empowerment Zones, the Community Reinvestment Act, the reform of welfare, and the Local Initiatives Support Corporation and the Enterprise Foundation, which shunt combinations of grants into inner-city communities: these have been mainly white creations. And so it is no wonder that the most influential treatment of the case for reparations written by a black person devotes itself almost entirely to the rhetoric of inflamed identity, to the revival of Malcolm X's bared teeth and upraised fist.

What kind of leading black thinker is one whose message for the black youth of America is that black success is marginal regardless of its prevalence; that we will find peace only by identifying with people of another continent who are largely alien to us; that the measure of our strength as a group is how articulately we can call for charity? Its popularity notwithstanding, it is hard for me to accept that *The Debt* is really representative of my brethren's thinking on the subject of their place in America. Surely black people in all walks of life are increasingly realizing that pity has never gotten a race anywhere, and that many groups in history have risen despite the power of prejudice.

It is instructive to compare Robinson with Bittker. Writing when the Black Power movement flourished, Bittker soberly based his prescriptions upon a conception of black Americans as a people holding a diversity of opinions. "Who is to decide," he asked, "whether a group that claims to be the vanguard is really only a body of stragglers because the army is moving in the opposite direction?" He came to the conclusion that reparations ought to be paid only to blacks who endured segregated schooling, this being in his view the only case that could be productively argued on a principled legal and moral basis. He rejected the distribution of payments on the basis of "blackness" alone, on the grounds that it would encourage a revival of the arbitrary conceptions of race that were used to justify slavery; and he distrusted the distribution of funds to any particular set of black organizations, on the grounds that it was difficult to know which groups could justly claim to represent all blacks. Bittker was struck by the diversity of African American life: "Among American blacks today, differences in economic status, geographical origin and current location, outlook, organizational ties, and educational background are powerful centrifugal forces that black nationalist groups have not succeeded in neutralizing."

Robinson, by contrast, offers an allegory in which all blacks are given a card listing twenty policies "favoring blacks": "instructions on the back of the card would oblige a bearer, as a matter of honor, to vote for the candidates who'd scored highest and against any who'd flunked." Robinson blithely assumes that the composition of "the card"—presumably an agenda made up of variations on the very handouts and set-asides that have so slowed the dissolution of "the color line" for decades—would be self-evident to all blacks. (What about the ones in the ugly glasses?) It is here that Robinson reveals himself and those of his ideological ilk to be precisely the "body of stragglers" to which Bittker referred.

The closest Robinson comes to acknowledging that there might be more than one legitimate way to think as a black American is in one of the oddest passages in his book, in which he dismisses the black radicals of decades ago:

> For reasons that were never clear to me, they elected to set themselves apart from those they presumed to lead by dressing and talking differently, using an unfamiliar idiom and cadence, leaving their voices up at the end of their sentences. . . . They seemed deeply suspicious, often with good reason, of those blacks who had received from white institutions a liberal arts education, which I think they viewed as rather closer to indoctrination.

Does Robinson really not see that those people were animated by exactly the ideology that animates his own book? Those people favored dashikis and exaggerated their black dialect out of the same estrangement from America and identification with Africa that Robinson sees as the salvation of the race. "Punic this, Pyrrhic that," dissing the black girl who deigns to learn Italian: those are precisely the separatist prejudices of the era of Stokely Carmichael.

One hundred years from now, the marvelous inevitability of interracial mixture will have created a deliriously miscegenated America where hundreds of millions of cafe au lait Tiger Woodses and Mariah Careys will be quite secure in knowing that American is "who they are." For those new Americans, ancient essentialist tracts such as *The Debt* will stand as curiosities. They will turn to historians to explain how it was that such a book was ever regarded as an urgent manifesto for the uplifting of a race. For now, however, pity is the only feeling that can be summoned for the man pictured on the cover of *The Debt*, this affluent and poised black American gentleman sitting grimly indignant that his government does not acknowledge his essence as a Mandingo tribesman, that the compensation that his race has received for almost forty years has come without a groveling apology, and that all black Americans, including himself, are eternally "lost" as a result.

JOHN H. McWHORTER *is a Manhattan Institute Senior Fellow in Public Policy. He studies various aspects of race and ethnicity. McWhorter's 2000 publication of* Losing the Race: Self-Sabotage in Black America *contends that sociopolitical misconceptions pervasive among black Americans are much greater hindrances to black advancement and interracial dialogue than white racism.*

... Or a Childish Illusion of Justice?: Reparations Enshrine Victimhood, Dishonoring Our Ancestors

My father was born in the last year of the nineteenth century. His father was very likely born into slavery, though there are no official records to confirm this. Still, from family accounts, I can plausibly argue that my grandfather was born a slave.

When I tell people this, I worry that I may seem conceited, like someone claiming a connection to royalty. The extreme experience of slavery—its commitment to broken-willed servitude—was so intense a crucible that it must have taken a kind of genius to survive it. In the jaws of slavery and segregation, blacks created a life-sustaining form of worship, rituals for every human initiation from childbirth to death, a rich folk mythology, a world-famous written literature, a complete cuisine, a truth-telling comic sensibility and, of course, some of the most glorious music the world has ever known.

Like the scion of an aristocratic family, I mention my grandfather to stand a little in the light of the black American genius. So my first objection to reparations for slavery is that it feels like selling our birthright for a pot of porridge. There is a profound esteem that comes to us from having overcome four centuries of oppression.

This esteem is an irreplaceable resource. In Richard Wright's *Black Boy*, a black elevator operator makes pocket money by letting white men kick him in the behind for a quarter. Maybe reparations are not quite this degrading, but when you trade on the past victimization of your own people, you trade honor for dollars. And this trading is only uglier when you are a mere descendent of those who suffered but nevertheless prevailed.

I believe the greatest problem black America has had over the past thirty years has been precisely a faith in reparational uplift—the idea that all the injustice we endured would somehow translate into the means of uplift. We fought for welfare programs that only subsidized human inertia, for cultural approaches to education that stagnated skill development in our young, and for affirmative-action programs that removed the incentive to excellence in our best and brightest.

Today 70 percent of all black children are born out of wedlock. Sixty-eight percent of all violent crime is committed by blacks, most often against other blacks. Sixty percent of black fourth-graders cannot read at grade level. And so on. When you fight for reparational uplift, you have to fit yourself into a victim-focused, protest identity that is at once angry and needy. You have to locate real transformative power in white society, and then manipulate white guilt by seducing it with neediness and threatening it with anger. And you must nurture in yourself, and pass on to your own children, a sense of aggrieved entitlement that sees black success as an impossibility without the intervention of white compassion.

The above statistics come far more from this crippling sense of entitlement than from racism. And now the demand for reparations is yet another demand for white responsibility when today's problem is a failure of black responsibility.

When you don't know how to go forward, you find an excuse to go backward. You tell yourself that if you can just get a little justice for past suffering, you will feel better about the challenges you face. So you make justice a condition of your going forward. But of course, there is no justice for past suffering, and to believe there is only guarantees more suffering.

The worst enemy black America faces today is not white racism but white guilt. This is what encourages us to invent new pleas rather than busy ourselves with the hard work of development. So willing are whites to treat us with deference that they are a hard mark to pass up. The entire civil rights establishment strategizes to keep us the wards of white guilt. If these groups had to rely on black money rather than white corporate funding, they would all go under tomorrow.

An honest black leadership would portray our victimization as only a condition we faced, and nurture a black identity around the ingenuity by which we overcame it. It would see reparations as a childish illusion of perfect justice. I can't be repaid for my grandfather. The point is that I owe him a great effort.

SHELBY STEELE *is a research fellow at the Hoover Institution who specializes in the study of race relations, multiculturalism, and affirmative action. He was appointed a Hoover Fellow in 1994. Author of the controversial bestseller* The Content of Our Character, *Steele is one of the most widely cited African American conservatives opposed to affirmative action.*

PART IV

*Reparations and
Grassroots Organizing*

CONRAD W. WORRILL

The National Black United Front and the Reparations Movement

If you beat a donkey every day, one day it will kick you.
—JAMAICAN PROVERB

Queen Mother Audley Moore championed reparations for over sixty years. She is considered the High Priestess of the Reparations Movement and formed the Reparations Committee of Descendants of United States Slaves, Inc., along with Dara Abubakari. In 1962, they delivered a petition to the United Nations demanding the United States be made to pay reparations.

Contributions to the reparations movement resurfaced through the leadership of the Honorable Elijah Muhammad and Malcolm X in the 1960s, making the demand for reparations through *Muhammad Speaks*, the print voice of the Nation of Islam. The Republic of New Africa made a reparations demand in 1968, demanding payment of $400 billion in damages for slavery.[1]

The National Coalition of Blacks for Reparations in America (N'COBRA) was organized in 1988 following in the tradition of Callie House. Since 1988, N'COBRA has developed a number of strategies designed to gain reparations for African people in America and to help advance international efforts to win reparations. Beginning in 1989, Congressman John Conyers introduced legislation in each Congress calling for the U.S. government to study the impact of slavery on Africans in America and the United States. This legislation is currently receiving wide support, primarily due to the work of N'COBRA.

Since the late 1980s, several organizations, including the December 12th

Movement, the Uhuru Movement, the Lost and Found Nation of Islam, the Republic of New Afrika (RNA), and the National Black United Front (NBUF), continue to organize around the demand for reparations. The Tulsa Race Riot Commission, under the leadership of Representative Donn Ross, has generated more interest in the movement. Since the late 1990s, attorney Deadria Farmer-Paellmann's research on the insurance companies that held policies on enslaved blacks in the 1850s has added to the reparations discussion over the last two years. Finally, the resolution on reparations sponsored by Alderman Dorothy Tillman in Chicago's city council received wide publicity and also generated a great deal of interest among black people in the United States regarding the demand for reparations. This visibility was further assisted by the publication of *The Debt* by Randall Robinson.

The reparations movement has moved from the realm of ideas pushed by a handful of intellectuals and activists to the masses of black people. This is an indication that African people have not lost memory of the historical atrocities inflicted upon them and that they will never forget or dismiss the continuation of this mistreatment by this country.

HISTORICAL BACKGROUND OF NBUF

NBUF was founded in 1980. It grew out of the spirit of the 1960s and 1970s when African people in this country were aggressively organizing around numerous issues. The activism of the Civil Rights movement and its challenges against *de jure* and *de facto* segregation was a spark that set off the mass movement of African people in America. The organization and mobilization of the Civil Rights movement provided the springboard for the emergence of the Black Power phase of our movement in the late 1960s and the renewed call for Black Nationalism and Pan-Africanism.

Through the disruptive tactics of the U.S. government via its counterintelligence programs, also known as COINTELPRO, the African Liberation Movement in America suffered serious setbacks. Many leading activists and organizers were arrested and convicted on false charges, some of whom remain locked up today as political prisoners. Others such as Malcolm X, Dr. King, Fred Hampton, and Mark Clark were assassinated or otherwise silenced with many people convinced that the U.S. government was complicit in their deaths.

By the late 1970s, the African Liberation Movement was in serious disarray. This stimulated numerous leading black activists, organizers, and leaders to convene a series of meetings. Two organizational meetings were held in Brooklyn in 1976 and 1977. The purpose was to address the ideological disunity among the various forces in the Black Movement and to formulate a united front. Many members of NBUF remember the all-day meetings held on the East Coast as an attempt toward national unity; however, the discrete commitments, points of view, and organizational interests were intransigent and in some cases unresolvable. The mistrust and apprehensions of the past years fueled by COINTELPRO scheming and various ideological rivalries lingered in the memories of most participants. However, a core group of participants in these meetings from around the country agreed that it was urgent that a call be made to convene the founding convention of the National Black United Front.

The meeting was held in Brooklyn, New York, at the Old Armory in June 1980. More than one thousand activists from thirty-four states and five foreign countries participated in this four-day convention. Rev. Herbert Daughtry was elected the interim national chairman, and we approved a draft of the constitution and bylaws. I succeeded Rev. Daughtry as chairman in 1985 and since then have continued to serve as national chairman.

At the second national convention in July of 1981, which was also held in Brooklyn, NBUF ratified a permanent constitution, bylaws, and leadership structure. NBUF chapters emerged across the country in Philadelphia, Atlanta, Raleigh, Greensboro, Mississippi, Houston, Dallas, Kansas City, St. Louis, Portland, Seattle, the Bay Area, Muskegon, Lansing, Detroit, New York, New Jersey, Milwaukee, Memphis, Chicago, and Washington, D.C. Nearly a quarter of a century later, most of these are still active.

NBUF has organized around the following set of principles:

- To struggle for self-determination, liberation, and power for African people in the United States.
- To work in common struggle with African liberation movements and African people throughout the world.
- To build a politically conscious, unified, committed, and effective African mass movement.
- To struggle to eliminate racism (including Zionism and apartheid), sexism (the oppression, exploitation, and inequality of women), monopoly capitalism, colonialism, neocolonialism, imperialism, and national oppression.
- To maintain strict political and financial independence of the National Black United Front.
- To build unity and common struggle with oppressed peoples in the United States and throughout the world, as long as the best interests of people of African descent are not contradicted or compromised.

- To continue to struggle to maximize the unity of the African Liberation Movement and people of African descent.
- To eliminate internal violence, character assassination, and self-destruction; to establish a viable process to arbitrate all major conflicts within the African Liberation Movement and the African community.
- To continue the political/cultural revolution to create a new vision and value system and a new man, woman, and child based on the common struggle around the needs of the African majority.

NBUF believes that in order for African people in America to become free, self-reliant, and independent, they must be organized. Therefore, we believe all African people should join an organization that works in the interest of African people. We believe that the National Black United Front is one such organization.

THE TRANSATLANTIC SLAVE TRADE: A CRIME AGAINST HUMANITY

The December 12th Movement International Secretariat, the International Association Against Torture, and North South XXI have official non-governmental organization (NGO) status with the United Nations. Over the last decade, these groups have committed much of their organizing efforts to participate in the United Nations Human Rights Commission by presenting numerous issues that impact African people in America. They have been NBUF's eyes and ears at the UN and were instrumental in encouraging the UN to hold the World Conference Against Racism (WCAR) in 2001.

Roger Wareham of the December 12th Movement recently wrote in an article circulated on the Internet, "Since 1997, when the UN agreed to hold this World Conference, the United States, Canada, and Western Europe (the WEO Group of countries) have done all they can to prevent it from succeeding."[2]

In the spring of 1998, at the African Group session during the meeting of the Commission on Human Rights in Geneva, a resolution was drafted identifying the Transatlantic Slave Trade as a crime against humanity. The United States, using all of its influence, succeeded in blocking the resolution. This did not stop, however, the momentum throughout the African World to push for this resolution to become an official position of the 2001 United Nations World Conference Against Racism.

At the African Regional Preparatory Conference for the World Conference Against Racism, held in Dakar, Senegal (January 22–24, 2001), the African ministers developed the historic Dakar Declaration. In their deliberations, they affirmed, in part, the following:

- The slave trade is a unique tragedy in the history of humanity, particularly against Africans, *a crime against humanity* that is unparalleled, not only in its abhorrent barbaric feature but also in terms of its enormous magnitude, its institutionalized nature, its transnational dimension, and especially its negation of the human nature of the victims.

- That the consequences of this tragedy, accentuated by those of colonialism and apartheid, have resulted in substantial and lasting economic, political, and cultural damage caused to the descendants of the victims, the perpetuation of the prejudice against Africans on the continent and people of African descent in the Diaspora.

- Strongly reaffirm that States that pursued racist policies or acts of racial discrimination, such as slavery, colonialism, and apartheid, should assume their full responsibilities and provide adequate reparations to those States, communities, and individuals who were victims of such racist policies or acts, regardless of when or by whom they were committed.[3]

International law supports the position that the enslavement of Africans was a *crime against humanity*. The Charter of the Nuremberg Tribunal defined crimes against humanity as: "Murder, extermination, enslavement, deportation, and other inhumane acts committed against any civilian population whether or not in violation of the domestic law of the country where perpetrated."[4]

The African Reparations Movement argues that historians and experts can show without difficulty how the invasion of African territories, the mass capture of Africans, the horrors of the Middle Passage, the "chattelization" of Africans in America, and the extermination of the language and culture of the transported Africans constituted violations of all these international laws.[5] For us the inarguable conclusion is that the *Transatlantic Slave Trade was a crime against humanity, and it is clear, African people throughout the world are owed reparations.*

CONCLUSION

Professor John Hope Franklin wrote in response to an advertisement by David Horowitz in the student newsletter of a major university:

Most living Americans do have a connection with slavery. They have inherited preferential advantage, if they are white, or the loathsome disadvantage if they are Black; and those positions are virtually as alive today as they were in the nineteenth century. The pattern of housing, the discrimination in employment, the resistance to equal opportunity in education, the racial profiling, the inequities in the administration of justice, the low expectation of Blacks in the discharge of duties assigned to them, the widespread belief that Blacks have physical prowess but little intellectual capacities, and the widespread opposition to affirmative action, as if that had not been enjoyed by whites for three centuries, all indicate that the vestiges of slavery are still with us.[6]

Franklin further states that until "we address what really happened during slavery to African people we will [continue to] suffer from the inability to confront the tragic legacies of slavery and deal with them in a forthright and constructive manner." Fundamentally, the Transatlantic Slave Trade and slavery were crimes against humanity and reparations are owed African people worldwide. This crime must be addressed if we are to move forward as a people.

Our esteemed ancestor Dr. John Henrik Clarke reminded us repeatedly, that:

[H]istory is the clock that people use to tell their political and cultural time of day. It is also a clock that they use to find themselves on the map of human geography.

The role of history in the final analysis is to tell a people where they have been and what they have been, where they are and what they are. Most importantly, the role of history is to tell a people where they still must go and what they still must be. To me and others the relationship of a people to their history is the same as the relationship of a child to its mother.[7]

Thus, we must look backward to go forward. It is the only way that we can repair ourselves and forge a future of our own making and a world safe for human habitation.

CONRAD W. WORRILL *is a professor at Northeastern Illinois University in Chicago and national chair of the National Black United Front. He is considered by many as being one of the best community organizers in the nation. The author of numerous articles on black nationalism and political organizing, Worrill is one of the founding members of the National Black United Front.*

ADJOA A. AIYETORO

The National Coalition of Blacks for Reparations in America (N'COBRA): Its Creation and Contribution to the Reparations Movement

A snake does not sting without cause.
—Oromo Proverb

MY ROAD TO N'COBRA

The concept of a national coalition of Africans in America organized for reparations flowed naturally from more than one hundred years of documented efforts to obtain compensation for enslaved Africans and their descendants. From an organizational perspective, the most inspiring effort historically was the Ex-Slave Mutual Relief, Bounty and Pension Association of the United States established in 1894 in Nashville, Tennessee. Rev. I. H. Dickerson, General Manager, and Mrs. Callie House, Assistant Secretary and National Promoter, were the leaders of this organization. The purpose of this organization was to "secure public sentiment in favor of a law to pension Ex-Slaves, being passed by Congress."[1] Through the efforts of the association, over 600,000 previously enslaved Africans joined this movement and petitioned the United States Congress "to pension every negro who was emancipated. And if the negro is dead give it to his child, if his child is dead give it to his grand child."[2] The United States government targeted Rev. Dickerson, Mrs. Callie House, and other leaders of this organization on politically motivated mail fraud charges and was successful in its attempt to derail the momentum of the organization.

This historical effort was a history many of those organizing the National

Coalition of Blacks for Reparations in America only learned after N'COBRA became a reality. The actual facts of the development of N'COBRA and the flow of events that led to its creation demonstrate the swirl of the ancestors' breath in moving us to resolve the damages done by this tragic crime against humanity, the Maafa, the Slave Trade, the institution of chattel slavery, and the continuing vestiges of slavery. This story can only be told by showing the fusion of the personal with the organizational.

One warm day in the early 1980s a large poster with a picture of Uncle Sam caught my full attention. The figure, pointing a finger directly at me, said: "Black people—The United States Government owes you 3 trillion dollars." It spoke about reparations for slavery and the brutality toward Black people at the hands of the United States government. It was my first introduction to reparations. I was called and I knew it. I obtained the poster and propped it up on the office wall where I worked and meditated on the meaning of its words.

Little did I know that this poster had given me the passion that would fill my life. I was committed to finding an organizational voice for reparations. I had chosen social justice activism in college in the 1960s, and escalated my activism as a social worker in the early 1970s. I had always worked through organizations such as the National Association of Black Social Workers (NABSW), the National Alliance Against Racist and Political Repression (NAARPR), the Black American Law Students Association (BALSA, now Black Law Students Association—BLSA), and the National Conference of Black Lawyers (NCBL).

At the time, active in the leadership of NAARPR and NCBL, I began to speak my newfound passion. This was surely the real remedy for the crime of the Transatlantic Slave Trade, centuries of chattel slavery, followed by more than one hundred years of structural impediments to our progress and rank violence perpetrated on our people by private and public entities. Surely this was the way to tie all the pieces of the puzzle together and to get at the source of our current condition as an African people in the United States.

Many in NAARPR saw reparations as a narrow race remedy that did not reach a broad spectrum of people. In both NAARPR and NCBL, many saw it as a nationalist issue and an unobtainable goal. Organizations at the time that had reparations as a part of their agenda included the African People's Socialist Party and its African National Reparations Organization, the Republic of New Afrika, the New Afrikan Peoples Organization and the Nation of Islam. Although I viewed myself as on the left, I also knew that reparations was an essential remedy and should not be dismissed because those who were identified as nationalists or revolutionary nationalists had been the primary supporters of them. I continued my lobbying in NCBL and aligned myself with Chokwe Lumumba, Chair of the New Afrikan Peoples Organization (NAPO) and Nkechi Taifa, then minister of justice of the Republic of New Afrika (RNA), also members of NCBL. Other NCBL members supported this issue, including my mentor and first executive director of NCBL, Haywood Burns.

Although there was resistance in the organization, we continued to talk about it at various meetings, and in the spring of 1987, David Addams, Associate Director of NCBL, asked if I would organize a panel on reparations for the September 1987 conference that was to be held at Harvard Law School in Cambridge, Massachusetts. This conference was designed to look at the Constitution of the United States, asking the question, What would the United States Constitution look like if it had been drafted to meet the needs of the enslaved Africans and their descendants? I immediately agreed to develop the panel and invited Richard America, noted Black economist, who has written extensively on this issue; Chokwe Lumumba; Imari Obadele, President of the Republic of New Afrika; and Nkechi Taifa to be panelists. The panel was riveting. Richard America outlined the economic basis for reparations, particularly the value of the unpaid labor of enslaved Africans. The other panelists spoke eloquently about the legal basis for a reparations claim, how the Fourteenth Amendment took away our right to self-determination and the legal, social, and political underpinnings of the reparations claim. The presentations of Lumumba, Obadele, and Taifa have been compiled in book form under the title *Reparations Yes!*[3]

THE CALL FOR N'COBRA

Imari Obadele, the president of the RNA, an avid organizer for reparations, seized the time. On August 21, 1987, several weeks prior to the propitious convening of NCBL's conference, and at the urging of Dorothy Benton Lewis, leader of the Black Reparations Commission, he issued a call to more than twenty-five organizations and individuals to come to Washington, D.C., and discuss building support for the armed struggle in Namibia, South Africa, Angola, and Mozambique and development of a definitive campaign for reparations. Five organizing meetings were held from September 1987 until April 1998. Vincent Godwin (now Kalonji Olusegun) chaired these meetings representing the convening organization, the Foreign Affairs Task Force of the RNA. Those invited to the organizing meetings included everyone from poets to politicians, from community activists to college professors.

The response to the call was overwhelming. Perhaps the House of Representatives passage of a bill on September 17, 1987, to pay reparations to Japanese Americans forced into internment camps during World War II placed a new urgency on this issue for many. As I walked into the meeting room I was engulfed in a room of primarily Black men, representing primarily Black Nationalist or Pan-Africanist organizations. I was where I never imagined I would be—this left-leaning social activist was deep in the nationalist community. I took a seat and continued on the path the poster had prophesied: I became an official part of the reparations movement and a player in the formation of the National Coalition of Blacks for Reparations in America (N'COBRA).

A committee for strategy, publicity, and political education was formed at the gathering. Over the course of the five organizational meetings, critical issues of focus, scope of work, and membership were discussed. After the first three meetings it was clear that the focus of the work would be the formation of a coalition to forward the movement for reparations for African descendants in the United States. A participant in the meetings and a member of the Pan Africanist Congress of South Africa, Dr. Nana Seshibe, supported the view that a vigorous struggle for reparations in the United States would aid the struggle of the people in southern Africa. The preliminary principles and plan of organization were modified in the fourth meeting to eliminate the focus on the armed struggle of the people of southern Africa. From the vantage point of 2002, three issues stand out as significant as we look at the reparations movement: (1) the actual focus of the work, (2) the anticipated membership, and (3) the choice of a name.

Focus of work

There was extensive discussion of the nature of the work of the to-be-formed organization. In some ways I was not sure everyone at the table believed that obtaining reparations was really possible. It was on the agenda of many of the organizations represented and in the rhetoric of many of the individuals present. Yet, I wondered if some lacked a firm conviction that reparations were obtainable.

Virtually everyone present agreed that reparations was an important issue, that it was important to maintain energy around reparations, and that cooperative work on the issue was essential. As one would expect, there was some concern expressed about the impact the formation of another group would have on the work of each organization's efforts in support of reparations. There was some organizational possessiveness; however, the overriding concern appeared to be whether we could take this all the way to victory, and if so, would it be recognizable as reparations.

The energy of the RNA and those of us who strongly supported increased organizing on this issue directed the discussion to focus on building a collaborative effort as we moved toward forming N'COBRA. It was agreed that the effort should be focused on obtaining reparations; it should be a coalition in which individuals and organizations would work without stripping them of their right to engage in independent work on reparations; and it would not endorse any one form of reparations—embracing all forms including land, money, repatriation and social justice and community-based programs. The question of endorsing any one form of reparations was critical since it was agreed that in order to be a broad-based coalition that was inclusive of all positions on the nature of reparations for African descendants, N'COBRA could not place greater value on one form of reparations over another. And, it was seen as critical to building the reparations movement that the message and the messengers of N'COBRA be one for unity, not highlighting political and ideological differences.

Defining the Anticipated Membership

The discussion revealed the disdain with which many members of the nationalist community viewed the left and the more mainstream civil rights groups. Stereotyping of nationalists by the left and mainstream was replicated by the nationalists and revolutionary nationalists, many of whom were suspicious of the agendas of the left and mainstream. They questioned whether they would embrace reparations, and if so, would they dilute its meaning to such a degree that what was called reparations was tepid compared to what was owed our people. The question resounded through the room, however: Do we want to win? We answered yes! The second question flowed naturally from the first: Who must be a part of this movement in order for us to win? The answer: As many Black folks as possible from all walks of life. It was conceded that, in fact, it was essential to mainstream this issue. It was agreed, therefore, that we would reach out to organizations not generally viewed as reparations supporters that had a human rights or civil rights agenda. This included left and mainstream organizations. Of course, the NAACP, the National Urban League, the National Bar Association, churches, and the various Black Greek-letter organizations were identified as mainstream organizations that must be reached if the average, everyday Black person was to become a supporter of reparations.

The question arose as to whether white people had a role in the reparations movement. A difficult discussion—even more difficult than the one involving the approach to integrationists. The difficulty was in not allowing it to be dismissed as a superfluous topic. Again, the question Do we want to win? loomed large. Yet, the next question was even more critical: Can we win without white support? We understood that white support was necessary. The discussion centered around whether we address this issue within the coalition that was forming and, if so, how?

A principled decision was made. The fact that white support was essential to victory was conceded by most. Yet, it was unanimously agreed that the most critical ingredient for victory was the development of broad-based support in the Black community and that the efforts of the coalition would be to organize African descendants around this issue. Support from white people would not be refused, yet the leadership on this issue must stay within the Black community. The principle of the *Nguzo Saba—Kujichagulia—*was raised up and embraced by everyone present: self-determination—we must speak for ourselves—and define this issue. Dorothy Benton Lewis, currently national cochair of N'COBRA, made a statement that I've carried since 1987: We must speak it everywhere we go. The discussion of membership led to the discussion of what we would call ourselves.

Defining Ourselves Through Our Name

This discussion brought up the importance of a name. A name reflects who you are; what you or your parents/founders thought of you. As we embraced the vision of a coalition and went through the often-heated discussion of our potential allies, naming the entity became a central issue. It was in the fourth meeting that it culminated into a firm decision. We had agreed we were an organization led by Black folks with the purpose of organizing Black folks to obtain reparations. We agreed we would be aggressive in our work: taking up the issue with vigor, organizing in the streets as well as the meeting rooms of Black organizations and the halls of Congress. We agreed we would raise the contradictions inherent in our current condition as a people and the agonizingly brutal facts of our ancestors' enslavement and post-slavery atrocities. We agreed white people would not have any central role in the coalition, yet could serve as supporters of our work. We agreed that although many of us supported reparations for Africans throughout the Diaspora and African nations raped by the slave trade, that our focus would be on reparations for African descendants in the United States. After a full, vigorous discussion we claimed the name: National Coalition of Blacks for Reparations in America. We also agreed to an acronym of N'COBRA, making distinct COBRA as the snake that is symbolized in ancient Egypt as Uraeus, the protector of the society. It is the symbol of our maturity from mental slavery to mastership, as Dr. Richard King's research indicates. We were formed: N'COBRA was a formidable voice for the demand for reparations for African peoples in the United States and for the righteousness of this demand.

Kalonji Olusegun (formerly known as Vince Godwin) designed N'COBRA's logo, which is composed of three *Adinkra*[4] symbols. The base of the symbol is our base: *Nkonsonkonson* (link or chain), symbolizing that those who share common blood relation are linked in life and death and never break apart. The middle portion of the symbol is our activity: *OwoForaAdobe* (snake climbing the palm tree), symbolizing performing the unusual or impossible. Finally, the top portion of the logo is our goal: *Biribi-Wo-Soro* (there is something in the heavens, let it reach me), symbolizing hope. As articulated by the organizing committee, the logo symbolizes that we are linked in life, and linked in death, that those who share common blood relation will never break apart and can perform the impossible to reach that which is in the heavens—Justice, Freedom, and Happiness.

STRUCTURING OURSELVES AND FORMULATING OUR WORK

The organizers agreed it was important to have organizational and individual memberships, consistent with the nature of the organizing committee. The primary goal of N'COBRA in 1988 and 1989 was to consolidate its organizational base. A committee worked over several months and developed the articles of incorporation and bylaws that were approved by the organizing committee. Of critical importance in the bylaws was the requirement of one male and one female cochair—asserting the importance of men and women in the leadership of the reparations movement. It developed and conducted a survey on reparations targeting Black organizations, seeking to educate people about reparations and obtain information about what people thought about reparations. It was in contact with Congressional Black Caucus Members, including Walter Fauntroy (D.C.), Merv Dymally (CA), a leading voice for the Japanese American bill, and John Conyers (MI). It urged that legislation on reparations for African descendants in the United States be introduced. Members of the organizing and executive committees wrote popular pieces to educate and mobilize the community and organize support for reparations. Articles about N'COBRA were carried in *The Final Call*, the Nation of Islam weekly newspaper, as well as in papers in Baton Rouge, Detroit, and Washington, D.C. N'COBRA had regular submissions to *Drumbeat*, a D.C.-based publication.

CRITICAL CONFLUENCE OF EVENTS

Much planning was going into the formation and development of N'COBRA. As with all significant occurrences, however, planning was enhanced by some events over which N'COBRA had no control and others over which N'COBRA members and supporters had influence. The attack on affirmative action had begun in the late 1970s with the decision in *Bakke*,* which began to undermine the barely ten-year-old affirmative-action policy and practice of private and public entities. This attack was continuing in the 1980s.

In 1988 the Japanese American reparations bill became law as the Civil Rights Redress Act of 1988. In 1989 Senator Bill Owens, a Massachusetts state senator, introduced Senate Bill 1621, cosponsored by Representatives Shirley Owens-Hicks and Byron Rushing. This bill was an act to provide for the payment of reparations by the Commonwealth of Massachusetts for slavery, the slave trade and invidious discrimination against the people of African descent born or residing in the United States of America. N'COBRA had been working with members of the reparations community in Detroit who had formed the African-American Reparations Committee. Cindy Owens was chair of the Detroit committee and was also the wife of Senator Bill Owens. The introduction of this bill gave a major uplift to N'COBRA and the reparations issue for several reasons. Senator Owens was passionate about this issue and participated willingly in meetings and public forums to inform the community about reparations and his bill. He was a charismatic speaker who had a good grasp of the historical, moral, and legal basis for reparations. It was also important to demystify slavery and the vestiges of slavery by showing that this heinous institution was not the sole province of the South. Rather, it helped establish the culpability of the entire country and the governments that supported the colonies, states, and the United States.

Senator Bill Owens was one of the major speakers at N'COBRA's first town hall meeting held in Washington, D.C., April 8, 1989. We were also honored by the presence and participation of Queen Mother Audley Moore, considered by most to be the mother of the contemporary reparations movement. The session was held at the Frank D. Reeves Center for Municipal Affairs and had a standing-room-only crowd. It was just what the newly formed organization needed: a current legislative effort for reparations tied to the historic demand for reparations. This demonstrated that it was not simply some wide-eyed, radical nationalists who believed in and worked for reparations.

Also in April 1989, the African-American Summit included support for reparations in its concluding document: "We call for reparations. If they are good enough for the Japanese-Americans and Native Americans, they are good for those of us who worked for hundreds of years unpaid, and who now need that capital for our own development in this country."[5]

---

*Regents of the University of California v. Bakke, 438 U.S. 265 (1978).

In May 1989, after the intense lobbying of members of the Detroit African-American Reparations Committee and N'COBRA members, led by Ray Jenkins, known as Reparations Ray because of his many years of work on this issue, the Detroit City Council unanimously approved a resolution introduced by Councilman Clyde Cleveland on April 14, 1989, to establish a $40 billion education fund for Black Americans descended from enslaved Africans. This was a great victory, although the value was largely symbolic since the source of the $40 billion was the federal government. The federal government, however, had not passed any legislation to appropriate monies for reparations for Black Americans descended from enslaved Africans. It was not until six months later, on November 20, 1989, that a bill that could lead to appropriations for such a fund was introduced in the House of Representatives by a Michigan congressman, John Conyers Jr.

In June 1989, N'COBRA received a letter from Congressman John Conyers asking it to review and comment on a legislative summary that he planned to put into the form of a bill and introduce in the House of Representatives. N'COBRA welcomed this opportunity and held a public meeting on July 8, 1989, at the Reeves Center to discuss and respond to Congressman Conyers's request. As with the April town hall meeting, the room was packed—standing room only. The discussion was passionate as longtime reparations activists along with the newly involved deliberated for several hours on the proposed provisions of the bill.

One of the primary objections to the bill was that it was a study bill. The official title of the bill was "A bill to acknowledge the fundamental injustice, cruelty, brutality, and inhumanity of slavery in the United States and the 13 American colonies between 1619 and 1865 and to establish a commission to examine the institution of slavery, subsequent de jure and de facto racial and economic discrimination against African Americans, and the impact [of] these forces on living African Americans, to make recommendations to the Congress on appropriate remedies, and for other purposes."[6] However, the bill digest indicated it was creating a Commission to Study Reparation Proposals for African Americans. Many avid reparations supporters were livid, grabbing hold of the word "study" and protesting heatedly that there was no need to study the question. They rested their position on the large volume of information available on the Transatlantic Slave Trade, the institution of chattel slavery, and the ongoing brutalization of and discrimination against African descendants that was well documented. Their position was that the legislation was far too weak and should be redrafted to demand reparations in specific forms.

Another primary objection was to the formation of the commission. The members of the commission were to be selected by the president of the United

States, the House of Representatives, and the president pro tempore of the Senate (i.e., the vice president of the United States). Many saw this as a death knell to any meaningful reparations, the assumption being that those appointed would not have the propensity to see the need for reparations or the commitment to embrace meaningful and substantive reparations proposals.

Cedric Hendricks, a member of Congressman Conyers's staff, explained the rationale for a commission to study reparation proposals and the proposed structure of the commission. He indicated that this bill was modeled after the Japanese American bill that had recently become law. It was Congressman Conyers's view that although a reparations bill for African Americans would have an uphill battle, perhaps even greater than the uphill battle faced by Japanese Americans, casting it within the format of the successful Japanese American bill gave it some strength—moral authority, if you will. The fact that this was yet another piece of legislation that made real the need for reparations was in itself exciting, building on the excitement generated by Senator Bill Owens's Massachusetts bill and the Detroit City Council resolution. That this was a national piece of legislation only added to the uplift and excitement. Yet, the substance of the bill made it starkly clear that this issue was being mainstreamed and, in doing so, it was taking on a tone that was reflective of the Congress of the United States, viewed by many as only representatives of the enemy.

The room was filled with emotion—full of the excitement of seeing a long-time marginalized issue coming out onto the national stage and the rage of many centuries of denial of our righteous claims for freedom, justice, and reparations. Voices were raised. There were many who wanted to go on, who wanted to give specific input into language and strategy. Yet, there was a small group of activists who objected to a congressional bill, particularly of this nature, with such force that it was difficult to go forward. Some of the most vocal opponents were members of the African National Reparations Organizations (ANRO), an organization of the African People's Socialist Party, which for a number of years in the 1980s conducted tribunals on the issue of reparations.

Underlying the objections to the bill were the concerns seen in the organizing meetings for the formation of N'COBRA—would the demand for reparations lose its stridency and substance, once embraced by not only mainstream Black people but also the mainstream in general? Perhaps there was also some organizational possessiveness—here comes a member of Congress, who is presenting his view of reparations rather than simply taking our demands to the Hill. And there were, of course, the continuing class divisions that were spoken of in the first organizing meetings that took the form of some not wanting to give in to the embrace of this issue by the Congress of the United States, albeit a member of the Congressional Black Caucus who was noted for his progressive advocacy. The fact that Congressman Conyers had been victorious in a twelve-year battle to get a federal holiday for Martin Luther King Jr. gave him little currency, particularly among the most vocally opposed to this bill. Rather than seeing him as a brilliant legislative strategist who placed the interests of Black people in the forefront, he was seen by some as simply a bourgeois, Black politician who was soft-pedaling our pain, our misery, and our conditions.

The leadership of N'COBRA prevailed in establishing order at the meeting. Some of those most hotly opposed to the legislation left in protest. The room was still overflowing with people who worked diligently for several hours to give input paragraph by paragraph. We provided Congressman Conyers's representatives with written recommendations immediately following the meeting. On November 17, 1989, we received a second discussion draft that was dated October 2, 1989. I provided comments to Congressman John Conyers as Director of the National Conference of Black Lawyers (NCBL), a founding organizational member of N'COBRA, noting that some of N'COBRA's July 8, 1989, recommendations were reflected in this redraft, yet recommending that a number of recommendations not reflected be included. There were four major recommendations reiterated by N'COBRA:

- That the text of the bill should indicate that some mention of the Middle Passage be made in the findings of the commission;
- that the terminology "Africans held as slaves" be used, denoting a condition, rather than "African slaves," which suggests an attitude as well as a position;
- that "brutalization of their bodies" be added in the appropriate section or some other phrase denoting the severe, demeaning, crippling, and frequently deadly physical force that was used on Africans held as slaves;
- that the commission's recommended remedies be broader than compensation and an apology since compensation is viewed by most as money or material goods and there may be recommendations that require nonmaterial remedies, such as requirements for some proportionality in elected offices to make up for the denial of fundamental voting rights, or to hold a plebiscite for those of African descent to choose their citizenship since United States citizenship was forced upon those freed Africans.

Although many of N'COBRA's recommendations were not incorporated in the bill that was introduced on November 20, 1989, as H.R. 3745, and has been reintroduced each year since then, most frequently as H.R. 40, the fact that N'COBRA was part of the discussion on this bill prior to its introduction placed N'COBRA in an important role as it worked to broaden the base for the demand for reparations for African descendants in the United States.

ORGANIZING FOR REPARATIONS AND INFLUENCING PUBLIC OPINION

The town hall meeting and public forum on legislative efforts were major components of N'COBRA's public education campaign. Wherever we went in 1989 and 1990, we took the message of the viability of obtaining reparations for African descendants in the United States. In September 1989, one of our leaders, Nkechi Taifa, moderated a Congressional Black Caucus brain trust on reparations hosted by Congressman Conyers. Two of N'COBRA's leadership and founders were panelists, Vince Godwin (now Kalonji Olusegun) and Dorothy Benton Lewis. This brain trust tossed us into the public eye, the local and state legislators and constituents of the Congressional Black Caucus who came to Washington, D.C., every year to rub shoulders with the powerful and influential. This was the very essence of mainstream Black America. The 1989 meeting has been followed each year with a forum on reparations hosted by Congressman Conyers, and in 2000 was cochaired by Congresswoman Carrie Meeks from Florida, an N'COBRA member. A member of N'COBRA has participated on virtually every panel and moderated the panel on at least two occasions.

On January 27, 1990, N'COBRA launched its 1990 campaign for reparations by holding a strategizing meeting at Watoto Shule, an independent African-centered school in Washington, D.C. The African American community in the metropolitan D.C. area was invited to attend. Ron Stroman from Congressman Conyers's office was present to describe the legislative process and field questions concerning the bill. This forum formally initiated N'COBRA's lobbying effort to push the Conyers bill. In addition, its purpose was to fuel the debate on reparations leading to N'COBRA's first annual conference in June 1990 in Washington, D.C.

Between January 1990 and June 1990, N'COBRA's primary activities were finalizing the draft bylaws and reaching out to the Black community nationally to participate in a major strategy discussion at its June 1990 conference. The conference participants evidenced what continues to be N'COBRA's face: a large group of persons identifying as nationalist, revolutionary nationalist, or Pan-Africanist with a smaller group of those who are mainstream or identify as being on the left. There was a conscious attempt on the part of N'COBRA to invite presenters who crossed ideological perspectives as well as to be reflective of our African heritage, again a continuing purpose of N'COBRA in all its annual meetings. Persons invited included people who had raised their voices on this issue yet were not a part of N'COBRA's organizing committee: Preston Wilcox, noted social worker from New York and active with the National Association of Black Social Workers; David Hall, Esq., Dean of Northeastern School of Law and in the leadership at one time of the Boston Chapter of NCBL; Stuart Adams (now Ajamu Sankofa), D.C. Coalition of Black Lesbians and Gays (now the D.C. Coalition of Black Lesbians, Gays and Transsexuals); and Queen Mother Audley Moore, whose spirit had been felt in all our organizing meetings providing the glue that held us together.

N'COBRA's public education and legislative agenda was formulated at this first annual conference—an agenda, with some modifications to fit the times, that continues to this day. It was agreed that N'COBRA would build support for H.R. 3745 in a number of ways, including hosting public education forums throughout the country, reaching out to Black organizations that have not spoken on this issue to obtain endorsements, and holding weekly education rallies on Capitol Hill.

By June 1990, N'COBRA had already become a leading voice in the reparations movement. Our principled efforts to be inclusive of all organizational voices that had been raising up this issue for some time was an asset for us since competitiveness among groups was minimized, although not eliminated. The African-American Reparations Committee and NCBL's involvement assisted greatly since Congressman Conyers and his staff had a relationship with both groups. Several articles in local papers in D.C., Detroit, and Baton Rouge, among other cities, spoke about the formation of N'COBRA and Congressman Conyers's legislative effort. In July 1990, N'COBRA hit the streets in D.C. and organized weekly rallies on Capitol Hill in support of H.R. 3745. These rallies were strategically scheduled for 12:00 noon to facilitate participation by our members and supporters as well as to catch the attention of the lunch crowd, including congresspersons and their staff. In July 1990, largely through the efforts of the Baton Rouge N'COBRA leadership, Peggy and Herbert Bookter, the legislature of the State of Louisiana passed a concurrent resolution calling on the Congress of the United States to pass H.R. 3745. In September 1990, Nkechi Taifa, as chair of the D.C. chapter of the National Conference of Black Lawyers, requested a similar resolution from the D.C. Council through the Honorable Wilhelmina J. Rolark, who was chair of the Committee on the Judiciary. The D.C. Council issued a resolution in response to this request.

Since 1990, N'COBRA members have been involved at various levels in obtaining resolutions from other state and local legislative bodies, including the City Council of Englewood, California; City Council of Cleveland, Ohio; Wayne County Commissioners; City Council of Detroit, Michigan; City Council of Chicago, Illinois; City Council of Nashville, Tennessee; and California state legislature. It has also been successful in reaching mainstream organizations, of both national and local character. This has been largely due to relationships some of its members have had with members of these legislative bodies and organizations, underscoring the importance of N'COBRA's resolve to have a broad-based membership, cutting across class and ideology. Therefore, N'COBRA has been successful in getting resolutions and support from organizations such as the NAACP, the general membership approving a resolution in 1990 at its annual convention, and now having as one of its primary legislative campaigns passage of H.R. 40; the National Bar Association, whose board of governors approved a resolution in 1991 in support of H.R. 40, and whose board of governors and members voted in 1999 to form a reparations task force; Delta Sigma Theta, Sigma Gamma Rho, National Association of Black Social Workers, National Conference of Black Political Scientists, and others.

In addition to articles appearing periodically since 1988 in local weekly and daily newspapers throughout the country on reparations for African Americans, the issue began appearing in television and radio programming. N'COBRA members have been tapped to participate on radio programs throughout the country, including in New York, Raleigh, Chicago, Washington, D.C., Philadelphia, Cleveland, Cincinnati, Baton Rouge, Miami, San Francisco and the Bay Area, and Detroit.

By the early 1990s the word was out: reparations was a claim that was being raised up in the African descendants' communities throughout the United States. It was furthered internationally by a conference held in Nigeria, hosted by longtime Nigerian reparations activist Bashorun M.L.O. Abiola in 1993. Ron Walters, noted political scientist and member of N'COBRA's advisory board, attended this meeting and made a presentation about it in N'COBRA's annual conference held in Baton Rouge, Louisiana. An N'COBRA chapter was formed in Kingston, Jamaica, in the early 1990s, and one in Accra, Ghana. The internationalization of the movement for reparations and the attention this received in the early 1990s only continued to build support for reparations throughout the Black community and in other communities throughout the United States. Even those who opposed it in these communities gave voice to the demand and helped make it a known commodity throughout the country.

ORGANIZING TO MEET THE GROWING MOVEMENT

In 1993, at its Baton Rouge convention, N'COBRA organized commissions to carry on the programmatic and outreach work of the organization under the leadership of the board of directors. Six commissions were formed on each of the following: internationalization, legal strategies, human resources, information and education, economic development, and membership and organizational development. Subsequently the Youth Commission was formed under the leadership of youth from Prairie View who were under the tutelage of Imari Obadele. In 2000 the Commission on Legislative Strategies was formed to continue the legislative work that had previously been subsumed under the Legal Action Strategies Commission. The Commission on Education was formed in 2001.

One in particular, the Commission on Economic Development, which has a strong base in the grassroots community, has developed economic development commissioners who are elected in their communities to serve for the purpose of, among other things, designing reparations proposals impacting the economics of the Black community that are submitted for approval to the N'COBRA board of directors and then circulated throughout the community and to legislative bodies. This commission presented proposals such as funding of Black colleges and universities as a form of reparations.

The Legal Strategies Commission has been active in the community primarily through its legislative work, which will now be done by the Commission on Legislation, as much as in its efforts to develop reparations litigation and obtain support for its legal fund to finance the litigation. It has been active in formulating resolutions and assisting in developing the efforts that have resulted in passage of these resolutions by local and state legislators as well as organizations described above. It has engaged in numerous activities to lobby for passage of H.R. 40, including a February 1997 Reparations Awareness Campaign organized by the D.C. chapter that resulted in two more congressional cosponsors of H.R. 40, and the Juneteenth 2000 lobbying effort, which also led to increased congressional cosponsors of H.R. 40 and all but two members of the Congressional Black Caucus being identified as cosponsors. In June 1997, the Commission on Legal Strategies formed a Reparations Litigation Committee that continues to work to develop and implement a litigation strategy for obtaining reparations. This committee is composed of lawyers, social scientists, and activists who are committed to develop a litigation strategy for reparations. A draft complaint has been prepared and reviewed by N'COBRA's board of directors. Filing of the complaint is imminent.

The Commission on Information and Education provides internal organs for the organization, periodically producing a newsletter. A separate editorial board that flows from this commission produces *ENCOBRA*, the information magazine for N'COBRA that is published annually in June. As the reparations movement has grown and become part of the discussions in the mainstream, more centrist community, there is a glaring need for N'COBRA to expand its dissemination of information to those who are not active in N'COBRA, not relying simply on the local and national media that may take an interest in the issue and may know of N'COBRA and include it in its programming. The current climate, as discussed below, requires that N'COBRA continue to grow in its expertise and develop mechanisms for creating public information that is targeted to the larger, nonmovement community.

The Commission on Youth, although started later than the other commissions, is developing strategies that are creating and building on youth interest in this issue. For example, it initiated a Day of Awareness on April 4, 2001, on which students did not go to class and engaged in activities to build support for reparations. The focus on youth in N'COBRA is not simply tied to the commission. Youth are involved throughout the organization, including serving as cochair of the Commission on Economic Development.

The development of the organizational structure to meet the demands of the growing movement is essential to N'COBRA remaining a viable, center-stage player in the movement for reparations. The release of Randall Robinson's book *The Debt: What America Owes to Blacks* in January 2000 made real the concerns raised at its founding of the possibility of losing leadership in this movement to more mainstream, high-profile leadership.

THE CHALLENGES OF BUILDING

In 1987, the founders of N'COBRA charged themselves with consolidating and building the movement for reparations to the point of victory. The fear of losing leadership of the movement and it becoming something we did not recognize was there at the first meetings. This fear was overcome by the vision of victory and the felt necessity for victory since we all knew too well the conditions of African descendants in the United States and the relationship of those conditions to the Transatlantic Slave Trade, the institution of chattel slavery, and the vestiges of that institution. Throughout the first twelve years of N'COBRA's existence there had been no real challenge to its leadership of the movement, although several mainstream groups came on board, taking up the clarion call for reparations, including the NAACP, which made it one of its primary legislative campaigns in 1999, and the National Bar Association, which created a reparations task force in 1999. Even the Black left came officially on board when the Black Radical Congress joined the reparations movement in 1998. Yet these groups to a large extent included N'COBRA in their work and recognized N'COBRA as an important leader in this movement.

WHERE DO WE GO FROM HERE?

The challenges that faced us in 1987 and 1988 when we were forming the coalition continue today. Yet, because of our organizational maturity and commitment to remaining a leading voice in the movement for reparations, we have ridden with the ebbs and flows of this movement. We continue to develop our program for winning reparations through public education, mobilizing and organizing African-descended people, working collaboratively with others, and developing public education, direct action, legislative and litigation strategies. Our challenges organizationally are the same challenges facing the success of the movement for reparations, since we are inextricably linked: to stay visible and increase that visibility, to posture ourselves to be a part of the discussions that will more than likely take place behind the scenes on the form reparations should take to assure that a package of reparations and not appeasement is developed, and to stay principled, demanding accountability to African descendants. Finally, we are an integral part of the continued internationalization of this issue. We are convinced more than ever that victory is on the horizon and will be ushered into manifestation by strong organizational work, unity, and the wings of the ancestors.

ADJOA A. AIYETORO *is chair of the Legal Strategies Commission and a founding member of the National Coalition of Blacks for Reparations in America. She has written extensively about reparations for Africans in the Maafa, including historical and legal analysis, and was a leader of the African and African Descendants Caucus, which successfully lobbied for reparations and to have the Maafa included in the final declaration document at the 2001 United Nations World Conference against Racism in Durban, South Africa.*

ROGER WAREHAM

The Popularization of the International Demand for Reparations for African People

A fish does not know it is wet.—Ghanaian Proverb

AUTHOR'S PREFACE: *The United Nations World Conference Against Contemporary Forms of Racism, Racial Discrimination, Xenophobia and Related Intolerance (WCAR, August 31–September 7, 2001) will be remembered for and defined by the struggle for reparations. The process of organizing for the WCAR was key in making the issue of reparations for African people,[1] in the Diaspora and on the African continent, an international one. This brief history of our experience in the UN will help illuminate the events leading up to the historic meeting.*

INTRODUCTION

The December 12th Movement is an organization of African people born in the struggle against racism in the United States. It takes its name from a statewide demonstration held in Newburgh, New York, on December 12, 1987. That rally was called to protest a tide of racist killings, beatings, and harassment that had been directed against Blacks and Latinos across New York State. The groups and individuals that came together that day called themselves the December 12th Coalition. Those that continued working together eventually became the December 12th Movement. The December 12th Movement has several constituent organizations. The one primarily responsible for international work is known as the December 12th Movement International Secretariat (IS). Roger Wareham, this essay's author, is a founding member of the December 12th Movement and a practicing attorney in the law firm of

Thomas, Wareham & Richards in New York City. He is also the international secretary-general of the International Association Against Torture, a non-governmental organization, which, along with the International Secretariat, has consultative status with the United Nations. As an organizational representative, he has been an annual participant at the United Nations in Geneva and New York since 1989.

> *Our problem is your problem. It is not a Negro problem, nor an American problem. This is a world problem for humanity. It is not a problem of civil rights, it is a problem of human rights.*
>
> —MALCOLM X, SPEAKING TO THE ORGANIZATION OF AFRICAN UNITY, 1964

One cannot understand how reparations for African people have become an issue of international dimensions without first paying homage to Malcolm X. Malcolm, following in the steps of Blyden, Garvey, Trotter, Du Bois, Patterson, and Robeson, among others, helped Black people see the important role the international arena plays in our domestic struggle for freedom and human rights.

The December 12th Movement International Secretariat [IS] first went to the United Nations Commission on Human Rights in 1989 as part of a group called Freedom Now, a campaign seeking amnesty for political prisoners and prisoners of war in the United States. Many of the IS members were veterans of the African Liberation Support Committee of the 1970s and had a wide range of experience in the international arena. We learned from this first experience in Geneva, Switzerland, that the international community's perception of the nature of Black people's existence within the United States was totally false. The reality had been deliberately skewed by United States government propaganda and by the omnipresent media images of the U.S. Olympic Dream Team, Michael Jackson, Oprah Winfrey, Colin Powell, and others. Their elite lifestyle was perceived as the Black norm rather than the exception. The need for reparations, much less the legal duty to provide them to Black folks in the United States, was not even being considered.

Countries who supported our struggle for freedom in the United States taught us that the only way we could change the misperception and replace it with the truth was to interact consistently with all the United Nations bodies that dealt with human rights, that is, the Commission on Human Rights, the Sub-Commission on the Promotion and Protection of Human Rights (Sub-Commission), the Third Committee of the General Assembly, World Conferences, *et cetera*. The IS took this advice seriously and committed ourselves to be everywhere in the United Nations raising the banner of the human rights violations suffered by African people, in the United States in particular and in the Diaspora in general. We did this over the years by presenting indisputable facts and an objective analysis that proved that African people around the world suffered from underdevelopment that was tied to the Transatlantic Slave Trade, slavery, and colonialism.

MORAL DUTY FOR AFFIRMATIVE ACTION

The Sub-Commission appointed Theo van Boven, a professor from the Netherlands, as a special rapporteur to conduct a study concerning the right to restitution, compensation, and rehabilitation for victims of gross violations of human rights and fundamental freedoms. Van Boven's 1993 report[2] found historical precedents for the reparations received by the Jewish victims of the Nazi Holocaust and the United States citizens of Japanese descent who were incarcerated by their own government during World War II. However, when it came to the issue of reparations for descendants of the victims of slavery, Professor van Boven stated, "[I]t would be difficult and complex to construe and uphold a legal obligation to pay compensation to the descendants of the victims of the slave trade and other early forms of slavery. . . . [I] agree that effective affirmative action is called for in appropriate circumstances as a moral duty."[3] The IS and LEVI [Lift Every Voice, Inc], another Black NGO (nongovernmental organization) from the United States, objected to van Boven's avoidance of the issue but were unable to change his position. We concluded that once again a different standard was being applied to African people. It reinforced a lesson that centuries of racist abuse have taught us, namely, that if we do not ensure that our interests are addressed and protected, we cannot depend upon anyone else to do so.

THE BATTLE TO HOLD A WORLD CONFERENCE AGAINST RACISM

At the United Nations World Conference on Human Rights held in Vienna in June 1993, the IS lobbied and made an oral and written intervention calling for a World Conference Against Racism. The United Nations ignored this call for another four years. In the interim it held a World Summit on Social Development (Copenhagen, Denmark) and the World Conference on Women (Beijing, China). In December 1997, the United Nations General Assembly reluctantly agreed to hold the World Conference Against Contemporary Forms of Racism, Racial Discrimination, Xenophobia and Related Intolerance. The United States and the other Western countries (whose United Nations geographic designation is the WEO Group, that is, Western European and Others [United States, Canada, Australia, and New Zealand]) vehemently opposed holding a World Conference Against Racism. This was not a surprise. Following the transition from apartheid in South Africa, the Western countries had initially tried to sweep the issue of racism off the agenda, saying it was a settled question. The IS and other NGOs fought this ploy and responded with a demand for a Special Rapporteur on Racism and Racial Discrimination.

The expansion of the title to include *Xenophobia* and *Related Intolerance* was part of the compromise reached to have a consensus call for the WCAR. The Western countries hoped that broadening the scope would dilute the focus on racism and racial discrimination. Rather than address the role of white supremacy and Eurocentrism in understanding present and historic inequities, the WCAR could conceivably spend more time looking at conflicts between Tutsis and Hutus, or allegations of sub-Saharan slavery. Even with this broadened scope, the United States was absolutely the last country in the United Nations to agree to hold the WCAR. This was just the first salvo in a four-year battle whose goal was to keep reparations off the agenda.

THE APPOINTMENT OF A SPECIAL RAPPORTEUR ON RACISM AND RACIAL DISCRIMINATION

The West's futile attempt to deny the continuing existence of racism as a major human rights violation led to a call for a special rapporteur on racism. The IS was in the middle of that fight. In 1993, the West agreed to appoint this rapporteur, only if his mandate was expanded to include Xenophobia and Related Intolerance. In 1994, Mr. Maurice Glele, a member of the Supreme Court of Benin, West Africa, was appointed as the Special Rapporteur on Contemporary Forms of Racism, Racial Discrimination, Xenophobia and Related Intolerance. Immediately, the IS successfully lobbied for him to visit the United States for his first investigation.

In October 1994, Mr. Glele came to the United States. The IS and the International Association Against Torture (AICT—its Spanish acronym) held an all-day, public tribunal on October 15, at Harlem's historic Abyssinian Baptist Church, where more than eighty victims of racism, individuals and representatives of organizations, testified directly on the situation in the United States. More than one hundred fifty other presentations were made and recorded on audio- and videotape for the rapporteur. In 1995, Mr. Glele issued a report[4] highly critical of the racial situation in the United States. The United States, disagreeing with the report's findings and recommendations, condemned it. Mr. Glele's report was unquestionably a factor in the negative United States response to the call for a WCAR.

The IS, along with other NGOs, continued its pressure to bring the United States to account for its human rights violations. In 1997, Bacre Wali Ndiaye, the UN Special Rapporteur on Extra-Judicial, Summary or Arbitrary Executions conducted an investigation. The IS and the AICT, along with the Center for Law and Social Justice of Medgar Evers College, sponsored a public meeting for the families of victims of police killings and of those who had been officially murdered by the government or were still on Death Row. His highly critical report[5] drew a similarly negative response from the United States government.

THE RESOLUTION CALLING FOR THE DECLARATION OF THE TRANSATLANTIC SLAVE TRADE AND SLAVERY AS A CRIME AGAINST HUMANITY

From the beginning of our participation at the United Nations, the IS had been working to implement Malcolm X's call for closer ties between Africans in the Diaspora and on the Continent. We saw the African countries not only as brothers and sisters but also as our natural allies in our struggle in the United States. Similarly, our location in the United States placed us in a position to help them move forward. We met with and lobbied the African countries regularly and over the years made presentations at meetings of the African Group (the United Nations' geographic designation) of countries. In these meetings we spoke of the issues we had in common, particularly the underdevelopment that we in the West suffered as oppressed nations and that they suffered as former colonies. As a reference point for understanding Africa's present situation we cited Walter Rodney's analysis in his classic work *How Europe Underdeveloped Africa*. The economic motive, the greed, and the racism that inspired and fueled the Transatlantic Slave Trade and slavery also stoked the fires of colonialism. African peoples' present situation, regardless of our geographic location, had the same source. We raised the issue of reparations as an issue for all Africans, not simply those of us in the United States. The form it would take might differ, but the fact that it was due could not be honestly denied. Our discussion with African countries was that their lack of development due to colonialism had to be recompensed by the former colonizers. Forgiveness of their debt was necessary but insufficient. Resources had to be provided for the actual development of their economy. Unconditional resources. No strings attached.

Nine years of lobbying with the African Group bore fruit at the 1998 session of the Commission on Human Rights (CHR). The African Group unanimously proposed a resolution calling for the Declaration of the Transatlantic Slave Trade and Slavery as a Crime Against Humanity. When word of the proposal got out it caused an uproar in the Western countries. Crimes against humanity have no statutes of limitations. The victims of crimes against humanity are due compensation from the criminals. The specter of paying reparations for the crimes they had committed and the unconscionable profits they had made over centuries shook the Western countries to their very roots.

Senegal was chairing the African Group when this resolution was proposed. In an ironic coincidence, President Clinton was on his way to Senegal to present the African Growth and Opportunity Act. Rumors have been consistent that President Clinton actually called Senegal's President Diouf from Air Force One to tell him that it would not be in Senegal's best interests to continue to push this resolution. The implications were clear—if you cross the United States, you suffer the consequences. The next day Senegal withdrew the resolution from consideration. The IS vowed that it would do all it could to resurrect the resolution at the WCAR.

THE THREE KEY ISSUES FOR
AFRICAN PEOPLE AT THE WCAR

The International Secretariat took the position that this World Conference was what Malcolm had in mind when he spoke of putting our situation before the world's court. This World Conference had to address the situation of African people. We knew from our prior experience in the United Nations and its world conferences that the most effective agenda is a narrow one. The more issues that you bring to the table, the less likely that any of them will be addressed. Early on, we identified those issues as:

1. Declaration of the Transatlantic Slave Trade and Slavery as a Crime Against Humanity;
2. Reparations for African people on the Continent and in the Diaspora;
3. The Economic Base of Racism.

These issues went to the economic root of and motivation for our kidnapping and enslavement; established international recognition of our humanity and the historically unprecedented crimes we had been subjected to;

provided for the forty acres and a mule that none of the original victims or their descendants/continuing victims had ever received; and explained why the ideology and practice of racism persists to this day. They were issues that spoke to African people on both sides of the Atlantic. They were issues that we could all unite on, despite the national particularities of racism that affected us. They were issues that, if addressed, would set an internationally recognized standard that would help all of us in our national struggle.

We met with NGOs of African people around the world and Africa to unite on lobbying for these three issues. The discussions with the African countries and NGOs led to the expansion of the first issue to include colonialism as a crime against humanity. This naturally followed, as colonialism, the Transatlantic Slave Trade, slavery, and the consequent underdevelopment of African peoples and countries were inextricably intertwined.

The importance of these issues has been testified to by the degree of resistance to them. No single set of issues had become such a line in the sand at any other world conference. At the First PrepCom, the WEO group, led by the United States, took the position that compensatory relief (i.e., reparations) should not even be discussed as a theme of the WCAR. They were just as adamant on the issues of the crimes against humanity and the economic base of racism. The latter is being played out in the struggle around how to characterize globalization. The "Developing" world wants to describe it accurately, that is, as a new name for an old game, namely, imperialism. The "Developed" countries want to tout its alleged benefits and to downplay, if not completely ignore, the fact that it has led to ever-increasing inequality in an already unequal world.

The WEO Group, the richest in the world, used money as a tool of disruption. It provided very little funding for the WCAR in comparison to the financial commitments it made for the four United Nations world gatherings of the 1990s.

In addition to the weapon of underfunding, the inventors of *Robert's Rules of Order* employed every parliamentary tactic to delay the process of preparation for the conference in the hope that the pressure of time would lead to concessions favorable to their position. They succeeded in making it necessary to convene an unscheduled Third Preparatory Meeting for the WCAR. Time and money limitations severely restricted NGO attendance at the last, crucial preparatory meeting.

To cover all their bases, during December 2000 the United States and Canada even went so far as to attend the GRULAC (Group of Latin American and Caribbean) countries' PrepCom rather than the WEO (their normal grouping) PrepCom. Their sole purpose in so doing was to ensure that the

GRULAC PrepCom (held in Santiago, Chile) did not come out with a strong declaration on the three issues. The United States and Canada accomplished their task by leading the GRULAC countries to believe that if they weakened the language on reparations and crimes against humanity the United States and Canada would agree to a consensus document. This lie was exposed at the last minute of the final session when both the United States and Canada admitted that they could not agree with the watered-down language that they themselves had suggested. There was not enough time at that point for the Santiago Declaration to be as strong as it might have been without the disruptive and disingenuous participation of the North American delegations.

At a public meeting in February 2001, the head of the United States delegation to the PrepComs admitted that the United States had sent a diplomatic delegation to attend the African PrepCom (in Dakar, Senegal) with the express purpose of having the Africans tone down their inflammatory language. The *language* she referred to was that concerning crimes against humanity, reparations and the economic motivations underlying the slave trade, slavery and colonialism. This time they were defeated. The African Group kicked them out of the meeting and produced a document that also included a program of action to implement reparations.

The West, and the United States in particular, employed yet another tactic, which was to state that insistence on the issues of crimes against humanity and reparations would derail the conference and thus make it a failure. This mantra was accompanied by threats of either nonattendance or the sending of a low-level delegation. In the end, the overt acts of pressure had their counterpart in the continuing use of covert threats like those that the United States employed in 1998 to sabotage the African Group resolution on the Transatlantic Slave Trade. One African Group diplomat, sounding discouraged, told us "the West is completely intransigent on this issue." We simply replied that "their intransigence is in defense of a crime against humanity. We must be even more intransigent in defense of our human rights."

IN THE UNITED STATES

In the United States, several organizations that have traditionally been involved in the struggle around reparations became participants in the international organizing. They include N'COBRA (National Coalition of Blacks for Reparations in America) and NBUF (the National Black United Front). Others, such as Fisk University's Race Relations Institute, were also active. One positive result of the WCAR was the increased organizing around and exposure of the issue of reparations. Meetings occurred all over the country and the word got out, along with a clearer understanding of the United Nations and the importance of work in international forums. This was particularly important because the United States did nothing to publicize this world conference. Contrast the $8 million it spent on the World Conference on Women with the $250,000 it committed to the WCAR. From early on, the IS knew that the mobilization for this conference would be at the grassroots level. And that analysis was proven true.

THE U.S. WALKOUT

Following the Third PrepCom, the WEO Group thought that it had successfully coerced the African Group to back down from the Dakar Declaration. The IS, along with NBUF and groups from Washington, D.C., and Atlanta, brought over a delegation of nearly four hundred Black people from around the United States, which we called the Durban 400. The strategy was to have an overwhelming force of lobbyists for our three issues. When the delegation reached Durban, it immediately began lobbying the African Group in particular and the other regional groupings in general. The work done by the Durban 400, along with that of the African and African Descendant NGO Caucus, combined with the presence of a delegation of the U.S. Congressional Black Caucus, provided crucial support to an African Group that was under siege by the Western countries.

Three days after the WCAR convened, the United States and Israel walked out. This was no surprise to the non-governmental organizations who had been involved with the preparatory meetings leading up to the WCAR, since veiled threats of the walkout, including promises to eliminate foreign aid to Africa if the reparations issue was pressed by delegates in Durban, had been continuous since the May 2000 PrepCom. Although the official reason given for the walkout was the debate over the Zionism-equals-racism issue, it was clear that the United States' *real* concern was the issue of crimes against humanity, reparations, and the ensuing flood of lawsuits that would emerge if the United States, along with the rest of WEO, supported these three critical issues. The U.S. walkout on the afternoon of September 3 followed a press conference convened that morning by the African Group, which restated its unequivocal support of the Dakar Declaration vis-a-vis the Transatlantic Slave Trade, slavery and colonialism as crimes against humanity.

CONSENSUS OR A VOTE?

Historically, United Nations world conferences produce documents by consensus, which results in a lowest-common-denominator declaration and program of action. This world conference more than any that preceded it reflected the realpolitik of the twenty-first century. On one side of the aisle in Durban stood the developing world and its brothers and sisters who reside in the developed world, but live a life of the Third World, as a collectivity of underdevelopment. On the other side, arrogant and unrepentant, was the developed world—beneficiaries of stolen labor and unpunished criminal acts, unwilling to admit its guilt yet fearful of the resistance of the victims. The former colonizers/enslavers versus the former colonized/enslaved.

Following the departure of the United States, the defense of crimes against humanity was left to the European Union, whose chair, Belgium, the architect of Patrice Lumumba's assassination, had a long colonial history drenched in African blood. Despite a valiant attempt to hold high the banner of white supremacy, the WEO Group was forced to reluctantly agree to consensus language that recognized the Transatlantic Slave Trade and slavery as crimes against humanity. African people had won a tremendous victory that set the basis for a higher level of activity in their national venues.

THE CRIMINALS NEVER GIVE UP

The events of September 11—just three days after the conference ended—temporarily refocused events away from the WCAR. But it did not deter the refusal of the WEO Group to accept its defeat. In an attempt to minimize the damage done to them by the events of the WCAR, the WEO Group fought, after the conference had ended, to have language eliminated or changed and paragraphs shifted in the final document. Their efforts at damage control delayed the issuance of the Durban Final Declaration and Programme of Action for nearly four months. As the perpetrators of crimes against humanity, the WEO Group, the self-proclaimed leaders of global civilization, remained unbowed and unrepentant.

CONCLUSION

The issue of reparations for African people is on the international front burner and will not go away. It is up to us to continue the pressure on the ground until our just and justifiable demands are met. Only by doing so will we have paid proper tribute to our ancestors.

ROGER WAREHAM *is partner in the law firm of Thomas, Wareham & Richards in Brooklyn. He is lead counsel in the historic Farmer-Paellmann lawsuit filed against corporations participating in the enslavement of Africans in the United States. He is also lead attorney for the December 12th Movement.*

PART V

Reparations and Intervention

TIM WISE

Debtor's Prison: Facing History and Its Consequences

> *If you burn the house you cannot hide the smoke.*
> —UGANDAN PROVERB

To read one's family history or hear it told, especially when that history has been compiled and presented by a member of said family, is an exercise in both irrelevance and farce. Irrelevant because the reader or listener knows, or should at least, that almost nothing he or she is being told or is reading is likely to be accurate, but rather, will be calculated to reflect only the most positive spin upon one's genealogy. Farcical because the intellectual gymnastics required to issue such a spin are just that, and demonstrate the depths to which some will descend in order to maintain a pretense of greatness, even amidst splendid mediocrity.

This is especially true for white Americans, of which I am one. Ask white folks from whence we come, and quite often one will hear an answer that in some way attempts to link (and occasionally succeeds in connecting) the white person in question to historical greatness. If English or Scottish, the story will typically lead back to royalty. That the Brits and Scots who came to the new world were mostly castaways and rejects—that's why they left—since most successful nobles don't relinquish their royal prerogatives and take to the high seas, never seems to faze white genealogical storytellers. It is as if one is just as close to the House of Stewart or the House of Windsor today, as one's family claims to have been a few hundred years ago. As if the Queen of England would merrily roll out the red carpet for one's kith and kin were they to strike out on a European vacation to the old homestead.

If one descends from a line of European "ethnics," as we like to call them, Italians, Russian Jews, or Irish, then the story is a bit different, and tends to resemble one or another version of the rags-to-riches fable. One's great-great whatever came here with nothing more than $13 and a ball of lint in his pocket. But through hard work and perseverance, he forged ahead and built a better life for his family. As with the epistles of nobility, these latter-day Horatio Alger tales are likewise steeped in a search for greatness, character, virtue, and ancestral honor.

Yet, at the same time that white Americans insist on celebrating the ostensible greatness of our ancestry, and in so doing glorifying the past, we show a remarkable propensity for acute historical amnesia when it comes to addressing the less noble of our ancestor's pursuits. We want the good, and we all but demand a retelling of it; but we ignore the bad, and insist all that was in the past. Indeed it was, as is all history, pretty much by definition. So was that lint-lined pocket at Ellis Island. But we can't stop talking about *that*.

The Civil War was most certainly in the past, and the Confederate States of America have been vanquished for 136 years as I write this. Yet, I still see Confederate battle flags flying over state capitols; and I still get cut off regularly in traffic by good old boys in trucks with bumper stickers that read "The South Was Right," and "If I'd Known This Would Happen, I'd Have Picked My Own Cotton." Still, white folks tell *black people* to get past the past, and to move on. Methinks they doth protest too much.

The issue of reparations for enslavement of Africans and the legacy of institutionalized racism is not new, and neither is white America's typically dismissive response to it. Our unwillingness to take the notion seriously and to recognize the ethical righteousness it embodies is symptomatic of the historical amnesia referenced above. For as James Baldwin explained in *Ebony* magazine in August 1965:

> [P]eople who imagine that history flatters them (as it does, indeed, since they wrote it) are impaled on their history like a butterfly on a pin and become incapable of seeing or changing themselves, or the world. . . . This is the place in which it seems to me most white Americans find themselves. Impaled. They are dimly, or vividly, aware that the history they have fed themselves is mainly a lie, but they do not know how to release themselves from it, and they suffer enormously from the resulting personal incoherence.

It is that incoherence, amnesia, blindness, and willed ignorance to which I now turn. For the issue of slavery and racism is not only a story of what happened to the victims, those African-descended persons who have fought tirelessly for justice and equity and freedom in this land. It is also a story about those who victimized, or collaborated, acquiesced, remained silent, or passively received the privileges of their white skin as the upside of another's misery.

And it is not only a story of how those persons profited from the pain of others; that tale has been told many times, and in this volume well enough. It is also a story of how those who profited so handsomely from that pain, in relative terms, lost something far more valuable in the process. It is a story about the loss not only of one's innocence, but also a healthy portion of one's humanity. Some relinquished it voluntarily; others did so on pain of rebuke and marginalization. But all paid for their privileges with the better part of their souls.

Now, in order to relieve the intergenerational psychic and spiritual burden purchased in this manner, it has become necessary to make recompense with a more lasting and useful currency. Please note that this debt requires payment not only to restore the victims, but indeed, to begin the healing process for the victors as well. It is an act not of atonement, which frankly can never likely be obtained, so great is the historical crime involved, but rather of coming clean. For whites, it involves the reclaiming of the souls jeopardized by that crime: the repurchasing of what was lost.

Of course, in order to initiate the repurchase transaction one must first begin to tell the truth. For as Baldwin also noted in his essay entitled "The Creative Dilemma," "whoever cannot tell himself the truth about his past is trapped in it, is immobilized in the prison of his undiscovered self."

Problem being, telling the truth about our past as a nation and a people is hardly at the top of most white folks' to-do list. Instead we avoid discussions of slavery and its legacy by insisting that we weren't there, and we never did anything to anyone. But it is for certain that *someone* was there, and *somebody* did something to someone else. And the somebodies in both instances are the ancestors of other somebodies still very much alive today.

So here we are, born into a legacy that we don't fully understand and for which we are not to blame, but which is ours nonetheless: our inheritance, as it were. And just like the gaudy piece of furniture left to you by your grandmother on the event of her passing, it is yours whether you wanted it or not. The only issue now is what to do with it. You can stick it in a corner if you choose; pretend it's not there. But every night before you go to bed, you'll catch a glimpse of it from the corner of your eye, and be reminded of its presence.

However, unlike the antique trunk or footstool, the legacy of racism is not something you can just take down to the city dump and burn. You can't sell it for a dollar in a yard sale just to get it out of the way. It's like a bad cold you can't shake. It is, for lack of a better way to put it, historical herpes. You can ignore it, scratch it, or maybe even treat the symptoms so as to minimize your discomfort, but it is *yours forever.*

Naturally, most white folks don't like to hear this. "What do you mean *our* legacy," they'll say. "Speak for yourself," comes the cry from others. "*I wasn't there. I didn't do it. I am innocent.*"

Innocence is an interesting concept, and in the mouths of persons born in the United States it is beyond interesting; it is, in fact, precious. More than precious, it is stunningly infantile. How anyone who finds themselves living on land that had to be stolen from others, and worked and plowed by still others, so as to make their own families' existence more comfortable could plead innocence at this late date is beyond the comprehension of the rational mind.

When a man such as Thomas Jefferson can be revered as a national hero, despite the fact that he saw no problem proclaiming from one side of his mouth that "all men are created equal," while from the other side insisting "the blacks are inferior to the whites in the endowments both of body and mind," the notion of innocence becomes comical to the point of distraction. When that same gentleman can ruminate about the greater beauty of whites, what with their "flowing hair," and "more elegant symmetry of form," while yet bedding Sally Hemings, whose form he must not have found too unappealing, the term becomes a perversity unworthy of comment.

What most white Americans fail to appreciate is that it makes no difference whatsoever if Thomas Jefferson and his fellow slave-owning founders are *our* direct ancestors or not. Either way, the system they put into place and the mindset that was cultivated so as to justify, rationalize, and maintain that system is very much their gift to all of us, or rather their curse. What's more, that curse not only damaged the objects of their oppression, but weakened the fiber of the white community as well.

Imagine, after all, what hoops the mind must jump through in order to ignore the obvious humanity of another person; to place that person on the same level as an ottoman. If human beings are born good—and I sincerely believe this to be the case—what stupefying process of conscience numbing must one endure so as to make owning another of God's creations acceptable?

My wife and I were recently blessed with the birth of our daughter, and as I look at her, I can't help but find myself wondering how one can take a child such as this, so small, so radically pure, so *free,* and in the one instance enslave that child, or in the other raise the child to eventually *own* the offspring of another. What tortuous illogic is required to make such a thing conceivable? And how entrenched must that illogic become so as to convince even those whites who *didn't* own slaves, and whose interests were hardly served by the system of elite domination, that they should be willing to fight *and even die* to maintain that system if need be?

What we must face is this: it was slavery itself that brought forth the excuses, the rationalizations, the justifications, and the pithy commentary about the "tainted blood" or bile of black folks. It was slavery that made necessary the mental game of Twister in the psyches of the Euro-American. And it is for this that we can say convincingly: no one is innocent.

How else, after all, could one smooth over the glaring contradictions between the talk of freedom on the one hand and the fact of bondage on the other? How else except by *lying* to oneself and others, and by making the spreading of that lie the most important task of one's society, could such a monumental fraud be kept in place? Without the acquiescence of a pliant mob—landless white peasants, white laborers, European immigrants, and so-called white trash—the trick would have failed. The illusion could never have been maintained. Because, you see, racism did not give birth to slavery, so much as slavery gave birth to racism. And it is that issue from the womb of the American slave system with which we are still entangled. To the extent all of us have been born into a nation where racism remains a social force, we all bear the scars of that delivery. Racism is no bastard child; its father was the chattel system. In its absence, there simply would have been no logical reason for the development of white supremacist ideology in the first place.

But let's not deal in abstractions. Let us take a look at what this system of white supremacy did to real flesh-and-blood Europeans. Both those who owned slaves and those who didn't even arrive here until decades after emancipation. For I insist both are implicated.

And I say this not as some exercise in finger-pointing, but rather as a preface to autobiography. For this is my story. I am both sides of the contradiction, representative of both the yin and yang of the white American condition: perpetrator and collaborator; descendant of slave owners and abolitionists; aristocrats and desperate immigrants; gentile and Jew; Yankee and Confederate; the blessed and the damned. But all of them white, and that is where my story begins. Were it only mine it would be of little import, and I wouldn't likely tell it. But it isn't, and so I will.

To thumb through the recently compiled hagiography that passes for the history of my mother's father's family, the McLeans of Western Scotland, is to know the true meaning of self-deception. It is also to stare the residual effects of racial oppression square in the face. There, one finds all the obligatory references to greatness: a paean to unfettered glory, without so much as a hint of modesty.

We learn that the McLeans' story is an important chapter from the American frontier and the settling of this great nation. We are invited to envision the family's ancestral castle, standing as a sentinel, and visible "through a mystical fog" covering the Isle of Mull. True enough about the mystical fog, the tale continues: our family became the Clan Mac Lean around 1250 (a truly proud period in European history, to be sure), and was one of the most prosperous clans in Scotland. We owned five islands, but lost our all with the defeat of some twit named Bonnie Prince Charlie in 1741. Our family motto: Virtue Wins Honor.

Fast-forward a few years to 1750, and the arrival of family patriarch Ephraim McLean in Pennsylvania. The McLeans, being as we learn a "hearty and athletic race of men, without any surplus flesh," kept on the move, settling in North Carolina in 1759, and then moving into Kentucky and Tennessee.

All the McLean men, it appears, were "gallant officers and devoted patriots," and all were imbued with the "characteristic Scotch traits of integrity and hospitality." One imagines the family to be perhaps contrasting these with the presumed characteristic Irish traits of drunkenness and perfidy, or those of some other lesser culture.

Ephraim, we learn, was granted a thousand acres of land in Nashville, as payment for relieving soldiers during the Revolutionary War, and another 11,000 acres in North Carolina simply for having been a property surveyor for the state. Just a few years off the boat, and Padre McLean was already accumulating land in places where Africans had been living and toiling for over 150 years without so much as the right to possess their own names.

Of course, land alone was not property enough for these proud men of Scotland. No indeed. Ephraim's son Samuel, my great-great-great-great-grandfather, "owned much land and slaves, and was a man of considerable means." This is explained in the McLean story with neither pride nor regret, but merely in the matter-of-fact style befitting those who are trying to be honest without confronting the implications of that honesty. Say it quickly, say it simply, and move on to something more appetizing. Sort of like acknowledging the passing of gas in a crowded room, but not admitting that you were the author.

Frankly, I find it hard to be so sanguine about the ownership of other persons. Especially because, contrary to what many white folks would have you believe, not all whites felt that Africans were subhuman. It is *not* the case that everybody back then felt that way, because they didn't. Certainly the Africans themselves didn't feel that way, so at best, those who try to pass off some uniform mentality of evil as an excuse for buying and selling flesh and blood are ultimately only considering white perceptions to have been important. But even then, many whites opposed slavery. Ardent white abolitionists, north and south, existed well back into the early 1700s. But rarely do we hear their stories. Unfortunately, those who believed and cultivated the lie of white superiority were stronger for two and a half centuries. So they got to write the history books, and they get to pass down with loving admiration the tales of their families' considerable means, without so much as batting a blue eye at the sight of such contorted indignity.

Indeed, the young abolitionist woman from the McLean family who, in the 1800s, convinced her parents to free their slaves merits hardly a mention in this official family history, and no mention at all of her antislavery tendencies. Apparently, this kind of bravery and resolve is not what some folks are thinking of when they intone, "virtue wins honor," though it seems as if such a character would know more about both concepts than any of the slave owners in the Clan McLean.

But truth be told, I know all too well why her story remains unmentioned. To tell it would reflect badly on the rest of the family. To acknowledge, let alone venerate, the commitment to freedom on the part of one family member is to contrast such a commitment to its utter absence on the part of all the rest. It is to bring the rest of the family up wanting, and pallid, and we couldn't have that.

Instead we have this: the 1844 will of Samuel McLean, which reads:

> I give and bequeath unto my loving wife, Elizabeth, my Negro woman named Dicey, to dispose of at her death as she may think proper, all my household and kitchen furniture, waggons, horses, cattle, hogs, sheep and stock of every kind *except as may be necessary to defray the expense of the first item above.*

In other words, sell whatever must be sold in order to hold on to the slave woman, for how might dear Elizabeth survive without her? Now read that again, and then explain how one can so seamlessly place farm animals, rocking chairs and Dicey in the same sentence as material to be inherited. But there is more.

> I also give the use and possession of, during her natural life, my two Negroes, Jerry and Silvey. To my daughter Sarah Amanda her choice of horses and two cows and calves, and if she marry in the lifetime of my

wife she is to enjoy and receive an equal share of the property from the
tillage, rent and use of the aforesaid 106 acres of land and Negroes
Jerry and Silvey that she may be the more certain of a more comfortable existence.

Furthermore, if Sarah were to marry before the death of her mother, she
and her husband were to remain on the property with Elizabeth so as to continue to benefit "from the land and Negroes." If, on the other hand, Mom were
to die before the wedding of Sarah, then the daughter was instructed to sell either the land or the slaves and split the proceeds among her siblings. Either
way, Dicey, Jerry, and Silvey would remain commodities to be sure. Choosing
freedom for them was not an option, for then the Clan McLean might have to
learn to do for self; might have to work. They might have to wash their own
clothes, and grow their own food, and nurse their own wounds, and make
their own beds, and suckle their own babies, and chop their own wood, and
wipe their own collective ass. And that would make them less "certain of a
more comfortable existence," so of course it was out of the question.

To his son, Samuel D. McLean, Sam Senior bequeathed "a Negro boy
named Sim," who would then be handed down like an armoire to his son
John. Then, according to family legend and in what can only be considered the
Margaret Mitchell version of the McLeans' history, "Sim went happily off to
the Civil War with his master." What's more, we even have *dialogue* for this
convenient plot twist, as Sim exclaims (and I'm sure this is a *direct quote*), "I've
taken care of Mr. John all his life and I'm not going to let him go off to war
without me."

For his loyalty, "Sim got a little farm to retire on because they (the
McLeans) knew he would not get a pension of any kind." No indeed, as property rarely receives the benefit of a 401(k) plan.

To his daughters, Sam McLean gave the slave woman Jenny and her child,
and the slave woman Manerva and her child, and in both cases "any further
increase": an interesting way to refer to future children. But we are to think
nothing of this subterfuge in the case of the children of Manerva or Jenny. We
have to keep telling ourselves that *they are not people*, no matter how much
they *look* like people. No matter that they share 99.9 percent of their genes
with us, and us with them. Pay no attention to what all your senses are telling
you. Pay no attention to the man behind the curtain, working overtime to
convince you to believe *him* and not your lying eyes.

Now I know what some are thinking as they read this: fine, they might say.
Perhaps your family *does* owe a debt. So then *you* should pay reparations

perhaps, but not me, and not most white folks. After all, they would insist, their families never owned or even heard of Dicey, Jerry, Silvey, Sim, Manerva, Jenny, their children, or further increase.

But such a self-assured entreaty misses the mark, and indeed misses it quite widely. For the government that allowed and even encouraged the McLeans to own those blacks listed above is the same government that favored even the nonslaving whites in matters of commerce, the legal system, and all other endeavors. The same government that enforced fugitive slave laws and allowed payment of rewards to nonslaving whites if they would capture and return the property of others, on those occasions when said property might decide life on the farm wasn't so pleasant. The same government that refused to consider blacks as citizens until roughly a century after it recognized the citizenship rights of European immigrants. The same government that allowed and even subsidized segregation, racial hierarchy, and apartheid for another century after slavery was crushed.

The same society that helped my family own slaves helped millions of other European-descended families to become overseers, or gain property under the Homestead Act. It made them *white* just like mine, and in so doing placed them as well as me above the lot of African-descended persons.

This is so even for those Europeans who arrived on these shores long after the fall of the chattel system; indeed, even for those Europeans who themselves faced substantial bias, hatred, and contempt from the betters of England, or Scotland, or other lands of Anglo-Saxony.

For example, my family on my father's father's side: Jews, from Russia, escaping from the Pale of Settlement to which the czar restricted all Hebrews in that place and in that time. Knowing nothing of whiteness before the trip to a strange and distant shore, it would not be long before my great-grandfather would learn its meaning. He would have little choice.

It began innocently enough, or so it seemed. To step onto Ellis Island and be told that because the immigration officials couldn't understand your accent, it would be necessary to give you a new name, to simply make one up. What was that, after all, which you were garbling? *Shuckleman, Sheckelman, Shuckman?* To hell with it, your name is now Wise. Not Weiss, but Wise. Whitened and sanitized for your protection: a Jewish Toby.

How exciting Jacob must have found this process. You come to a new country and you get a new name! But little did he realize and little did most Jews who have similar stories to tell that this was but the down payment on the process of becoming white; only the beginning of the process of deculturation. Had it stopped there it might have been tolerable. But of course it wouldn't. This was the inheritance waiting for him, handed down by a system he had neither created, nor been involved with in any way, but which was his nonetheless.

As Jacob entered New York and later found his way south to Tennessee, he no doubt worked hard. And he no doubt faced the slings and arrows of anti-Semitism, as did most Jews at that time. And he also, no doubt, learned a number of vital lessons about this place. First, that if one wanted to make it, and be accepted in the United States, one had to assimilate to a dominant cultural norm. One had to blend in and make others comfortable with the thought of your existence.

That this might prove difficult for someone with a strange way of talking, and praying, and eating would have struck him as fairly self-evident. But having made the journey, what could he do about it now? There was no turning back. And so he sucked it up, as did most Jews, most Irish, and most Italians. He began the slow process of transformation. Don't seem *too Jewish*. Don't teach your children the language of their forebears, nor the customs. Don't talk about the old country. Put all that behind and become a new man, an American, and more to the point, a white American.

If one did this, one might yet inherit the kingdom of white supremacy. If one resisted, one might well be crushed, remain ghettoized, and ostracized, and marginal. It doesn't take much historical knowledge or common sense to imagine that most chose the path of least resistance. Having done so, they were able to obtain jobs that were off-limits to African Americans, whose presence here predated theirs by three centuries. As long as they hid their religious tradition and tried to pass as millions did, they could avoid the restrictive covenants that sought to place them on the nonwhite side of the color line along with blacks when it came to housing. They could go to schools off-limits to blacks. While there might be restrictive quotas that limited the number of Jews able to attend Yale and Princeton, for blacks there was a quota of exactly none, so that even repression took on a more limited feel for outsiders with the proper melanin deficiency.

Please understand, my great-grandfather's decision to take the bait and accept the bargain offered him is neither surprising nor particularly indicative of his character one way or the other. It is a bargain most anyone probably would have taken and would still take today. The problem is not that he took it, but rather that he was put in a position where doing so became an option, and more than that, a virtual certainty. Any society that requires of its members a song and dance so as to make its rulers more comfortable is a society filled with court jesters, but not too many human beings worthy of the title. Any place that requires one to stifle one's cultural traditions and *go along* as the price of their ticket is a place that is charging too much for the matinee.

To see the cost of becoming white, which became necessary because of a slave system that needed a category for those who could not be owned, one need only stare into the eyes of one's Jewish grandfather, as he stumbles, incapable of passing down any story, any fable, *anything at all* about his father, or grandfather, or mother, or grandmother, or the place from which you come. And why couldn't he do that? Because to be able to tell such stories would require that he had been taught them. And for him to have been taught them would have required that his father be willing to do so. And for his father to have been so willing would have meant that he had been able to resist the pull and lure of whiteness. And to do that was unthinkable.

So your grandfather joins his ancestors, about whom neither you nor oddly enough even he knows much of anything. With him goes your connection to the past, and you are left, in my case, with one-half of a set of gold candlesticks, the only items smuggled out of Russia on that ship your papa boarded in 1910. And you don't even know the story behind the candlesticks. You only know that they are all you have left.

All, that is, except for your white skin. And though that may provide you with innumerable benefits, it is hardly better than the candles in the candlesticks at keeping you warm at night. Because you know their true price: you know how much your family paid for them, and for your name. And you begin to realize that the balance sheet is terribly out of balance, not only for the direct victims of slavery and apartheid, but for you as well. You have been played, and tricked, and cheapened, and made to give up that which you were in order to become something you were not. You have become a white American, and now can neither bear nor fully even understand just what that entails.

Slavery and the racism that resulted from it as a mechanism of justification are to blame for this. It has become a pox upon the house of anyone living in this place. There is no escaping it, or pretending that it isn't there. The souls of too many millions gone are crying out for recompense, and for justice. And those souls are not merely the souls of Dicey, and Jerry, and Silvey, and Sim, and Manerva, and Jenny, and their children, and further increase. They are the souls of Ephraim and Samuel McLean and their increase as well. One need only look at their pictures: faded, tattered, but quite telling. Not a smile on even one of their faces, but rather, the kind of severe looks that indicate an unhappiness, an emptiness, that is nothing if not predictable for those who have had to circumvent their better natures to live out a lie. It is the visual manifestation of the word *numb*.

And so too are they the souls of Jacob Wise and his increase, who had to hit the reset button on their own lives and behave the way others needed them to, in order that they might find a place in this land.

Now one might think that if whites were also harmed by slavery and the system of racial supremacy that has been its legacy, then reparations make little sense. After all, how can whites be asked to pay for something that on balance diminished them as well? But it is not whites per se that are being asked to pay, but rather the government that elevated those whites over others and in so doing diminished them in relation to the human beings they might otherwise have become. The fact that the government receives its money from the people is, of course, true. And any reparative program or investment—and it is that being called for, not personal checks or money orders to individual black persons—will in fact be paid for by *all* Americans, white, black, and otherwise. As it should be. Not merely as a way to make right the wrong done by certain individuals to other individuals, but also as a way to offer compensation for a system that did wrong by us all.

To heal these wounds requires a proper accounting. It requires the payment of a debt, the interest on which is still building. To be sure, one can choose to ignore the nagging reminders of that debt. As with any other debt, one can ignore the phone calls from the collection agents. One can pretend not to have received their notices in the mail. One can even, if one is especially crafty, send in a check to cover one's debt and forget to sign it, thereby delaying things further while appearing to have been operating in good faith. One can do any and all of these things, and yet the debt still looms. It is the thing that shackles the debtor and the debtee forever and prevents either from tasting true freedom.

Though there is no such thing as debtor's prison in the formal sense today, there are steellike bars that we erect quite independent of the justice system. One can be incarcerated in these prisons, no less than if found guilty of a crime and punished. One can be enslaved and held captive to one's *own* history, and usually will be, for precisely so long as one refuses to face up to it.

TIM WISE *is one of the most sought after speakers on the college circuit specializing in racism and its consequences. A social justice activist for the past two decades, Wise has spoken to over 75,000 people in forty-six states, on over 275 college campuses, and to hundreds of community groups. His work in Louisiana was instrumental in derailing the candidacy of David Duke for the U.S. Senate and for the office of governor of Louisiana. He is the author of* Little White Lies: The Truth About Affirmative Action and "Reverse Discrimination."

JEWEL CRAWFORD, WADE W. NOBLES, AND
JOY DeGRUY LEARY

Reparations and Health Care for African Americans: Repairing the Damage from the Legacy of Slavery

Serious diseases require great remedies.
—Haitian Proverb

INTRODUCTION

This chapter is written with an underlying premise: that the debt owed to African Americans by those benefiting from our enslavement can never be fully repaid. The authenticity of this premise is exemplified in the context of the injury, death, and disease suffered by Africa's children historically and up to the present day as a result of the Transatlantic Slave Trade and its ever-present accomplice, racism.

The status of one's health is inextricably linked to the quality of one's life. The Transatlantic Slave Trade, slavery, and racism have left an indelible scar on the bodies and the minds of generations of displaced Africans whose forefathers persevered despite the sting of the lash. For it was only the inner strength of these predecessors, deeply rooted in an unwavering faith in the Almighty, that empowered them to withstand hardships and guarantee their people's survival in a hostile, foreign land.

The review of the horrific history of the health and health care of Africans in America in the context of the reparations movement forces the questions: How should the beneficiaries of the slave trade settle a debt owed to an entire race of people? How does one begin to repair the emotional, psychological, and physical damage inflicted by kidnapping millions of members of African families, transporting them in chains across the Atlantic as they lay in their excrement, and for those who survived the trip to the New World, subjecting them to unimaginable horrors: scattering their family members, never to be rejoined again; persistently raping their women, such that almost all of their descendants are tainted with the blood of their captors; and then lynching, torturing, and castrating their men, who dared to defy their brutal treatment. It is the persistence of the unresolved, ongoing trauma of racism that has impacted the health of the descendants of captured Africans right up to this very day. It is in the context of this most heinous crime against humanity that we review racial discrimination in health care and what remedial actions would amount to fair, just, and partial compensation for the descendants of those Africans who lived and died so exploited for their labor.

U ntil very recently, very little scholarship has directly related slavery's institutional effects on Black health. Although Black slave health data are sketchy, a definite pattern of poor Black health outcome survives. African Americans' health would be shaped directly by slavery for the first 246 years of their experience in North America and indirectly by its sequelae to the present day.[1]

When we recall the sordid events of the past, and just compensation thereof, it is clear that reparations cannot be reduced to a giant welfare check to assuage the guilty consciences of the perpetrators' heirs as they continue to enjoy the spoils of their unjust enrichment. And while perhaps a sincere, symbolic, apologetic gesture, on the part of some, a check cut and dispensed to the descendants of the victims of the Transatlantic Slave Trade—now free to spend and consume in a (most likely) white-owned establishment—making a few purchases or paying off a few bills would hardly settle the gory four-hundred-year score.

The root of the word "reparations" is repair, which is quite apropos in the case of the victims and descendants of the Transatlantic Slave Trade. For, as we shall see below, much damage has been done. For many the damage has been irreversible, as the millions of African skeletons (were they to be dredged up from the bottom of the ocean, where they jumped or were thrown overboard during the notorious Middle Passage) would attest to. Or perhaps, on quiet starlit nights, if we could listen to the whispering of millions of ghosts hovering over unmarked graves throughout the land, we could hear their stories. It may be a mass grave from whence comes the murmuring, dug in the aftermath of the carnage of the bloody put-down of a slave rebellion. Or it may simply be the scattered plots now overgrown with weeds of those who were beaten, overworked, tortured, maimed, exposed to the elements and left for dead during the slave era and the vicious period of institutionalized racist discrimination that persists more than one hundred years following emancipation.

HISTORICAL BACKGROUND

The scholar and historian of ancient African history Basil Davidson articulated the sentiment of many of Africa's children when he stated that the coming of the Europeans in the fifteenth and sixteenth centuries was the worst thing to happen to Black Africa.[2] From the years A.D. 1000 to the 1500s, western Africa and later central Africa were the scene of a number of flourishing trading empires. The African kingdom of Ghana was famous for its opulent gold and salt trade. The rise of Mali followed, controlling the gold trade routes, and included the famous cities of Timbuktu and Gao. Timbuktu was renowned for its great mosques, royal palace, and most of all for its prestigious university, which was the foremost academic center for African scholarship. In the late 1400s arose the mighty empire of Songhai, followed by the kingdoms of Luba and Lunda in central Africa, important for their trade in copper and salt. The Kongo kingdom and the Great Zimbabwe exemplified other wealthy African civilizations that flourished before the invasion of the African coast and its interior by the Europeans.[3] It is against this backdrop of a vigorous African tradition of intellectual development, cultural achievements, and trading economy, where Africans themselves were the beneficiaries of their own rich and vast natural resources, that the Transatlantic Slave Trade began.

HEALTH CONDITIONS IN SUB-SAHARAN AFRICA PRIOR TO THE TRANSATLANTIC SLAVE TRADE

The tropical environment of Africa is home to a host of disease-producing insects and parasites that transmit maladies such as malaria, yellow fever, and sleeping sickness. While these diseases had a significant impact on sub-Saharan African infant mortality and death rates, the indigenous people of Africa developed specific immunities to local diseases. As more and more Africans were captured for slavery, mortality rates soared due to a number of factors. It was a long and treacherous journey from the African interior to the coast, wherein the captives, many of whom were skilled craftsmen and laborers, had to travel shackled and on foot. Members of various African tribes

from different regions with their own specific immunities and disease suscep-
tibilities were crowded together in holding pens and slave dungeons while
awaiting transport onto slave ships. This situation allowed for the exchange of
germs unfamiliar to Africans from diverse areas and to which they had no immunity. Compounding the problem was the exposure to an entirely new host
of deadly European diseases foreign to Africans, such as tuberculosis, syphilis,
smallpox, and measles. All of these circumstances resulted in a 50 percent
mortality rate for captives even before leaving the African coast.[4]

THE MIDDLE PASSAGE

*Black Africa sacrificed 40 to 100 million souls to the slave trade; 15 to 25 million
survived. . . .*[5]

There is little recorded in human history to compare with the sheer hor-
ror of the Middle Passage. Human beings were chained together, and then
piled on top of each other, where they had to lie and sleep in their own waste
as well as that of the persons crowded next to them for weeks on end. A vicious cycle of disease ensued as African people huddled together crying,
screaming, vomiting, and defecating uncontrollably. Along this human chain
of misery, some were dead and some alive, the waft of rotting bodies adding
to the stench. There was no escape from disease. The captives suffered from
dysentery, diarrhea, eye infections, malaria, malnutrition, scurvy, worms,
yaws, and typhoid fever. Slaves also suffered from friction sores, ulcers, injuries and wounds resulting from accidents, fights, and whippings. By far the
greatest killer disease during the Middle Passage was the bloody flux: amebic
dysentery. Many succumbed to this bloody diarrhea, which alone resulted in
a mortality rate of anywhere between 20 and 80 percent.[6] The longer the journey, the more the human cargo was to die in route. One can only imagine the
state of mental health for those trapped in this living nightmare. Panic, anxiety, and hysteria prevailed. Pure rage alternated with a deep collective depression, manifesting in mutiny and onboard rebellions, the most well known of
which was led by the famous Cinque. For various reasons—too much cargo,
too little food, to eliminate evidence of being a slave ship—African people
were oftentimes thrown by the crew into the shark-infested waters. Believing
that their souls would return to the place of their birth, long lines of chained
captives sometimes jumped overboard together, committing group suicide,
and mothers threw their babies overboard. The three hundred years of the
Transatlantic Slave Trade amounted to a massive *Maafa* (system of death and
destruction beyond human comprehension and convention) unparalleled in

the annals of history. The notorious Middle Passage destroyed the lives of somewhere in the range of 25 million to 75 million innocent human beings.[7]

THE LANDING OF THE SLAVE SHIPS: LIFE FOR CAPTURED AFRICANS IN NORTH AMERICA

Only the strongest of the captured Africans survived the Middle Passage. Their uncanny ability to survive the harshest of conditions was again put to the test in the weeks and months that followed arrival in the New World. This critical phase of adjustment to a life of slavery in North America was called the breaking-in period. It lasted for up to three years. The breaking-in period was traumatic both physically and psychologically. Physically, there was the adjustment to a much colder climate, without adequate clothing to keep warm. Exposure to the elements in and of itself was enough to induce illness, but in addition, the new arrivals were exposed to an entirely new set of germs and diseases that African people had little or no resistance to. The auction block experience, where one was stripped naked and inspected like cattle, was only surpassed in cruelty and inhumanity by the overwork and beatings that were typical of plantation life. While white indentured servants also were forced to work on plantations in the 1700s, racism was legally sanctioned in statutes prohibiting public beatings of white indentured servants. In 1705 the law stated: In addition to the prohibition against publicly whipping a naked white Christian servant, masters were admonished to find and provide for their servants wholesome and competent diet, clothing and lodging, by the discretion of the county court. In contrast, with regard to enslaved Africans: Masters were allowed to feed, clothe, and nurse their slaves in whatever manner they saw fit.[8] Psychologically, it was during the time of the arrival of enslaved Africans in North America that the formal destruction of the Africans' culture, language, and family life took place. If not already separated, families were torn apart—husbands from wives, children from their parents and siblings. It was common practice for them to be sold to different plantations in different states, never to be reunited again. Practicing African religions was prohibited, and speaking in their mother tongue, their native African language, was forbidden. The drum, a focal point of African life, was outlawed for communication. Depression, suicide, anxiety, rage, and psychosis, combined with tuberculosis, pneumonia, influenza, typhoid, yellow fever, and cold injuries combined to produce a 30 to 50 percent mortality during the breaking-in period. Housing on the plantation in the slaves' quarters generally consisted of poorly ventilated cabins with leaky roofs and dirt or clay floors. The average cabin was fifteen square feet, housing five to seven people. Mattresses were rare; sleeping on the dirt floor was not. Due to crowding in the living quarters, conditions such as respiratory infections spread rapidly throughout the cabin. Accidental and fatal fires were common in the cabins in winter when large fires were built in an attempt to stay warm.[9]

Clothing was generally limited to a ration of one to three outfits per year, with one pair of shoes a year on the average. Presumably, it was difficult to launder the clothes if one had only one outfit that was likely worn, especially in the winter months, for twenty-four hours a day.

While there was some variation in the food rations received, overall nutrition was inadequate. Some survived on a diet of cornmeal and molasses, while for others the diet was high in fat and calories—with pork scraps used commonly as a protein source. Generally the typical diet was lacking in some of the necessary vitamins and nutrients. Some slaves, however, were able to have their own vegetable gardens and could also fish and hunt to supplement their diets.[10]

Childbearing during slavery became a prominent issue with the official end of the African slave trade in 1808. The fertility of Black women, who were referred to by slaveholders as breeders, was crucial to the economy of the United States, as its growth was based on slave labor. Black women were subjected to every type of sexual abuse imaginable. They were expected to be sexually available to the master, his friends and relatives, or Black male "studs" assigned to impregnate them regardless of whether they had a relationship or not. This was at the discretion of the owners, some of whom allowed their Black slaves some semblance of marriage and family. Those owners were the exception, however.

High rates of maternal and infant mortality reflected the overwork and poor nutrition of enslaved pregnant women, who often worked in the fields up until delivery and returned to work shortly afterward. This allowed very little time for breast-feeding. In addition, midwives attending deliveries were poorly trained.

Once purchased by a slaveholder, enslaved Black people could be used for any purpose their owners saw fit, including, in the case of some white science researchers, for experimentation. J. Marion Sims, M.D. (1813–1883), considered the Father of Gynecology in medical schools throughout the United States, purchased Black female slaves to perform experimental operations on them or to use them for teaching purposes. During presumed didactic sessions, he exposed their genitals to the public. He admitted to operating on one slave woman thirty agonizing times before achieving the results he desired. Sims performed surgical procedures on these women without using anesthesia, as, purportedly, Black people did not feel pain.

Enslaved African people did not passively accept this barbaric treatment. There were numerous instances of slaves doing violence to the master class. Poisoning was always feared, and perhaps some planters felt a real need for an official taster. As early as 1761 the Charleston *Gazette* remarked that the "Negroes have begun the hellish act of poisoning." Arsenic and other similar compounds were used. Where they were not available, slaves are known to have resorted to mixing ground glass in the gravy of their masters' tables. Numerous slaves were convicted of murdering their masters and overseers, but some escaped.[11] Accounts of runaways and slave rebellions were frequent headliners in the news periodicals of the slave era. The most well known slave revolts took place in New York in 1712, Stono, South Carolina, in 1739, in southern Louisiana in 1811, and Southampton County, Virginia, in 1831. Other famous slave revolt conspiracies included those whose chief strategists were Gabriel Prosser in Richmond, Virginia, and Denmark Vesey in Charleston, South Carolina. The specter of organized bands of Black slaves marching over the countryside from plantation to plantation was impressed upon the minds of the white South with the slave rebellions that occurred in the Pointe Coupee section of Louisiana in January of 1811. This triggered a white civilian mass migration into New Orleans.[12]

Systematic terrorism and the lash were the means by which slavery was maintained. Punishment, however, for participation in rebellions and slave revolts became even more brutal. *Floggings, castrations, amputations, nose splittings, and a variety of mutilations added to the slaves' already high injury burdens related to agricultural and industrial work and punishments associated with the workplace.* (Italics supplied.)

The Civil War and the signing of the Emancipation Proclamation proved to be a mixed blessing for Africans in the United States as former slaves were now completely destitute and homeless, without a support system of any kind. In the first few years following the Union victory in the Civil War, Northerners introduced legislation guaranteeing rights to ex-slaves, ratifying the Thirteenth, Fourteenth, and Fifteenth Amendments to the Constitution in spite of virulent opposition from white Southerners. In 1865, Congress established the Freedmen's Bureau, a program that provided food for destitute ex-slaves, emphasized education, and helped develop medical programs. It was during this time that former slaves were promised forty acres and a mule. The failure of the government to keep its promise to the ex-slaves, now referred to as freedmen, was one of a long litany of shattered hopes of government support for the recently emancipated Black population as Northerners' enthusiasm for solving Black problems waned a few years after the end of the Civil War. The Freedmen's Bureau did manage to stay functional long enough to make a significant contribution to the education of Blacks. It was during this time that many of the historically Black colleges were established, along with the medical schools for Blacks. By 1900 there were ten Black medical schools, nine nursing schools, and two schools of pharmacy and dentistry. Though these numbers were not nearly adequate to care for the health of the ex-slave populations, now in dire straits as epidemics of poor health spread, these new schools dramatically increased the number of Black health professionals to care for Black patients. Rejected by white organizations, African American health professionals began to organize themselves. The National Medical Association was established for Black physicians in 1895. It was not long, however, before the few small gains in politics and health care were rolled back. By 1870 most of the Freedmen's Bureau programs were closed. By the 1890s Ku Klux Klan activity—that is, lynching—was at its peak, and racial apartheid was legally instituted in the South via the segregation codes. The rights of Blacks were essentially wiped out. By 1920 only two of the ten Black medical schools remained open (Howard and Meharry), the others having been rendered unnecessary by the infamous Flexner Report of 1910. Black doctors comprised 2 percent of the physicians in the United States at the turn of the century. At the time of this writing (2002), they comprise approximately 3 percent of all physicians. The system of racial apartheid and segregation in health care persisted legally until the Civil Rights era of the 1960s. The Flexner Report said that it was unnecessary to train many Black physicians since whites would not go to them anyway.

In view of the present racial disparities in health care, one might ask, What has really changed? Drs. W. Michael Byrd and Linda Clayton have painstakingly described the slave health deficit in their brilliant documentary of the history of African American health care: *An American Health Dilemma*. Despite the gains made over the last hundred-plus years of the African American struggle for equality, the slave health deficit remains.

PSYCHOLOGY AND PSYCHIATRY'S ASSAULT ON THE MENTAL HEALTH OF BLACK PEOPLE

The mental health of a people is an essential aspect of their human wellness and health. The consequence of white supremacy (racism) and its requisite negation and nullification of African people and things African (ideas, philosophy, history, traditions, et cetera) have resulted in more everlasting damage than the whip or the physical chains of bondage. In fact, in a very real way the physical damage and destruction of Black life was paralleled, if not precipitated, by an insidious assault on the psychological value of African human beingness. Next to anthropology, psychology and psychiatry have served as the fundamental disciplines and intellectual tools used to justify the dehumanization of African people. Psychology and psychiatry have been and remain critical instruments in the falsification and denigration of the image and meaning of African people. The mental assault on African people that resulted in a worldwide perception and treatment of African people as inferior and less than human was designed and conducted by white supremacists under the cloak of science. What, in fact, was created was an intellectual atmosphere that was intentionally designed to be destructive and detrimental to the mental health and well-being of all African people.

Beginning almost simultaneously with the introduction of Africans into the American colonies, psychiatry and psychology have assaulted the very humanity of African people. Dr. Thomas Szasz, professor emeritus of psychiatry, wrote in his book *The Manufacture of Madness* (1997) that people like Dr. Benjamin Rush, the father of American psychiatry, made Black people medically safe domestics, while simultaneously calling for their isolation sexually because Black people were carriers of a dreaded hereditary disease. The formula was almost sacrosanct. Because the psychological community had determined that African people were human defects and carriers of mental illness, white society was, as a medical prophylaxis, justified in medically protecting itself by dominating and killing African people.

In her introductory remarks to "Psychiatry's Betrayal," Jan Eastgate, the international president of the Citizens Commission on Human Rights, notes that psychiatry exploits man's desperate need for a workable solution to resolve spiritual travail. It has consciously used pseudo-scientific terminology and experimentation to secure a fraudulently obtained position of authority on the subject of the human psyche. It is a problem masquerading as a solution. Eastgate further notes that, based on documentation and evidence amassed by the Citizens Commission on Human Rights, there has been a calculated psychiatric attack on Blacks and other racial minorities.

The history of psychology and psychiatry's relation to African people suggests that the field of mental health has been more than just a problem masquerading as a solution. Both psychologists and psychiatrists have been amongst the lead perpetrators in the negation, nullification, denigration, and dehumanization of African people.

Starting as early as 1797, Dr. Benjamin Rush declared that the color of Black people was caused by a rare, congenital disease called negritude which, in turn, was a derivative of leprosy. He added that the only evidence of a cure was when the skin color turned white. In effect the cure for the illness of being Black was to become white. Rush used the disease of negritude to justify segregation.

Dr. Edward Jarvis, a specialist in mental disorders, used the 1840 census, which purportedly proved that Blacks living under "unnatural conditions of freedom" in the north were more prone to insanity, to conclude that slavery shielded Black people from some of the liabilities and dangers of active self-direction.

Publishing in the *New Orleans and Surgical Journal* in 1851, Dr. Samuel A. Cartwright penned an article entitled "Report on the Diseases and Physical Peculiarities of the Negro Race." In this article he claimed to have discovered or identified two mental diseases peculiar to Black people that justified the enslavement of African people. The two diseases he called Drapetomania and Dysaesthesia Aethiops. Cartwright claimed that Drapetomania caused Black people to have an uncontrollable urge to run away from their masters. The psychiatric treatment for this disease was to whip the devil out of them. Dysaesthesia Aethiops affected the mind and the body and was evidenced by disobedience, answering disrespectfully, and refusing to work. The psychological cure or therapeutic intervention was to force the patient to undertake extremely difficult and hard labor, which sent vitalized blood to the brain to give liberty to the mind. One can not even estimate the degree of brutality and destruction unleashed against enslaved African people in the guise of curing their diseased minds.

The Sanctioning of the Destruction of Lives Unworthy to Be Lived, by psychiatrist Alfred Hoche and jurist Karl Binding in 1920, and *The Principles of Human Heredity and Racial Hygiene*, by Fritz Lenz, Erwin Bauer, and Eugen Fischer in 1924, were books that set the tone for the Jewish Holocaust but also reinforced racial hatred toward black people. Both books—legitimized by the science of psychology—added to the large body of "scientific" literature that already proclaimed the genetic inferiority of black people.

Continuing this attack was no less a luminary than Sir Francis Galton. In support of white supremacy, Galton believed that judicious mating would give the more suitable races a better chance of dominating or prevailing over the less suitable races. Predictably, for Galton the least suitable of the less suitable was the African. In fact, in his book *Narrative of an Explorer in Tropical South Africa* (1889), he unashamedly asserts regarding African people, specifically the Damara people, that "these savages court slavery . . . you engage them as servants, and you find that he considers himself your property, so you become [reluctantly, he implies] the owner of a slave." Africans, he concludes, have no independence about them and follow their masters like compliant puppies. This assault was further aided by the work of G. Stanley Hall. As the first president of the American Psychological Association (APA), Hall theorized that Africans, Indians, and Chinese were members of adolescent races and in a stage of incomplete growth. Hall's thinking obviously influenced the development of professional psychology in America. His notion of adolescent races legitimized and gave rationale for Western psychology's role and responsibility to intervene and save the adolescent races from the liabilities and dangers of freedom. Dr. Eugen Fischer added to this devious and mean-spirited atmosphere of Black inferiority and mental deficiency. Fischer, the Director of the Kaiser Wilhelm Institute of Anthropology, Human Heredity and Eugenics, used psychology to justify the annihilation of Negro [sic] children.[13] Fischer claimed that Africans were devoid of value and useless for any employment other than manual labor and servitude.

In the historiography of America, Germany has been sacrificed on the altar of righteousness for its bigotry. However, what are less publicized are white American doctors' admiration, approval, and agreement with Germany's 1933 sterilization laws, calling for the sterilization of Jews and colored German children. In fact, from 1902 up until 1931 (two years before the enactment of Germany's sterilization program), over 15,000 sterilizations were performed in the United States. The director of the United States Birth Control League, Dr. Lorthrop Stoddard, applauded Germany's efforts at cleaning up its "race problem" and noted that Germany's sterilization laws weeded out the worst strains (Jews and coloreds) in the German stock. Stoddard's thinking had influenced and most assuredly continues to influence the idea and practice of population control in America. Evidence of this thinking is found in the establishment of Planned Parenthood of America. Planned Parenthood's founder, Margaret Sanger, in fact, had a plan (circa 1939) to exterminate the Negro population by convincing Black preachers to support birth control.

Paul Popenoe and Roswell Jenkins, in their 1918 book, *Applied Eugenics*, propagated the idea that black intelligence was determined by the amount of white blood present in the person.

Dr. Popenoe asserted that the lighter the Black person, the higher his IQ, and the darker the Black person, the lower his IQ. Popenoe concluded that the Negroes' low mental estate was irremediable, and since the Black people were eugenically inferior to white people, all of the Negroes must take into account the fundamental fact of their inferiority. The barbaric treatment (rape, murder, castration, brutality) of African people is chronicled throughout psychiatry. Not so long ago, around the time of the 1954 Supreme Court decision outlawing segregation, Black prisoners in New Orleans were routinely used for psychological experiments that involved surgically implanting electrodes into their brains. In 2002, *Psychiatry's Betrayal*, published by the Citizen's Commission on Human Rights, noted the total disregard and disrespect for African life in a comment made by the directors of these experiments, Dr. Robert Heath from Tulane University and Dr. Harry Bailey from Australia, who boasted twenty years after the experiments that they used black people because it was "cheaper to use Niggers than cats, because they were every-where and cheap experimental animals."

Dr. Heath had also been funded by the CIA to carry out drug experiments and tested LSD and bulbocapnine on African American prisoners at Louisiana State Prison. The National Institute of Mental Health Addiction Research Center in Kentucky for a ten-year period administered LSD and the experimental drug BZ (100 times more powerful than LSD) to African American men. The founding director of the National Institute of Mental Health (NIMH), Dr. Robert Felix, was also involved in the CIA-funded LSD experiments on African Americans in Kentucky. Is it not strange that in the Community Mental Health Centers funded by NIMH, 45 to 55 percent of the people using these centers received powerful mind-altering (psychotropic) drugs and/or tranquilizers? It is now admitted that these same psychiatric drugs create violent and aggressive behaviors. Statistics also show that after admission to a CMHC, patient arrests for criminal activities were twice as high as those from the general population.

Given this history, it is not surprising that in the wake of urban rebellions during the Civil Rights movement, Dr. Ernst Rodin, head of the Neurology Department at Lafayette Clinic in Detroit, would suggest that medical technology should be applied to solve the problems of riots in Black ghettos. Citing psychiatry's premise that children of limited intelligence often turn to violence if treated as equals, Rodin equated young dumb males who riot to oxen and noted that the castrated ox will pull his plow and human eunuchs are not given to physical violence.[14] Psychiatry had simply moved from whipping the devil out of them to psychosurgery and castration.

Finally, in 1983 the Duke University Medical Center in North Carolina was given almost three-quarters of a million dollars to specifically research the aggressive behavior patterns of African American children living in high-crime-risk areas. Less than ten years ago, Dr. Frederick Goodwin, a psychiatrist and director of the Institute overseeing NIMH, continued the assault and without any hesitation compared African American youth living in inner cities with hyperaggressive and hypersexual monkeys in a jungle. The psychiatric oppression of African American people is unparalleled. The same NIMH that supported drug and psychosurgical experiments on African American people in the 1950s and 1960s attempted in the 1990s to launch a National Violence Initiative, which would have used psychiatrists to determine through biological markers which African American and Hispanic children were likely to develop criminal behavior. And the beat goes on.[15]

The stench of psychiatry's and psychology's racist underpinnings continues with the publishing and popularization of racist attacks in the guise of scientific discourse, like Murray and Herrnstein's *The Bell Curve* (1994), in which they simply echo the historical lie that African Americans are intellectually inferior to whites, genetically disabled, and unable to cope with the demands of contemporary American society. Psychology's and psychiatry's assault on African American people has been and continues to be unrelenting. The consequent rupture in the psyche and human spirit of African people has never been addressed or fully understood.

To fully appreciate the devastation and destruction experienced by Black people in America, one must start with a critical examination of slavery and white supremacy (racism) and move forward to explication of the contemporary systems of racial negation and nullification in America.

THE PSYCHOLOGICAL EFFECTS OF SLAVERY

In his book *Breaking the Chains of Psychological Slavery,* psychologist Dr. Na'im Akbar writes: *The slavery that feeds on the mind, invading the soul of man, destroying his loyalties to himself and establishing allegiance to forces which destroy him, is an even worse form of capture.*

While some descendants of enslaved Africans have been more susceptible to this mentality than others, overall most Black people in the United States were affected by this mental conditioning to some extent. This mentality, described herein, in which the descendants of enslaved Africans manifest self-destructive behavior, self-hatred of themselves and their race, an overall inferiority complex, and become ardent subscribers to the ideology of white supremacy, is by no means coincidental or simply a side effect from a legacy of chattel slavery. Centuries before Dr. Akbar described psychological slavery in his book *Breaking the Chains of Psychological Slavery,* white slave trainer Willie Lynch delineated a strategy for mind control of enslaved Africans. In 1712, the Barbados slave owner was invited to Virginia to teach his method of slave control to plantation owners. There on the banks of the James River he said:

> ... I have outlined a number of differences among the slaves; and I take these differences and make them bigger. I use fear, distrust, and envy for control purposes. Take this simple little list of differences, and think about them. On the top of my list is age but it is only there because it starts with A. The second is color or shade. Then there is intelligence, size, sex, size of plantation, status on plantation, attitude of owners, whether the slaves live in the valley, on a hill, east, west, north,

south, have fine or coarse hair or are tall or short. Now that you have a list of differences, I shall give you an outline of action. . . . you must pitch the old Black against the young Black. . . . you must use the dark skin slaves against the light skin slaves and the light skin slaves against the dark skin slaves. You must also have your white servants and overseers distrust all Blacks. But it is necessary that your slaves trust and depend on us. They must love, respect and trust only us.

Gentlemen, these kits are your keys to control. Use them. Have your wives and children use them. Never miss an opportunity. My plan is guaranteed, and the good thing is that if used intensely for one year, the slaves themselves will remain perpetually distrustful.

African American revolutionary Malcolm X aptly described the success of Willie Lynch's slave training kit in describing the lingering house Negro vs. field Negro phenomenon, which persists to the present day. Malcolm stated:

Back during slavery, when Black people like me talked to slaves, they didn't kill 'em, they sent some old house Negro along behind him to undo what he said. You have to read the history of slavery to understand this.

There were two kinds of Negroes. There was that old house Negro and the field Negro. And the house Negro always looked out for his master. When the field Negroes got too much out of line, he held them back in check. He put 'em back on the plantation.

The house Negro could afford to do that because he lived better than the field Negro. He ate better, he dressed better, and he lived in a better house. He lived right up next to his master—in the attic or the basement. He ate the same food his master ate and wore his same clothes. And he could talk just like his master—good diction. And he loved his master more than his master loved himself. That's why he didn't want his master hurt.

If the master got sick, he'd say, What's the matter boss, we sick? When the master's house caught on fire he'd try and put the fire out. He didn't want his master's house burned. He never wanted his master's property threatened. And he was more defensive of it than the master was. That was the house Negro.

But then you had some field Negroes, who lived in huts, had nothing to lose. They wore the worst kind of clothes. They ate the worst food. And they caught hell. They felt the sting of the lash. They hated the master. Oh yes, they did. If the master got sick, they prayed that the master died. If the master's house caught on fire, they'd pray for a strong wind to come along. This was the difference between the two. And today you still have house Negroes and field Negroes.[16]

Malcolm was equally adept at articulating the problem of self-hate ingrained in the African American psyche. On February 16, 1965, speaking to a packed church in Rochester, New York, he said:

They made us think that Africa was a land of jungles, a land of animals, a land of cannibals and savages. It was a hateful image. . . . They were so successful in projecting this negative image of Africa . . . we looked upon Africa as a hateful place. . . . Why? Because those who oppress us know you can't make a person hate the root without making them hate the tree. You can't hate your origin and not end up hating yourself . . . you can't make us hate Africa without making us hate ourselves. And they did this very skillfully. . . . We ended up with twenty-two million Black people here in America who hated everything about us that was African. We hated the African characteristics. . . . We hated our hair. We hated our nose, the shape of our nose, and the shape of our lips, the color of our skin. . . . Our color became a chain, a psychological chain. Our blood—African blood—became a psychological chain, a prison, because we were ashamed of it. . . . We felt trapped because our skin was Black. We felt trapped because we had African blood in our veins. . . . This is how you imprisoned us. . . . Because we hated our African blood we felt inadequate, we felt inferior, we felt helpless. But a change has come about. In us.[17]

This ingrained inferiority complex resulting from the legacy of slavery has manifested in a myriad of pathological behaviors familiar in the African American community. African American psychiatrist Dr. Alvin Poussaint, in his book *Why Blacks Kill Blacks*, remarks:

Why are we caught up in this destruction? What has society done to our minds? It is an ugly fact that the American cultural experience has taught us that crime and violence is a way to success and manhood. Crime data indicate that Americans value guns and other destructive weapons. The whole frontier cowboy mentality sanctions and teaches violence. Television and movie folklore reinforces the popular conception that problems can and must be solved by violence. It is a rare occasion when the good guys do not triumph over the bad guys either by maiming or killing them.

Americans respect violence and often will not respond to the just demands of its citizens unless they are accompanied by violence.... Consequently, some of us have come to feel that the quickest way to solve any problem, personal or social, is through an impulsive act of violence.[18]

One of the most devastating effects of slavery for African Americans was the "damaging" of the family. Many of the present family problems suffered by African Americans are rooted in slavery. During slavery, the Black man was valued as a stud and a workhorse. The Black woman's value to the slave master was as a breeder or sexual receptacle capable of having many healthy children. Healthy children that the white slave master could sell or work to death to make money for himself.

The white man in the culture of slavery forbade marriage, a sacred rite in almost all human cultures. William Goodell in 1853 described the futility of slave marriages as viewed by slaveholders:

The obligations of marriage are evidently inconsistent with the conditions of slavery, and cannot be performed *by a slave. (In marriage)* the husband promises to protect his wife and provide for her. The wife promises to be the helpmeet of her husband. They mutually promise to live with and cherish each other, until parted by death. But what can such promises by slaves mean? The legal relation of master and slave render them void! It forbids the slave to protect even himself. It clothes his master with authority to bid him to inflict deadly blows on the woman he has sworn to protect. It prohibits his possession of any property wherewith to sustain her.... It gives master unlimited control and full possession of her own person, and forbids her, on pain of death, to resist him, if he drags her to his bed! It severs the plighted pair at the will of their masters, occasionally or forever.[19]

Permission for Africans to marry had to be granted by the slave master, many of whom, as demonstrated in the above passage, simply did not see the point. When marriage did occur between enslaved Africans, knowledge of the traditional African sacred rite had been diminished and only retentions of the deeper meaning of the ceremony remained. For example, in many cases the deeper meaning of marriage was replaced by simply jumping over a broomstick. This act signified that the two individuals had entered into a new state (oneness or togetherness) of being. Jumping over the broom literally swept away their old separate state and swept them into a sacred union. However, while the sacred meaning of marriage was being modified and retained, the physical ability to protect the union was shattered by slavery. Among the chief culprits reducing the stability of African marriages was forced separation. Sale, inheritance of people, and the mobility of slave owners broke up many families.

Traditional rites-of-passage ceremonies, an integral part of many African cultures, wherein the roles of the man and husband, woman and wife were explained to young people, were unknown to African descendants born into slavery. Despite the best efforts of the slave masters to destroy the Black family, poignant accounts remain of the extreme measures some African people took to maintain family ties. This included repeated trips by some brave Africans on the Underground Railroad (with its life-and-death risk) to liberate family members still held in bondage. The most well known liberator of her people was Harriet Tubman, who returned to the South repeatedly to free her family members along with others.

Another legacy of white supremacy that has been extremely detrimental to the mental health of African Americans is the image of God, His host of angels, Jesus, Mary, and all biblical and religious characters as white people. During slavery the practice of traditional African religions was brutally suppressed. The white slave master's distorted version of Christianity, redesigned with the maintenance of the slave system in mind, was forcefully imposed on the enslaved African. This imagery, of the divine as a white man, is consistent with the delusional syndrome of white supremacy with which many whites are afflicted.

For African Americans, if one is to believe that human beings are created in God's image and the image of God resembles the people brutally oppressing your people, what is one to think? This imagery, of God and the heavenly host as white, has had an extremely adverse effect on the psyche of people of African descent, as it has served to reinforce the inferiority complex, which results from (amongst other things) an image of God that does not look like oneself.

POST-TRAUMATIC SLAVE SYNDROME (PTSS): AN EXPLANATORY THEORY

Key Concepts

PTSS theory states that African Americans sustained traumatic psychological and emotional injury as a direct result of slavery and continue to be injured by traumas caused by the larger society's policies of inequality, racism, and oppression. Another dimension of this type of injury occurred as a result of the destruction of the African culture (i.e., belief systems, customs, and values). Even prior to arriving in the Americas, African families were often torn apart and people were separated from their various groups or tribes. *Relationships with each other*, which was the foundation of their historical survival as well as their primary axiological construct or "value system," were threatened and destroyed. Due to their resilience, Africans survived the sudden violent disruption of family and home, as is demonstrated in the reconstruction (albeit fragmented) of their cultural values and customs. However, as a result of exceptionally harsh treatment at the hands of their captors, they sustained severe, immeasurable psychological and emotional as well as physical damage. For example, recent unearthing of a slave cemetery in Washington, D.C., revealed startling evidence that slaves suffered physical injuries that have been rarely seen in recent history. These injuries were the result of extreme exertion and physical deprivation.

Additionally, the institution of chattel slavery forced Africans to integrate the ethnic ideology of their captors into their own psyche, which led to what can be described as a "cultural dissonance," a feeling of disharmony and psychological conflict resulting in a loss of identity and self-esteem. Thus, multitudes of African Americans were forced to function within a system that was in conflict with their own traditional customs, values, and needs.

In light of the cumulative and continued stressors experienced by African American communities, viewing such stress in light of post-traumatic stress disorder (PTSD), as well as socially learned maladaptive behavior, is appropriate. The direct traumas associated with slavery occurred long ago. If one considers differential dose exposure (differences in the amount of stress individuals were exposed to) and differences in individual psychology, it is difficult to determine the degree of severity of PTSD in those who were affected. Although it is likely that PTSD was prevalent, there is currently no evidence that would suggest that those Africans who may have suffered from PTSD in the past received any formal treatment. The lack of any therapeutic intervention for millions of captive, displaced Africans who likely suffered from PTSD during or after the advent of slavery suggests that PTSD remained a perennial problem among African Americans. It is also overwhemingly clear that new traumas continued to plague African Americans long after slavery was officially ended.

PTSS Theory

PTSS theory proposes that while the remarkable resilience of the African allowed for the perpetuation of a modified and improvised version of African culture, varying levels of both clinically induced and socially acquired residual stress-related illness was passed along through the generations. Transmission occurred on several levels:

- the family level;
- the community level; and
- the societal level

The most common trauma involved a serious threat or harm to one's life or physical integrity; a serious threat or harm to one's children, spouse, or other close relatives and friends; sudden destruction of one's home or community; or seeing another person who has recently been, or is being, seriously injured or killed as the result of an accident or physical violence. In some cases the trauma may be learning about a serious threat to a close friend or relative, e.g., that one's child has been kidnapped, tortured, or killed. The stressor is usually experienced with intense fear, terror, and helplessness. The disorder is considered to be more severe and will last longer when the stressor is of human design.

All of the above-mentioned traumas were the common and perpetual experience of African slaves and their African-American descendants. A less severe form of the violence and abuse continued after slavery officially ended with peonage, Black Codes, Convict Leasing, lynchings, beatings, threats to life and property, the rise of the Klan, Jim Crow segregation, the death of Emmett Till, the race riots of the 1960s, the 1989 beating death of Mullageta Sera (an Ethiopian man) by white-supremacist skinheads, the proliferation of white supremacist groups, the near election of an ex-Klansman to governor for the state of Louisiana, the 1991 police beating of Rodney King, the 1999 dragging death of James Byrd in Jasper, Texas, by four white youths, the police shooting death of Amadou Diallo in 1999, and the 2002 police beating of sixteen-year-old Donovan Jackson-Chavis, a special-education, hearing-impaired youth. All these events remind African Americans that the trauma has never really ceased and that it is likely to continue if there is no intervention.

It is not plausible that after more than two centuries of relentless oppression and brutal violence, slavery simply ended, leaving no traces of its psychological impact upon the generations that followed. The evidence of the transmission of the psychological effects is represented in racial economic disparities, consumerism, higher rates of morbidity and mortality for many diseases and decreasing overall life expectancy. Unfortunately the consequences of human degradation, oppression, and injury cannot be neatly swept away.

These effects manifest like a recurring nightmare in the heart of America's prodigy and in her institutions.

Multigenerational Transmission of Trauma

The Family

The first mode of transmisson was through the family. According to Comer, 1980:

> Each society has a vital interest in the indoctrination of the infants who form its new recruits. It lives only through its members, and its culture is its heart, which it must keep pulsating. Without it, its members are rootless and lost . . . they must be so raised that the culture exists in them and they can transmit it to the next generation. It is a task that every society largely delegates, even though unwittingly, to an agent—the family.

The African American Community

The second mode of transmission is through the African American community. During slavery the community was a suppressed and marginalized group. Today the African American community is made up of individuals and families who collectively share different levels of anxiety and adaptive survival behaviors resulting from prior generations of African Americans who suffered from PTSD and received no treatment.

The community serves to reinforce both positive and negative behaviors through the socialization process. For example, in the 1940s, according to Comer (1980), families frequently destroyed any signs of aggression in their children, particularly the male children. It was an acceptable and expected practice in African American communities to beat unruly boys severely so that they would never make the mistake of standing their ground with a white person in authority. While this practice was clearly the result of the hostile and oppresive environment in which African American families lived, it resulted in an assault on the collective psyche of the group as a whole.

The Larger Society

The third method of transmission is through the larger society. The interaction with the larger society has occurred at all the levels of transmission, adding consistent and enduring trauma. This occurs through policies of continuing inequality, discrimination, and scarcity of resources, coupled with crass materialism and a mass communication system that allow everyone to see the stark disparity between the "haves" and the "have-nots." New policies such as reversals in affirmative action act as setbacks, reopening old wounds, and making it increasingly difficult for young black people to fulfill the societal expectation or individual and group aspiration.

A CLINICAL VIEW

Post-traumatic stress disorder (PTSD) has generated profound interest. Many psychological journals, articles, and books have been written with elaborate details of the symptoms, causes, and treatment of this disorder. Individuals and groups said to suffer from this disorder include victims of rape, war veterans, Holocaust survivors and their children, victims of incest, heart attack victims, natural disaster survivors, victims of severe accidents, etc. Absent from this list, however, are enslaved Africans and their descendants.

According to the *Diagnostic and Statistical Manual of Mental Disorders, Third Edition, Revised (DSM IV)*, PTSD is described as being a reaction to a distressing event that may have occurred months or years before.

THE HEALTH OF AFRICAN AMERICANS IN THE TWENTY-FIRST CENTURY

The persistence of the health deficit due to slavery documented in Table 1, is a racial comparison of health data through the early 1990s from volume I of *An American Health Dilemma*, reproduced on the following page. A cursory perusal of the table reveals an excess of morbidity and mortality in the life of African Americans in almost all categories of assessment of well-being and quality-of-life indicators. The following is a discussion of three of the several major factors impacting poor health outcomes.

1. Lack of Health Insurance Coverage

From 1998 to 1999, 21.2 percent of African Americans had no health insurance of any kind. This is approximately double the 11 percent rate of uninsured for the white, non-Hispanic population. A lack of health insurance is one of the major deterrents to accessing health care. Persons lacking in health insurance are more likely to delay in seeking medical care for health problems. Delays in seeking care reduce the likelihood of achieving an early diagnosis for impending health problems and contribute significantly to adverse health outcomes.[20]

TABLE I: THE PERSISTENT SLAVE HEALTH DEFICIT*

Health Parameters	*Black/White Comparison*
Infant Mortality Rate (IMR) (death in the first year of life per 1,000)	Black IMR of 17.7 per 1,000 live births is 2.2× the white rate
Longevity	Blacks live five to seven years less than whites
Death Rate (deaths per 100,000 persons)	Black death rate of 783 per 100,000 is 1.6× higher than the white rate
Death Rate ages 25–44 (per 100,000 persons)	Black rate of 374 per 100,000 is 2.5× the white rate
Selected leading causes of death (National Vital Statistics Systems)	Blacks lead in 14 of 16 categories
Prevalence Rates	
Diabetes	33% more common in Blacks
Hypertension	Blacks: 34% have hypertension Whites: 20% have hypertension
Heart Disease	1.5× more common in Blacks
Stroke Deaths	Black Rate 1.93× Higher than whites
Cancer Incidence	Cancer has increased in Blacks 27% Cancer has increased in whites 12%
Cancer Mortality Increase (since 1995)	Blacks, 50% increase Whites, 10% increase
HIV Infection	Black male rate 3× white rate Black female rate 9× white rate

*Racial Comparison of Health Data through the early 1990s, Byrd and Clayton, *An American Health Dilemma*, Volume I (New York: Routledge, 2000), pp. 29–33.

2. Environmental Racism

There are approximately 40,000 hazardous waste sites in the United States. Many of these are landfills containing hazardous chemicals, which contaminate the surrounding air, soil, and water. Of the twenty most hazardous chemicals found in landfills by the Environmental Protection Agency, over half are carcinogens. Other toxic substances detected at these sites include chemicals that can potentially cause birth defects; neurological, renal, and liver impairment; and disorders of other bodily systems. One class of chemicals found at hazardous waste sites is known as the endocrine disrupters. They are so called because they have the potential to interfere with the function of the endocrine (glandular) system and may play a role in the development of certain types of cancers, such as breast cancer. In the southeast, approximately 75 percent of all landfills are located in close proximity to African American communities. In addition, there are 3 million to 4 million children in the United States who live within a one-mile radius of at least one hazardous waste site.[21] African American children and adults are disproportionately affected by this problem. African Americans are known to have higher rates of certain types of cancer. Tragically, childhood cancers such as leukemia and neurological malignancies have increased in incidence in children in recent years. Environmental factors are implicated in playing a significant role in the development of a number of these pathological conditions. The weight of evidence indicates that living in close proximity to sites containing toxic and hazardous waste can have an adverse effect on health.

The incidence rate of asthma in the United States has increased dramatically in recent years. Asthma is 26 percent more prevalent in African American children and they are six times more likely to die from asthma than their white counterparts. Again the weight of evidence implicates environmental factors, such as emissions from oil refineries, industrial incinerators, and vehicles, as contributing to the increased rates of asthma attacks.

3. Racism in Health Care and the Lack of Adequate Numbers of African American Health Providers

Several studies have documented less-favorable treatment by some white physicians toward African Americans.[22] Studies indicate that racist attitudes on the part of white health care providers manifest as less-than-optimal health care for Blacks. This disparity in treatment results in less-than-thorough workups for conditions such as stroke and heart disease with subsequently missed diagnoses, which translate to missed opportunities to prevent a catastrophic event. At worst, for African American patients, this racist attitude can result in permanent health impairment or death. This factor, combined with cultural insensitivity on the part of white health care providers, contributes to higher rates of morbidity and mortality in the African American population.

The percentage of African American physicians, 3 percent, has changed very little since the turn of the century. The result has been inadequate numbers of African American physicians to care for its population. Many African American patients often do not have the option of seeing a physician of their own race because there are none available.

THE LEADING CAUSES OF DEATH IN THE AFRICAN AMERICAN COMMUNITY: HIV/AIDS

Every day, seven people contract HIV. Of those, three are African American, according to Dr. Beny Primm, Executive Director of the Addiction Research and Treatment Corporation. African Americans are less likely to know their HIV status, get treatment, and be prescribed and take combination drug therapies for the disease, Dr. Primm said during the Congressional Black Caucus (CBC) 1998 Spring Health Braintrust.

Earl Ofari Hutchinson in a June 11, 2001, column entitled *Blacks Must Confront AIDS: Fear and Denial*[23] notes that the face of AIDS has changed from gay white males to poor inner-city Blacks. There is a state of emergency regarding the AIDS crisis in the African American community. While representing approximately 12 percent of the total U.S. population, African Americans make up almost 37 percent of all AIDS cases. Almost two-thirds (62 percent) of all women reported with AIDS are African American. African American women are rapidly becoming the new face of AIDS. AIDS is now the number-one killer of African American men between the ages of twenty-five and forty-four and is the second leading cause of death among African American women the same age.

Of the 641,086 AIDS cases reported to the Centers for Disease Control and Prevention, Blacks and Hispanics accounted for 53 percent of the total. Substance abuses, along with unsafe sexual practices, are major risk factors for HIV/AIDS.[24]

Cardiovascular Disease

Heart disease, stroke, and hypertension are the major killers in the African American community. Risk factors for cardiovascular disease include smoking, obesity, genetic factors, stress, dietary factors, lack of exercise, and environmental factors. The majority of these deaths are preventable with proper treatment, medication, and lifestyle changes.

Cancer

Cancer is the second leading cause of death in the United States. African Americans experience the highest incidence and mortality rates from cancer of any population group in the United States.[25] The highest incidence of cancer occurs with cancer of the lung, breast, prostate, colon and rectum, esophagus, and oral cavity. National Cancer Institute Statistics from 1989 indicate a 20 to 40 percent higher mortality rate in blacks as compared to the general population.

Risk factors include cigarette smoking, alcoholism, poor diet/nutrition, occupational and environmental exposures, and low socioeconomic status.

Other Major Health Issues in the African American Community

1. Obesity
2. Drug and alcohol abuse
3. Diabetes
4. Infant mortality
5. Violence: suicide and homicide
6. Lead poisoning in children
7. Homelessness
8. Sickle cell anemia
9. Mental health
10. Care of the elderly and disabled
11. Stress management

THE USE OF REPARATIONS AND REMEDIAL ACTIONS TO ADDRESS THE HEALTH CRISIS IN THE AFRICAN AMERICAN COMMUNITY

The health problems of African Americans are deeply rooted in the history of slavery. Equity and the sense of fairness should compel one to recognize that minimal reparations should be equivalent to the replacement of the loss incurred during the three hundred years of chattel enslavement. Hence, one could be justified in calling for a minimum of three hundred years of free health care, housing, sustenance, and education for every African American in the United States. In specific regards to health, reparations to repair the damage resulting from the persistent slave health deficit should be directed toward the development of programs to provide health education, disease prevention, diagnosis, intervention, treatment, and follow-up for mental and physical disorders, as well as care for the elderly. In addition to using reparations funding for the training of many more health professionals, legions of lay persons (who should be residents of the communities that they serve) should be highly trained as community outreach workers to assess and follow the mental and physical health care needs of individual members of the communities under nursing supervision. The following programs are proposed with the express condition that they are both psychoculturally congruent with the highest values and traditions of African people and are sociopolitically empowering for the African American community:

1. Universal health coverage

Remedial actions must first and foremost provide for free health care for prevention, screening, diagnosis, and treatment of all physical and mental disorders as well as providing for holistic wellness and preventive care. Patients should be free to choose their own practitioner and modality of treatment; both alternative and mainstream medicine should be covered.

2. Massive and comprehensive AIDS screening and prevention programming

- Prevention programs should be oriented toward behavioral change (via cultural realignment, cognitive restructuring, and character development) and the establishment of health and healing lifestyle rituals.
- Outreach workers are needed to assist community residents in obtaining AIDS screening.
- Mobile screening units should be conveniently located throughout African American communities to give residents information in a confidential and expedient manner.
- All treatment should be easily accessible and completely covered by universal health insurance.
- Records should be kept on the percentage of community residents screened with a goal of 100 percent screening for the target population.

3. Extensive and comprehensive health education programs for prevention of cardiovascular disease

- Programs would include blood pressure screening stations located conveniently throughout African American neighborhoods, diet and exercise programs, weight loss counseling, and smoking cessation programs.

4. Cancer prevention programs

- Smoking cessation programs should be marketed to communities and made widely accessible.

- Community outreach workers should assist residents in keeping up with all cancer screening exams, including pap smears, mammograms, prostate exams, etc.
- Community review boards should be involved in the development of all environmental health policies in communities.
- Cancer prevention programs, support programs, and other health education and support measures should be widely available, including pertinent information on nutrition, lifestyle, and stress management.

5. Optimal psychological health programs

Many African Americans feel the need for public acknowledgment of the many harsh manifestations of racism that have impacted their lives, including discriminatory practices, racial profiling, unjust imprisonment, and mental and physical abuse. A collective catharsis is needed to help alleviate the emotional and psychological pain that many African Americans and their families have endured. A truth and reconciliation commission, similar to the one established in South Africa with the dismantling of apartheid, is much needed to bring what has been festering below the surface out into the light so that healing can take place. In addition, psychological counseling designed by African American mental health professionals such as the Association of Black Psychologists should be made available to African Americans free of charge. This is necessary in order for African Americans to address the psychological damage, which manifests as an inferiority complex, and other negative states of mental and emotional health, that is, to address the state for which the term has been coined: post-traumatic slave syndrome. Other programs and concepts to be put in place would include those listed below:

- To promote a positive self-image, cultural programs should be instituted to teach African and African American history emphasizing contributions made by people of African descent to both children and adults.
- Images of Jesus Christ shown as a white man should be strongly discouraged in Black churches.
- To address the difficulties faced in forming lasting family units, whole families as well as couples should be encouraged to attend classes on roles and responsibilities in traditional African family systems, and marital counseling would point out problems in relationships related to the legacy of slavery.

- Parenting classes should be relative to the *Maafan* forces that impact and negatively affect child development and interpersonal relationships.

6. Remedial education programs

To decrease the incidence of poverty, which is a risk factor for a variety of diseases, both adult education and after-school programs should be created offering testing and remedial and supplementary education for those desiring job training or wishing to close educational gaps. Engaging youngsters in constructive activities after school will develop their talents and keep them out of trouble. This will greatly decrease the incidence of the myriad of problems resulting from teen idleness, including teen pregnancy, substance abuse, school failure, and juvenile delinquency.

7. Programs for ex-inmates of penal institutions

In order to reduce the incidence of recidivism, violence, and crime, and develop ex-inmates into productive members of society, halfway houses, transitional homes, counseling and adult education, and job training programs are needed for those exiting from the prison system. These programs need to be psychoculturally congruent with the reality of being African inmates in an anti-African society.

8. Drug and alcohol rehabilitation

In many instances, African Americans have turned to drugs and alcohol as a result of their inability to cope with the impact of racism and the effects of the legacy of slavery.

Drug and alcohol abuse must be decriminalized and put into the medical model where affected persons, both users and those who sell drugs in the community, can undergo psychological counseling, job training, as well as drug and alcohol rehabilitation as needed.

Both inpatient and outpatient programs are needed on a massive scale.

9. Maternal–child health programs

- Community outreach workers should follow up newborns and their mothers for the infant's first year of life. This should include a home inspection for potential health hazards.

- Mothers should receive a stipend for staying home with their children the first two to three years of life.
- Afrocentric preschools should be free of charge and available to all preschool-age children.

10. Programs to address the shortage of Black health care professionals

Programs for recruiting, mentoring, and tutoring potential health professional students should be promoted starting in elementary school. These programs should include field trips and be fun for students.

11. Cultural competency programs should be made mandatory for all health professionals. They should:

- Increase sensitivity to cultural and racial differences.
- Be designed for participants to discern internalized racist tendencies so that they can be recognized and rectified.

12. Lifestyle-change programs

To help combat widespread obesity, neighborhood teams should implement programs to promote group exercise programs with neighbors, assist with shopping and food choices for those who need it, facilitate psychological counseling for those who have emotional issues such as sexual abuse, and teach meditation and other stress-management techniques.

13. Eliminating homelessness

Funding should be made available to provide low-income housing to families and individuals. Counseling should be available through these programs regarding job training, mental health issues, and family issues such as domestic violence. Referrals should be made as needed for drug and/or alcohol rehabilitation or for treatment for mental health disorders.

14. Elder care

African American elderly, who have experienced the worst racial discrimination (as they lived through the apartheid years of overt segregation), should be able to secure placement in a care facility of their choice should they become in need of care. African American assisted-living facilities, or nursing homes or other types of clean and pleasant environments, should be made available and accessible to them if family members cannot care for them. For those who wish to remain in their own homes, home assistance should be provided in consultation with the family and other social service agencies. If possible a family member of his or her choice should be hired to look after the elder. In these facilities, rules must be enforced that ensure that the elder will be treated with the utmost respect. They should be receiving these services, in addition to any prescribed medications, free of charge.

Overall, there is a need to support existing—and where needed, create—African American think tanks that are charged with the task of developing ideas and programs to deal with the four hundred years of racism and its impact on the mental and physical health of the African American community.

A community health index, which would include all indicators of quality of life, should be developed for communities and used to follow the progress of programs implemented. In addition, it would allow for comparisons of communities on a standardized basis, enabling one to learn from successful programs and duplicate them where needed.

CONCLUSION

Historically, the accumulation of profits from the capitalist economic system of the United States is deeply rooted in the exploitation of people of African ancestry. This process of exploitation began with the kidnap of African people and their transport to America during the three hundred years of the Transatlantic Slave Trade. The loss of life that occurred during the notorious Middle Passage amounted to nothing less than genocide—a *Maafa*, unparalleled in the history of mankind. White America was able to amass tremendous wealth over the centuries of cruel exploitation of the children of Africa. White Americans, the heirs to this bloody fortune, have continued to benefit from this unjust enrichment and continue to enjoy white-skinned privilege. Racism, the twin brother of American slavery, has been internalized, legitimized, and institutionalized by white America with the full sanction of the United States government. The white American twins of slavery and racism, with their concomitant violence, terror, and repression, have caused immeasurable pain, suffering, and death to African Americans. And while the United States government gives lip service to freedom and democracy, these racist practices stubbornly persist to this very day.

The health of African Americans has been and continues to be irreparably harmed. Premature and excess loss of life has resulted from a legacy of slavery and racial discrimination at all levels of society and has continued to go unchecked without the aggressive intervention of the full forces of the United States government to stop it. We charge genocide: genocide for which there is no statute of limitations. We can never be repaid for the atrocities committed against us. Apologies, however sincere, are not enough. At the very least a concerted effort to repair the damage must be undertaken. Indeed, as has been done repeatedly in the case of other groups of people that have suffered unjustly, and in keeping with human decency and international law, a massive effort to repair the mental and physical health of African American people must commence. A well-planned and comprehensive program to optimize the health of African American people is the minimum needed.

WADE W. NOBLES *is Professor of Black Studies at San Francisco State University and holds a Ph.D. in psychology from Stanford. He is considered one of the founding members of African Psychology and is a leading advocate of African-centered treatment for victims of racism.*

JEWEL CRAWFORD *is a physician who works at the Centers for Disease Control and specializes in the impact of racism on the health of African American children. Her M.D. is from Howard University.*

JOY DeGRUY LEARY *has developed a culturally based educational model for working with children and adults of color. She has consulted widely on her theory of posttraumatic slave syndrome, and currently teaches social work at Portland State University.*

HAKI R. MADHUBUTI

The United States' Debt Owed
to Black People*

Debt is owed to African people for centuries of unpaid
   forced labor, suffering,
     death in the tens of millions and the systematic
seasoning and
victimization of an entire race of people.

Debt is owed for the willful and brutal separation of
   African people from their
   land, mothers from children, husbands from wives
   and families, children
   from fathers and mothers and a whole people
from their African land, culture and
consciousness that defined them and gave them
substance and
memory.

Debt is owed for redefining all African people, women,
men and children as

*Excerpted from *Tough Notes: A Healing Call for Creating Exceptional Black Men*, 2002

slaves and sub-humans not deserving of salvation,
universal love,
kindness, human consideration, education and fair
compensation for all
forms of labor.

Debt is owed for the designation of African people as
property, three-fifth
humans whose only use is to slave from sun-up to
sun-down, for slave owners, therefore
reduced African people to the category of animals
accorded less care than that of pigs,
sheep, cattle and dogs.

Debt is owed for the inhuman treatment of Africans
forced to slave for the sole
benefit of Europeans, Americans and others for left
over food, clothes
and sleeping space and whose worth and value
was determined by their
production in building wealth for Europe, America
and their people.
Debt is owed for the brutal, ordered, encouraged and
unrelenting rape of
African women by white human (aka slave) traders
and owners, thereby producing a
nation of half-black, half-white children whose
color, status, history and consciousness
branded them forever as bastards, thereby creating
a color and class and cultural
consciousness that to this day continues to rip the
hearts out of African
people maintaining a vicious circle of self-hatred, self-destruction and
denial in them.

Debt is owed for centuries of ruthless, planned and
destructive looting and
wholesale theft of Africa's people, land and mineral
wealth for the sole
purpose of creating wealth for Europe, the United

States and their
people.

Debt is owed to Black people for centuries of merciless
treatment, mendacious
reordering of the historical record and torturous
psychic damage.

Debt is owed for systematically stealing the cultural
memory from African people, for
the denaturing and renaming them thereby creating
a people unaware of
themselves and whose history and person is now
synonymous with
slavery and slave.

The successful creation of the "Negro" people in the United States is the tortuous American tragedy. This white supremacist metaphor started in this land with the ethnic cleansing and genocide of American indigenous people, renamed Indians and/or Native Americans. Thereby, let it be stated forcefully and without doubt or hesitation that the United States was founded and developed on two holocausts, that of the indigenous people and that of the enslaved Africans. Now, today, in this new century and millennium it is documented, confirmed and agreed upon by all thinking and well meaning people that Debt is owed to Native Americans and Black people forced:

To laugh when there is nothing funny,
to smile when they are in pain,
to demean themselves on stage, in film, television and
videos,
to dance when their hearts hurt,
to accept delusion as truth,
to lie to their children in the face of contradiction,
to pray to a God that does not look like them,
to pay compounded interest on the wealth they
created,
to sell their souls for acceptance in a fairy tale,
to mortgage their spirits for another people's history,
to support white people's affirmative action with centuries
of Black labor and taxes,

to create America's music and be denied the fortunes
made for others,
to see clearly and act as if they are blind,
to act stupid in the eyes of a fearful rulership,
to say yes when they really mean No!
to go into battle to maim or kill other whites
and non-whites
for the benefit of whites.

What does America owe Native American Indians
and Black people?
What is the current worth of America? or
Count the stars in all of the galaxies and multiply in
dollars by 100 billion,
For a reflective start.

HAKI R. MADHUBUTI *is considered one of the greatest writers on Black national-*
ism of the twentieth century. He founded Third World Press, the oldest continuously
operating Black-owned publishing house in the United States. He is the author of
numerous books and poems, including Don't Cry, Scream *and* Black Men: Obso-
lete, Single, Dangerous?

MOLLY SECOURS

Riding the Reparations
Bandwagon

Hold my baby like you hold yours, said the woman.
—OROMO PROVERB

I admit it. I, a middle-aged white woman, have jumped on the reparations "bandwagon." When I was asked several years ago if I thought reparations to African descendants of the slave trade were in order, I stammered and shuffled my feet and said, "it depends." A safe response for a liberally minded white woman who had absolutely no idea what I was really being asked. Like so many other whites—and people of color—I thought reparations was only (and all) about money. I thought reparations meant someone tossing out an arbitrary number and someone else (the government) writing a check, and that was the end of it.

It is exactly this kind of misinformation that causes most white people who hear and/or read about the reparations movement to get nervous and defensive, and respond with one of an infinite number of objections. It seems the most common states of mind in which whites live when it comes to dealing with racial issues are denial and deflection. It's a truly symbiotic relationship. Denial prevents us from getting anywhere near the issue and deflection allows us to stay cloistered in denial.

The following section briefly illustrates some devices used by those who would rather not talk about race, reparations, and white privilege. They are effective conversation stoppers that successfully prevent meaningful dialogue from taking place. If you are a person of color, the list may be all-too-familiar territory. You have probably heard some of these so many times that it seems a waste of paper. The only benefit may be in validating and confirming any frustrations you've experienced when attempting to discuss racial disparities with a white person.

If you are classified as white, you will also recognize them and may react with embarrassment or anger. Either one can be productive. It means simply that your mind is open. Here is *The Big Five List of White Deflections*.

THE BIG FIVE

1. My family didn't own slaves.

This assertion usually prevents an intelligent discussion about reparations from ever getting off the ground. The result (and often the intent) keeps at arm's length any responsibility for current-day racial injustices. It comes in several forms but basically the same thing is being said: "I didn't do it!" What follows is usually a vigorous defense explaining how not all whites are perpetrators and should not be punished for being white. In other words: "I didn't ask to be white and I resent feeling like I'm supposed to apologize for things my ancestors did that have nothing to do with me." What is implied in this deflection is that whites who were *not* slaveholding plantation owners did not benefit from slavery.

Let's be honest. Just because my ancestors didn't own slaves does not mean that I don't benefit from the legacy of slavery. In fact, as a white woman I have enjoyed first-class privileges my entire life and I do not come from an economically privileged family. I have never forgotten the shame I felt as a teenager when my mother was forced to pay for groceries with food stamps after our family disintegrated because of divorce. Regardless of my economic status, I was (and am now) the recipient of numerous invisible advantages every day simply because I am white. I never knew what those advantages were or meant until I started working with those who have dealt with discrimination and oppression all of their lives.

I thought nothing of it when, as a young and perky twenty-four-year-old, I held a high-level position in an international computer software company. I was the youngest executive (and one of very few females) to hold a position of power in this particular company. I had no formal education at the time nor did I have an impressive resume to draw upon. A white male recognized my potential and gave me an opportunity. Although I am most appreciative for his faith in my talent (and his respect), I am no longer naive enough to believe that I got there solely on merit. I didn't. I am a card-carrying member of the

White American Fast Track to Success. Recently, when I shared this story with a well-educated and successful black woman in Washington, D.C., she quipped that the chances of an opportunity like mine happening to a black woman (with equal education or better) were about as good as winning the lottery twice—in the same state!

As a white person, understanding the insidious nature of racism has been an effort. I needed to educate myself, because racism had no bearing on my everyday life. I couldn't see racism because it simply didn't exist in my world. I had to open my mind and heart to understand that racial discrimination is a stain on the fabric of American culture that has never been wiped out. Indeed, it has always been a part of the social fabric of this country. I'm not referring to just the obvious kinds of racial injustices but to all the subtleties that prohibit people of color from walking through the world with the same ease and privilege that most whites enjoy.

I recently traveled to Europe for an international conference. Among the group with which I traveled, I was the only white person. I watched as my luggage zipped through Customs while my traveling companion was sent into a long line to undergo the scrutiny of a Customs agent. I witnessed several instances in which waitresses who spoke horribly broken English were abrupt and impatient with my associate but turned to offer me service with a smile—in perfectly good English. There was even a moment when I caught myself wondering if maybe she (my associate) was giving off some kind of "vibe" and these strangers were picking up on it.

And at that instant, it hit me. I was about to blame the victim because it was more comfortable than admitting that the nice smiley woman behind the counter—who greeted me so cordially—was exhibiting racist behaviors. That I was the beneficiary of such hospitality within ten seconds of her humiliation was perhaps not a shock, but certainly a rather sobering moment in my journey.

To most privileged people this situation is difficult to conceive of and may not even seem like much to endure. But try to imagine these encounters twenty-four hours a day, 365 days a year, and then imagine what kind of psychological, social, economic, and spiritual consequences might come as a result. When I am immersed in my "all-white world," I don't notice these things unless I commit to doing so. It is this commitment that gives me hope of overcoming my racist tendencies.

2. I'm not a racist.

For whites prone to guilt or who have an investment in being perceived as nonracist there are many forms in which this objection is reflected. Often when whites claim to be nonracist it is followed with an accounting of how many friends of theirs are people of color, how many black people they've dated, and how they've moved into neighborhoods in which blacks live.

Many whites claim that it is impossible for them to be racist because of their Christian faith. Others will cite the black people they admire: Oprah, Michael, Whoopi, Colin, Stevie, or Denzel, to prove their multicultural perspective. What is significant is that the famous people listed are recognizable by their *first* names (a holdover from the slaveholding era?) and that whites repeatedly refer to them as evidence of their antiracist tendencies. Never mind that these people are highly visible, wealthy individuals who have made it in a "white world" and do not pose a significant threat. If these black people can make it without a handout then everyone can! The implication is that these successful blacks—who are held up as exceptions—are also used to prove that not all African Americans suffer, nor do they deserve compensation.

3. Reparations is only going to divide us more.

If I only had a dollar for every time someone insisted that talking about reparations will only serve to "further divide the races" and therefore should be abandoned. I can think of nothing more divisive than what has already been done to African Americans and indigenous peoples in this country. Are white objectors honestly convinced that race relations are so good now that we don't want to jeopardize our progress?

This deflection is often infused with fear, fatigue, and anger. Many fear that digging up all this "stuff" will create more hard feelings. Some whites claim they are tired of being blamed for everything and are sick of being labeled the problem. Others seem weary and just want to know "why can't we all just get along?" Reinterpreted: If we just don't talk about it, the problems will go away.

4. Blacks receive preferential treatment. What about me?

This deflection often emerges from those who feel their own hardships place them outside the realm of more privileged whites or those who just cannot see their own advantage. Very often these individuals do not want to be tagged as racists and resent the notion that they have benefited from the system of white supremacy. When the subject of reparations emerges, these people will tell you how deprived they were growing up and how they worked for everything they own.

I once received an angry e-mail from an attorney responding to an article

I had written about white privilege. He railed at me about how he did not have advantage over anyone and went on to decry the evils of affirmative action and talked about people "pulling themselves up by their bootstraps." He strongly objected to any financial compensation that might be "doled out" to blacks in the form of reparations. I was intrigued by the details of his personal struggle until I noticed the Roman numeral III after his name. As it turns out this man is a third-generation attorney employed at his father's law firm (formerly his grandfather's), and yet he insisted that he earned his way in the world—unassisted.

It is a social paradox that this objection, which many whites have concerning blacks receiving preferential treatment, is never viewed *negatively*. As we read in the newspapers and see on the news, blacks and other people of color are *preferred* targets of the police and law enforcement officials. People of color are recipients of preferential treatment in our criminal justice system, as is evidenced by the fact that they receive harsher sentences for similar crimes committed by whites.

A cursory study of the criminal justice system supports these allegations of preferential treatment in black and white. When we look at the figures of who is in prison and we realize that there were few prisons in America during slavery, we must ask unthinkable questions. Why are American prisons one of the fastest-growing businesses in this country? Why are they being privatized? Has our prison system replaced slavery? Is it possible that black men and women are deemed less threatening to whites when they are held captive in controlled environments? Does the justice system reflect an unconscious conviction of white society, that prison is where people of color belong? Have we returned to the era of "convict leasing" as recently reported by Doug Blackmon in the *Wall Street Journal*, when African American prisoners were leased to private corporations? All of these complicated questions are relevant to the issue of reparations and deserve to be answered honestly and objectively.

5. Slavery is over. Let's move on and forget about the past!

It is often difficult for whites to connect past events with present realities when it comes to the social conditions of African Americans. They claim that they cannot understand the benefit of digging up the sins of our forefathers. They don't believe that an honest review of our ancestors' governmental policies and practices might enlighten us as to the source of current injustices. They simply do not accept that it could be productive to dredge up something that has been over (technically) for one hundred forty years.

This is probably one of the weakest deflections of all. Anyone involved in personal transformation will admit the value and necessity of addressing family-of-origin issues. That is, in order to move forward, we must review where we have been. This is one reason why a national study on reparations is critical to understanding our racial past.

WHAT IS REPARATIONS?

First of all, reparations is not a recent notion or something born out of the 1960s Civil Rights movement. Reparations advocates (black and white) have been around for well over one hundred years. Reparations is not a noun, it's a verb. It is a *process* of exploring the damages to the descendants of those who were kidnapped from Africa and enslaved throughout the world for two hundred fifty years. It involves dialoguing about potential compensatory remedies.

But reparations is not just about investigating the damages to the victims. It is about taking an agonizing look at corporations,[1] and state, local, and national governments that participated in the slave trade and who are prosperous today because of that participation. Who would argue that receiving an inheritance from a relative affects the kind of life one might have as a result of that inheritance (remember Mr. Lawyerman III)? Well, African Americans have inherited a legacy of injustices inflicted upon their ancestors and are living with them today in a multitude of forms. For those who are miseducated about American history, it is tempting to view a few civil rights laws and affirmative-action programs of the last thirty years as remedies for three hundred fifty years of hatred and mistreatment.

As Congressman John Conyers argued in February 2001 at the Race Relations Institute Conference on Reparations in Nashville, those who are promoting reparations on the governmental level are only requesting a *study* of the damages, which so far the government has refused to support. This does not mean that studies are not being done—just that the government refuses to finance them or acknowledge their existence.

It is illuminating that many Americans do not object to countless billions of dollars spent on strategic air missile defense systems that may or may not work and even more billions on studies that examine why they fail in protecting us from an enemy—who may or may not even exist. Yet many white Americans vehemently object to a study of reparations—something that could specifically identify the injustices heaped upon our African brothers and sisters and the lingering effects.

A study of reparations would be an honest analysis of our history, and something that would help promote racial healing and justice for blacks and whites alike. The damages being sought are not limited to money. And realistically, there is not a dollar figure large enough that could adequately compensate all those who have incurred the devastating effects of the slave trade.

What many reparations protesters overlook is that a national study does not mean that compensation should, can, or will only take the form of individual checks written to African Americans. It is also premature to say whether dollars would be allocated for social programs or to individuals or even in what form financial compensation might be made.

What Congressman Conyers and his supporters are advocating is a government-funded study that would assess the impact of enslavement on African Americans. That is all. It is a necessary step discussed earlier in this book by John Van Dyke and others who write about compensatory measures for crimes against humanity; reparations for the incarceration of the Japanese during World War II were initialized by a similar study, which resulted in $1.2 billion being given to the families of those impacted by incarceration. It is clear that those in Congress who oppose such a study might feel that it would open up the proverbial floodgates for government-sponsored reparations for African Americans.

REPARATIONS AND HEALING

Many in our culture have embraced Alcoholics Anonymous's (AA) twelve-step program in which substance abusers methodically address their destructive behaviors. The transformational success stories of those committed to the program are well documented and place AA at the top of the list of effective treatment programs for alcoholics. In this twelve-step approach, a number of key steps focus on confronting wrongful acts, making amends, and publicly committing to rehabilitation. These steps—or plans of action, if you will—are designed not only to address the behaviors of the addict but also the effects on those victimized by their addiction. Through the process, many addicts say that they discover the extent of their selfishness and the deleterious effects their destructive behaviors have had on others. Before recovery, most addicts claim to be so self-centered that they are often oblivious to everyone's needs but their own. Quite often addicts will tell you that the healing doesn't begin until the denial phase is over and they have progressed to Step Four: make a searching and fearless moral inventory of ourselves.

Step Four is the phase when most addicts/alcoholics admit all the wrongs committed against others and contract to make amends to those whom they have offended. After making a list of the wrongs, the person is charged with the responsibility of making atonement. Most often this isn't just about saying "sorry." It requires taking an action that will counteract the offenses, if possible. Usually it is a face-to-face apology to the offended—unless, of course, they are dead or live far away. The next step is a commitment to insuring that past offenses are not repeated.

In a sense that is what the reparations movement is: a process of recovery for those in denial about our addiction and dependency on white privilege. Whites have inherited this affliction and are clearly steeped in denial about the devastation of slavery and its legacy. For those drunk with power and privilege, addressing the "inner white supremacist" can be quite sobering and fraught with anger and frustration. But like AA members say, it's "one day at a time—one step at a time."

I am not suggesting that by tackling Step Four and saying "I'm sorry," deeply imbedded problems in our racist society will be absolved. That would minimize the magnitude of suffering and the damage that continues to be inflicted upon people of color. However, we must look to models for transformation and healing that already exist and then expand them. Legislation prohibiting discrimination cannot resolve the problem because the laws don't address the deeper issues of white legacy, denial, social systems, and institutional racism. The *system* of white supremacy is the problem. And white dependency on a system that provides advantage and privilege is what continues to perpetuate the policies and practices that help keep the addiction alive. Clearly, we are still in the denial phase; we remain in Step Four.

A reparations study presents an enormous opportunity for healing. In discussion workshops and racial dialogues I've participated in and facilitated, it is clear that the process of discovery is the only road to racial healing for the United States. Learning about and acknowledging another's oppression does not take anything away from me. If anything, I benefit enormously from revisiting history and making connections between the history and contemporary social ills.

I recently wrote an opinion piece on the need for reparations that was published in a local newspaper and an Internet magazine. As the article hurled through cyberspace, I was inundated with e-mails from both blacks and whites from around the world within hours of its publication. One black man wrote, "Thank you for taking the time and energy to publish your refreshing thoughts and feelings so that I, and others, might find brief respite in our bitter quest for justice." He said my words helped because he has to continually remind himself "all light-complexioned people are not bad."

I was struck by his honesty and moved by his struggle for peace. I was also saddened that for many people of color, a white person who shows compassion and speaks candidly about race is a rarity. In this instance, there was no money exchanged, no bloodshed—just words of acknowledgment, which for him seemed to offer consolation, hope, and maybe even a little peace.

Hearing the truth about racism is difficult at best, and violence provoking at worst. We white Americans have been digesting fictional sound bites about race for so long that we can't recognize truth when it is presented to us. Most of us want so much to believe that everything is okay with America that we deny African Americans the right to an accurate telling of history. And every time we deny our true history, we obscure the connection between the slave trade and the residual effects that are so prevalent today. Again, steeped in denial, we prevent ourselves from getting anywhere near solutions because we don't want to acknowledge the depth of the problem.

Our social compromise has been to allocate African Americans one month a year in which to be included in the educational system. Every February we smugly promote an endless number of programs focusing on the struggles, victories, and contributions of African Americans. By no means am I suggesting that the contribution of blacks should not be lifted up. What I am saying is that African American history is not a "separate" history. Black history is *all* of our history and should be included in every educational institution in America—all year round. The separation occurs between blacks and whites because the truth of our history is wrought with conflict, contradiction, and oppression. Yes, it is uncomfortable, but it is our history.

An official study of reparations could be the greatest source of healing for everyone involved simply because it's not just about blame. It is about accountability and truth and responsibility. A reparations study will allow whites and blacks to explore the lingering effects of American slavery together, thereby allowing us to ameliorate the crux of the injury. Instead of applying more Band-Aids, which only inhibit healing, we expose the wounds to the benefits of the fresh air of truth.

After hearing some preliminary figures of how much the U.S. government profited from taxing the slave trade, it is laughable to suggest that there is not enough money in this country to compensate the victims of racism. Personally, I believe that a check of some kind must be written. But again, it is only *part* of the solution.

Angry when we hear the words "reparations" or "compensatory remedies," we need to understand where the anger originates. It's not the words that cause such a violent reaction in us; it is something much deeper. Our own unwanted inheritance of fear, hatred, and anger has done a number on whites. It has made us arrogant. Fortunately for many, arrogance is only a by-product of ignorance. And that is something that can be remedied if we can stop blaming the victims long enough to search out the unpopular truth. It is also arrogance that prompts whites to dismiss (out of hand) the idea of reparations. We have been in the driver's seat for so long that we can't abide the notion that someone else might have another route worth exploring. As one of my African American friends once said to me, "y'all always have to be the experts." I'm afraid she's right. White people tend to think we always have the solutions—even when we haven't honestly explored all aspects of the situation. Perhaps, like Alice Walker said, whites need to keep silent for a hundred years and simply listen.

THE LANGUAGE OF RACISM AND WHITE PRIVILEGE

On an average day I don't think about being classified as white. I don't have to. I walk around oblivious of my race and ethnicity unless someone else brings it to my attention, which rarely happens. I'm white. Most of my friends are white. Most of the institutions I frequent are white. Most of my coworkers and all my superiors at my place of employment are white. I've actually experimented with being mindful of my race and observing how it colors my world for an entire day. At some point in the day—often very early—I inevitably just forget and vow to try again the next day.

When I ask an African descendant what it means to be black in white America, there is usually little hesitation—given that my intentions are trusted to be sincere. Sometimes an uncomfortable laugh is followed by a low sigh, like someone attempting to exhale a couple hundred years of pain—but there is never a shortage of insightful answers. When I first decided to write about whiteness and privilege, the page sat blank for the longest time. Even though I've spent the last several years focusing much time and energy exploring racism and privilege, the question of what it means to be white remained elusive. Addressing my own white privilege turned out to be pretty stressful too. I soon discovered it was only stressful when I was thinking about it. If it got too unnerving, I could just set it aside until I felt like dealing with it again. This was my first lesson in white privilege. I could *choose* whether to deal with it or not. People of color have no such choices since they are constantly reminded of what Derrick Bell calls their faces being at the bottom of the well. The advantages of being classified white are simply infinite.

How many times my friends and I have entertained each other with our *Traffic Ticket Evasion* stories. One incident is particularly memorable where the message of white privilege is clear. About twenty years ago, I was driving home after a Thanksgiving party at midnight and was stopped after turning left at a red light without using my blinker or my headlights. I was also speeding, slightly inebriated, and my tags were expired. For my multiple offenses and extremely bad judgment, I received a wagging finger and a warning by the policeman "to be more careful." Instead of arresting me, he insisted on *escorting* me home, which was two blocks away. After watching me turn safely into my driveway, he waved good-bye and wished me Happy Thanksgiving!

In retrospect, although I appreciate his compassion for a stupid adolescent, I am clear that my skin color was my good fortune rather than my youthful charm or dumb luck. Many whites I know have similar stories of close encounters with the law—and most have similar endings.

Unfortunately, that is not the case with most African Americans, Latinos, and other friends and associates of color. Whenever I've shared this story with my black friends they just shake their heads and look at me as though we inhabit separate planets. I think this might explain the sinking feeling I get whenever I observe a person of color being pulled over to the side of the road. I know in my gut that the chances of their being extended the same courtesies and treatment I've experienced are slim to none. If whites and blacks could have open, productive, and truthful dialogues about such privilege, the discussion of reparations and compensatory measures would follow naturally. I've been told by more than one person of color that just hearing a white person admit their privilege helps reaffirm that they are not crazy. Whites rarely recognize our resistance to discussing racial issues. We are rarely challenged to do this. We are so unaware of our privileges that it seems ludicrous when someone suggests that we have advantages because of our racial classification. Instead we repeat reactionary rhetoric about civil rights laws, affirmative action, and the free ride African Americans are getting through social programs. It is our best and our worst defense.

RACIST LANGUAGE

Finally, it is so easy to suggest that perhaps we are being too sensitive when it comes to our language. Whites hate it when we are told that our language is racist. We counter that political correctness is oppressing our right to free speech. We want the freedom to express whatever is on our minds—even if it's offensive and rife with racist "undertones."

Recently, I appeared on a television talk show to discuss racism. The other guest was the founding member of a prominent rap group and a well-known magazine journalist. He is from New York and after the television taping, our white male host approached him and jovially asked when he was heading back to the "jungle"? After a moment, my friend composed himself and said, "Oh, you are referring to New York." Even if one is not thinking in explicit racial terms at the time, this kind of *unconscious* racist language has the effect of reinforcing racially loaded imagery of crime and danger that our culture identifies with inner-city blacks. *Thinking racially* is endemic to being white in America. Even the widely used term "Caucasian" alludes to the racial supremacy of whites. It was anthropologist Johann Bluemenbach two hundred years ago who anointed the word "Caucasian" to describe "the most advanced and superior race among mankind." I have asked both whites and blacks what it means and the reply is usually that it means white people. Both groups are usually oblivious of its origins in Europe's racist past.

By *not* dealing with racist language, we unwittingly perpetuate ideologies and behaviors usually ascribed to those whom we identify as obvious racists. Many whites, if asked how they feel about racism, will climb an invisible soapbox and preach about the evils of discrimination by condemning sheet-wearing Klansmen and other white supremacist groups without hesitation. *What most whites fail to recognize is how our own behaviors and language help uphold the system of white supremacy on a daily basis.*

The language used in many history texts obscures and abandons certain truths that leave us wallowing in rather shallow historical waters. Because of this, it is difficult for whites to gain perspective on what white advantage means and to understand how potent the language of race is when recounting history. We cry blasphemy when confronted with controversial truths about our colonial ancestors. It is an awakening to learn that Columbus *didn't discover* America and read about his bloody relationship with the Arawaks of the Caribbean. Their loving greeting to Columbus on the islands where they had lived for thousands of years was met with rape, exploitation, murder, and ultimately extermination by the disoriented Europeans who saw them as primitive Indians ripe for exploitation.

So what is the harm? you may ask. It is this: By perpetuating the myth that Columbus *discovered* America, we teach our children to overlook those who have been marginalized, disenfranchised, and often murdered. We teach them to honor the conquerors and overlook the victims, just as it may sting to admit that Thomas Jefferson—the man who symbolizes freedom and liberty for all—actually owned well over one hundred slaves and freed only a handful at the time of his death—his own children. Our historical language betrays the truth that Thomas Jefferson had sex with a slave girl named Sally Hemings and had several children by her. Although the sanitized version of the story that we finally admit to depicts a very loving relationship between slave master and slave, the disquieting truth is this: A man of enormous power and wealth had sex with a young girl who was fifteen years old and his legal property. In most civilized societies, this is called rape. For us to call it anything else is disingenuous and dangerous. For an enslaved black woman who was not even considered to be a complete human, there could be no such thing as consensual sex with a white master. The great imbalance of power that dominated that relationship prohibits any reasonable thinking person to suggest otherwise. The familiar stories of Christopher Columbus and Thomas Jefferson are but two examples of how white supremacist language perpetuates lies and mythologies about our nation's history and further promotes a distorted view of relationships between the powerful and the powerless.

African Americans often have but one story to tell about their ancestors' journey to America. And before you rush to the defense of old cousin Harry who came during the potato famine, I recognize that some Europeans did endure hardships in coming to America. Certainly there were Europeans who were alleged to have escaped religious persecution, debt, or famine, and some were even indentured servants. But whatever the reason or circumstances, most Europeans arrived *voluntarily*. They were not kidnapped, separated from their land and families, stripped of their identities, culture, and language and then chained to the bottom of a ship and sold into slavery. That legacy is reserved for African Americans.

There are no shortcuts or easy solutions when it comes to truth and healing. Dialoguing about the past is essential if people of color and whites are ever going to recover from the devastating effects of the slave trade. From the past we will learn how not to repeat the sins of our fathers. Reparations could provide a framework for that dialogue and from that discussion would come the compensatory remedies and the healing. What will that look like? That is what a reparations study is for.

The reparations bandwagon is really a movement to reestablish ourselves racially in a world that has not allowed us to embrace difficult truths. No one knows where the movement will take us but it must be a journey taken by both black and white to a place where we all need to be.

MOLLY SECOURS *is a freelance journalist and film producer who documents the effects of racism on a global level. Based in Nashville, she has written extensively on the issue of white privilege and its consequences for both victim and perpetrator. She was a delegate to the 2001 United Nations World Conference Against Racism in South Africa.*

YAA ASANTEWA NZINGHA

Reparations + Education = The Pass to Freedom

> *White folks never teach us to read nor write. They were*
> *afraid the slaves would write their own pass and go over*
> *to a free country.*
> —VICTORIA ADAMS, SOUTH CAROLINA SLAVE

MY OWN STORY

As an educator for over twenty-five years in the public school system of the United States, I feel Victoria Adams's chilling statement about white folks' fear of black folks writing their own pass to freedom remains true. It is essential for the ruling elite to take the mind of the African, and keep the body. In order to live peacefully with the very people you enslaved, lynched, beat, burned, castrated, stripped of their religion, language, sense of family, culture, identity, et cetera, you must find a way to control their thoughts. If African people in America understand who they are, where they come from, and what was taken away from them, they would have no choice but to revolt. Imagine 40 million black folks armed with the truth still living in the land of the very people who enslaved them. America, therefore, must continue to *miseducate* the African and keep him docile. As Dr. John Henrik Clarke states in his book *Notes for an Afrikan World Revolution,* "Powerful people never educate powerless people in the strategy of taking their power away from them."[1]

Teaching a slave to read and write was forbidden, under heavy penalties. Lungford Lane describes his experience as a slave child who grew up on a plantation near Raleigh, North Carolina:

My father was a slave to a near neighbor. The apartment where I was born and where I spent my childhood and youth was called the kitchen, situated some fifteen or twenty rods from the great house. Here the house servants lodged and lived, and here the meals were prepared for the people in the mansion. My infancy was spent upon the floor, in a rough cradle, or sometimes in my mother's arms. During my early boyhood, playing with the other boys and girls, colored and white, in the yard, and occasionally doing such little matters of labor as one of so young years could, I knew no difference between myself and the white children; nor did they seem to know any in turn. When I began to work, I discovered the difference between myself and my master's white children. They began to order me about, and were told to do so by my master and mistress. I found too that they had learned to read, while I was not permitted to have a book in my hand. To be in possession of anything written or printed, was regarded as an offense.[2]

Education was not only restricted in the South, but also in the North.

In 1832 Miss Prudence Crandall undertook to open a private boarding-school for young colored girls, [in] Canterbury, Connecticut. The enterprise was denounced in advance, by the people of this place, in a public meeting. When the term opened, with fifteen or twenty girls from Philadelphia, Boston, New York, and Providence, storekeepers, butchers, milkmen, and farmers, with one consent, refused to sell provisions to the school, and supplies had to be brought from expensive distances. The scholars were insulted in the street; the doors-steps and doors were besmeared with filth, and the well filled with the same; the village doctor refused to visit the sick pupils, and the trustees of the church forbade them to set foot in their building. The house was assaulted by a mob with clubs and iron bars, they broke the glass of the windows and terrified the inmates. Finally, the State Legislature passed an act making this school an illegal enterprise, and under this act Miss Crandall was imprisoned in the county jail.[3]

Even in the twenty-first century, some teachers are still victimized when they attempt to use an African-centered curriculum for the children they teach. Similar to conditions under slavery, many teachers find themselves teaching behind closed doors, whispering, gesturing, looking over their shoulders, hovering in corners, checking the hallways, and constantly worrying about their job security when it comes to teaching a curriculum that stimulates, motivates, and awakens the minds of African children. Teachers who make a conscious choice to openly instill high self-esteem and academic greatness in children of African descent are dealt with severely. African educators, on all levels, who are African-centered in their approach are in jeopardy of losing their jobs or ridiculed for teaching the truth about the state of Africans. Laurence C. Jones, founder of Piney Woods School in Mississippi in the early 1900s, was constantly threatened by local white supremacists who attempted to thwart his instructing black children to read. At the university level, Dr. Leonard Jeffries of City University of New York in the early 1990s was also excoriated for statements concerning the role of ancient Africans in science, religion, and the arts.

During my thirty years as an educator I have experienced isolation, hostility, harassment, demotions, suspensions, and termination because of the role I played in successfully educating children of African descent. My most devastating blow came when I was terminated from my teaching position at Junior High School 113, in Brooklyn, New York, for teaching children of African descent to refer to themselves as Africans rather than Americans. While covering my theater class, a Jewish teacher became troubled when he asked a thirteen-year-old black male student to do a skit where he would play the role of a thief to impress young black females. The young man responded to the teacher by saying, "My drama teacher, Miss Nzingha, teaches us, that as Africans we shouldn't fall into the low standards America has set for us, but should have the high standards of African people. Therefore, I do not feel I should have to act out the part of a thief." The teacher felt telling children to call themselves *Africans* not only lowered their self-esteem, but also taught them hopelessness. He took his concerns to the principal, an African female, who sided with him, and after years of praising my curriculum and teaching style, ordered me to discontinue my teaching methods and no longer include anything in my curriculum that dealt with issues of rejection, racism, peer conflict, personal conflict, or the atrocities of slavery. She took the role of an *overseer* and restricted the educational process of the children I taught. The white male teacher's fear was the same as the slave masters'. He saw the young black male writing his own pass to freedom. My incident triggered a host of attacks on other teachers in the same school who attempted to teach from the African perspective or support my position. Lauristine Gomes, an outstanding language-arts teacher (who also taught from an African perspective) and the union representative at Junior High School 113 in Brooklyn, New York, joined my struggle and strongly voiced her objections. She was dealt with severely by the administration. Ms. Gomes was banned from the school and also terminated.

A NATIONAL TREND

My case is by no means an isolated one. In Brooklyn two educators did a joint project on the effects of slavery on black people. Students were given quotes such as "I am an African," and "If it were not for chattel slavery I would still be home in Africa" to research, write and report on. The quotes, along with other statements, were mounted on red, black, and green paper and put on walls outside the classroom. The teachers were ordered to take the quotes down and discontinue the lesson. The teachers refused. The vice principal of the school ripped the quotes from the wall, leaving an appalling display of torn and tattered red, black, and green paper. The teachers were immediately given letters of reprimand. I have received e-mails and letters from all over the country describing similar situations. A teacher from Atlanta wrote, "I lost my job at Morehouse [College] because I demanded that students develop into African Men. I think I understand your struggle. It may do your spirit good to know that our struggle has led to a creation of a full-time African-centered middle school that is home-based now two years running."

DELETERIOUS EFFECTS OF ENSLAVEMENT IN EDUCATING AFRICAN PEOPLE

Such harassment has a devastating and psychological effect on the teachers, students, and parents forced to witness this modern-day lynching. It symbolizes the breaking process described in the document *Let's Make a Slave* in the Black Arcade Liberation Library. The document describes the social process of man breaking and slave making. It describes the rationale and results of the Anglo-Saxon's ideas and methods of insuring the master-slave relationship.

> When it comes to breaking the uncivilized nigger, use the same process (as for a horse), but vary the degree and set up the pressure so as to do a complete reversal of the mind. Take the meanest and most restless nigger, strip him of his clothes in front of the remaining male niggers, the female, and the nigger infant, tar and feather him, tie each leg to a different horse in opposite directions, set him a fire and beat both horses to pull him apart in front of the remaining niggers. The next step is to take a bullwhip and beat the remaining nigger male to the point of death in front of the female and the infant. Don't kill him, but put the fear of God in him, for he can be useful for future breeding.[4]

Great damage has been done to African people in America through the denial of an education that fosters cultural consciousness and academic excellence. My experience as an educator has taught me to attempt to teach children of African descent the basics; reading, writing, and arithmetic are useless if the child's mind is filled up with images that lower his/her self-esteem and promote an inferiority complex. Many children of African descent believe they are the undisciplined, unintelligent, ignorant lowlifes that the racist media drunk with stereotyping force upon them. This propaganda has been imbedded in their minds such that when a child of African descent speaks and acts in a proper manner, they are teased by their peers and accused of acting white. On several occasions I have asked students why they feel proper behavior is representative of Europeans. The responses vary. Some will say, because black people are loud and disrespectful. Others say they curse, even though they themselves don't behave in that manner. I then challenge them by saying, "On my many trips to Africa, I have visited African schools with African children and have never witnessed loudness, curses, or disrespect. So, obviously this behavior you are referring to is certainly not the behavior of an African child. Therefore, if we did not bring the behavior with us from Africa, where was this behavior learned?" After some discussion, it became obvious to the students that this is not *African* behavior that they are claiming, but *European* behavior. Simply watch Europeans at a football, baseball, or basketball game and you will see the loudness, the curses, and the disrespect that has been falsely instilled in the African child as their behavior. As Marimba Ani states in her book *Yurugu*:

> Now that slavery, as an institution, had ended, the attempt to dehumanize Africans on the part of the European would have to be continued using other methods. It was important to the system of white supremacy that (1) white people continually reinforce their European consciousness at the expense of the African image, i.e., through our degradation, and (2) that the Africans continued to act like slaves of a new sort and indeed become what Europeans portrayed them to be. It was during this period that a Euro-American-controlled media began its long career as one of the most effective weapons used to ensure the exploitation and dependency of people of African descent.[5]

MIND CONTROL AND EDUCATING AFRICAN CHILDREN

In June of 1998, a document was presented to the U.S. Supreme Court on coercive persuasion in the case of *Wollersheim v. Church of Scientology.* The

Wollersheim case was being reviewed by the court on issues involving abuse in this area. In the document, coercion was defined as "to restrain or contain by force." The following seven tactics were described as being used in a coercive persuasion program:

Tactic 1. The individual is prepared for thought reform through increased suggestibility and/or softening up, specifically through hypnotic or other suggestibility-increasing techniques, such as extended audio, visual, verbal, or tactile fixation drills.

Tactic 2. Using rewards and punishments, efforts are made to establish considerable control over a person's social environment, time, and sources of social support.

Tactic 3. Disconfirming information and nonsupporting opinions are prohibited in group communication.

Tactic 4. Frequent and intense attempts are made to cause a person to reevaluate the most central aspects of his or her experience of self and prior conduct in negative ways. Efforts are designed to destabilize and undermine the subject's basic consciousness, reality awareness, worldview, emotional control, and defense mechanisms as well as getting them to reinterpret their life's history, and adopt a new version of causality.

Tactic 5. Intense and frequent attempts are made to undermine a person's confidence in himself and his judgment, creating a sense of powerlessness.

Tactic 6. Nonphysical punishments are used, such as intense humiliation, loss of privilege, social isolation, for creating strong aversive emotional arousals, etc.

Tactic 7. Certain secular psychological threats [force] are used or are present: that failure to adopt the approved attitude, belief, or consequent behavior will lead to severe punishment or dire consequence (e.g., physical or mental illness, the reappearance of a prior physical illness, drug dependence, economic collapse, social failure, divorce, disintegration, failure to find a mate, etc.).

The similarities found in the Wollersheim case and how African children are taught are compelling and disturbing. Mind-control tactics, instilled in African people through the use of the media and other brainwashing institutions such as public schools, European religions, and social and corporate organizations all teach African children they should imitate European culture at the expense of African culture. Students are taught at an early age to praise white slave owners such as George Washington and Thomas Jefferson and to admire Abraham Lincoln for the act of freeing slaves for moral reasons rather than political ones. They are drilled repeatedly on the greatness of European people and European accomplishments with little if any mention of the greatness of African people and their accomplishments, encouraging them to want to identify with and become a part of a culture that has openly oppressed them. Students who reject these thoughts are often labeled as troublemakers and punished. Similar experiences occur in religions. Religious figures, such as Jesus, are depicted as white men with blue eyes and long straight blond hair. African people are taught to worship and bow down to these images that have no resemblance to them. Their own religions are made to appear foolish and barbaric. Social and corporate organizations also play a major role in destroying the minds of African people. Success in these areas is often dependent on one's ability to adopt the attitude and beliefs of the organization. Many Africans are made to believe that their names, hair, clothing, speech patterns, and physical appearance are unacceptable and are rewarded if they are willing to make changes and assimilate. African people in American society have been subjected to tactics similar to the ones Wollersheim said the Church of Scientology used to brainwash its adherents to submit to its authority. As Carter G. Woodson stated:

> When you control a man's thinking you don't have to worry about his actions. You do not have to tell him not to stand here and go yonder. He will find his proper place and will stay in it. You do not need to send him to the back door. He will go without being told. In fact, if there is no back door, he will cut one for his special benefit. His education makes it necessary.[6]

Our challenge as a people is to undo the brainwashing process and restore mental wholeness to African people. We must *reafricanize* those who have been *deafricanized*. As long as our children and our people remain ignorant of who they are and where they come from, they will never achieve the academic excellence history has proven to naturally exist in African people. This can be done by significant reparations being allocated to education in the K–12 curriculum.

THE DESTRUCTION OF SELF-ESTEEM IN
AFRICAN CHILDREN

In April 2001, the results of the National Assessment of Educational Progress were released. The writer Michael Kelly commented in the article, "We are looking at another lost generation." After reading the article, my mind pondered, lost for whom? Certainly not whites and Asians. Michael Kelly goes on to say:

> The NAEP confirmed that we are continuing to create a nation radically divided along meritocratic class lines. The top level is held by a tiny, hyper-schooled and highly competent overclass to which, in an information-based economy, accrues a vast disproportion of the nation's jobs, wealth, status and power. This class is disproportionately composed of Whites and Asians. The second class, numerically broad, is made up of people who are educated enough to work in blue, pink, and low to medium white-collar jobs that often pay too little to support a family, or offer no security and hold little hope of advancement to the upper tiers. The third class, also broad, comprises people who are functionally illiterate or close to it, and for whom life holds at best the joyous prospects of bike messengering, table busing, weed pulling, hamburger flipping, and broom pushing—episodically relieved by unemployment and descendence into deep poverty. Members of this lowest class are disproportionately Black or Hispanic.[7]

The study said that 40 percent of whites and Asians tested at or above proficient (able to understand and draw inferences from a text). The percentage of blacks at or above this level: 12 percent. Only 16 percent of Hispanics and 17 percent of American Indians tested at this level. Even among the so-called lowest class, children of African descent tested the lowest. Either children of African descent are inherently less intelligent than whites and Asians (which we know is not true), or they have been deprived of something whites and Asians have not been deprived of—*cultural consciousness*, a sense of self. European and Asian children know who they are and where they came from. They know their religion, their homelands, and their names. Amos Wilson states in his book *The Falsification of Afrikan Consciousness*:

> We have a school system that is based upon the psychology of white children and white people. We are trying to educate our children in that system; they are bound to fail. The very structure of the educa-

tional system itself is based upon a white model and therefore it has a built-in failure mechanism for us, one way or the other. We must develop a psychology of our children based upon our history and experience. It is only then that our pedagogical and educational approaches will be in line with their personalities. Only then can we move our children forward to fulfill our needs and desires as a people.[8]

I recall asking an Asian boy, new to my seventh-grade class in Brooklyn, New York, his name. He replied, "Ayo." I then asked him the origin of his name; he said it was an Asian name. He also told me he was an Asian. I responded, "Were you born in Asia?" He replied, "No, I was born in Brooklyn." I had a similar situation with an Asian child in Milwaukee named Kao. When asked the same questions, his answers were identical to Ayo's. He also told me he was an Asian, although he was born in Milwaukee, Wisconsin. The black students in both classrooms saw nothing unusual about Asian children born in America calling themselves Asians, but had a difficult time referring to themselves as Africans. Many students of African descent have a distorted view of Africa. They see Africa as barbaric and uncivilized. When I returned from one of my trips to Africa, several of the black children eagerly asked me, "Do Africans wear clothes? Do they drive cars, live in houses?" All ideas that had been instilled in them can be attributed to their miseducation in America. As the late Malcolm X said, "America has taught us to hate ourselves. America has taught us to hate our skin. America has taught us to hate our hair. America has taught us to hate our nose. America has taught us to hate Africa." Malcolm made it clear that you can't hate the root of the tree and not hate the tree. Both black and white educators, who underestimate the intelligence of children of African descent, further bring about these feelings of inferiority. I have heard black educators talk about black children as if they themselves were not black, using statements such as, "Look how you behave. That's why no one wants to teach black children," or "I need to get a job in the suburbs where the kids are civilized." In Milwaukee, a white teacher showed black students a picture of apes on the cover of a magazine and told them that these were their ancestors. In the February 1999 issue of *Emerge*, there was a disturbing article called "A Romantic Take on Slavery?" written by Gregory Wright, a contracting officer for the federal government and a freelance writer living in Erial, New Jersey, with his family. He told the story of his seven-year-old daughter coming home from school and saying, "I wish I could have been George Washington's slave. He was kind and gentle in his role as a master." Mr. Wright expressed that his daughter had been taught a romanticized version of slavery that discounts the atrocities suffered through America's history by her own African

American ancestors; she had been drawn to this conclusion by the (teacher's? school's? society's?) desire not to sully the reputation of a very prominent figure in our country's history.

He further stated:

> No matter how well a slave owner allegedly treated his slaves, the overriding factor is that they were, in fact, slaves. People in bondage. Oppressed. Treated not as humans, but as property. Less than cattle. Whipped, beaten, malnourished, raped, mentally degraded and emotionally humiliated. Families destroyed. Identities stolen. Slavery cannot be minimized. It cannot be romanticized. Any attempt to find some slavery-related moral consciousness in the hearts, minds or psyches of slave owners is an insult to those of us who are descendants of slaves.[9]

In 1849, James C. W. Pennington, the minister of a Presbyterian Church in New York City, echoed Gregory Wright's feelings when he published a narrative of his life that revealed the astonishing news that he was a fugitive slave and former blacksmith from Maryland. In his account of his life, Pennington offers the following reflections on the impact of slavery upon slave children:

> My feelings are always outraged when I hear [ministers] speak of "kind masters,"—"Christian masters,"—"the mildest form of slavery,"—"well fed and clothed slaves," as extenuations of slavery; I am satisfied they either mean to pervert the truth, or they do not know what they say.[10]

EDUCATION FOR *REPAIR*

Just recently, I was in a drugstore in Brooklyn, New York, and a student was using the copying machine to make a copy of a term paper she wrote entitled "Elizabethan England: Crime and Punishment." The paper described cruel crime and punishment of the people in that era. I asked the young student if she chose the topic. She replied, "No, my teacher chose it." I then asked could she have chosen her own topic. She said no, they had to write on the topic the teacher gave them. I asked her did every child have the same topic. She replied no, they all had different topics. I then asked if any of the topics the teacher gave dealt with black people. She said no, none. The next question was how many students were in her classroom. She replied, "Thirty-two." I followed with how many of the thirty-two were black. She said, "At least ninety-five percent." I said, "Ninety-five percent of the students are black, and not one of the topics you had to choose from dealt with black people? What race is your teacher?" She replied, "She is white." Such classroom situations are not unique in public schools in America. Unfortunately, they are quite normal. There is serious need for educational institutions that not only instill strong academic skills, but also *repair* the damage done to the minds of African people during the *Maafa*. As historian John Henrik Clarke states in Marimba Ani's book *Yurugu*, "We must instill will into the African mind to reclaim itself." This will not be an easy task given the critical condition of African minds as a result of four hundred years of slavery. Repairing all the damage may not be possible. A number of African people in America have terminal psychological illnesses, their condition so advanced it is irreversible, as African psychologists Na'im Akbar and Wade Nobles have said. But many can be rescued with the help of reparations. Reparations should be used to finance educational institutions that restore cultural consciousness, wholeness, and academic excellence to people of African descent. These institutions should be totally controlled by African people and staffed with scholars committed to the reafricanization of Africans. Kwame Agyei and Akua Nson Akoto in their book *The Sankofa Movement* state the following objectives they have successfully used in the reafrikanization process:

> To reconstruct the full history of the Afrikan world within an Afrikan centered framework. To identify, analyze and record the core values, spiritual-moral systems, socio-economic systems, philosophy and political systems of traditional Afrika and Afrikan antiquity.
>
> 1. To reestablish direct linkages with the spiritual and physical environment, people and traditions of traditional Afrika where efforts to sustain the purity of the culture have been successful.
> 2. To identify the personal and collective self (family, and community) in the continuum of Afrikan history and Afrikan culture.
> 3. To study, analyze, record, interpret and teach the languages and scripts of Afrikan antiquity and tradition.
> 4. To identify, establish, and/or support individuals and institutions devoted to the recovery and perpetuation of the traditions, knowledge and languages of Afrikan antiquity and tradition.[11]

The above objectives are merely suggestions. Curriculum developers, coordinators, and staff should establish definitive objectives for an education for reparation.

A CRITICAL QUESTION

It is of extreme importance that the staff of African-centered schools be meticulously screened. Too many African-centered institutions fall short of their goals and objectives because educators slipped through who do not truly believe in the *reafrikanization* process.

In my opinion, there is one crucial question that should be asked in the interviewing process for anyone interested in teaching African children, What do you call yourself? If the answer is anything but an African, I believe that person suffers from what Amos Wilson calls social amnesia and should not be considered for classroom duty. He states in *The Falsification of Afrikan Consciousness*:

> When we suffer from social amnesia, we identify with abstractions: I am not Black; I am not Afrikan; I am a human being. I am an American. Sterile Abstract identity, What is that? Who is that? What does that stand for? What does it mean? It's empty, and people who identify themselves with these abstractions are also empty and experience their lives as empty, as people who have no feelings. I love the challenge of being Afrikan in today's world; it's wonderful. We should eat this kind of challenge for breakfast.[12]

My experience has taught me that educators who love the challenge of being an African in most cases also love the challenge of *deeuropeanizing* black minds. *Education for Reparation* must begin with strong African-thinking teachers and will be critical to the success of institutions professing to teach cultural consciousness, wholeness, and academic excellence. What we call ourselves will determine who we become. Be mindful of those teachers who insist on referring to themselves and their students as African Americans. They are generally not comfortable with being an African and will defend "African American" vehemently, claiming they are both. Ironically, you seldom see the African side of these individuals. As we develop our own institutions we cannot and should not ignore students of African descent in the public school system. Reparation money should not only be available to reeducate public school teachers in the mind-healing process of African people, but also to give financial assistance to educators and scholars who have suffered financially and socially for attempting to provide African-centered education to children. Educators in public school institutions should not be fearful of developing a curriculum that will begin to make the African child whole. This curriculum should include, regardless of age, an extensive study of the pamphlets *The Willie*

Lynch Letter and *Let's Make a Slave,* two documents showing the psychology used during slavery to take the minds of African people. Lynch, a slave owner from the West Indies, gives American slave owners a foolproof method for controlling their slaves, which he guarantees will control the slaves for at least three hundred years, and that the slaves, after receiving this indoctrination, will carry on and become self-refueling and self-regenerating for hundreds of years, maybe thousands. He outlines a number of differences among the slaves and pits them against each other according to these differences. He suggests you use fear, distrust, and envy for control purposes. Pitch the old black male against the young black male and the young black male against the old black male. You must use the dark-skin slaves versus the light-skin slaves and the light-skin slaves versus the dark-skin slaves. You must use the female versus the male and the male versus the female. You must also have your white servants and overseers distrust all blacks, but it is necessary that your slaves trust and depend on them. They must love, respect, and trust only them. As an educator, I thoroughly discuss the Willie Lynch method with my students and have them examine how his method has been indoctrinated into today's society, just as Lynch predicted. In a separate lesson I discuss the cardinal principles of slave making from the pamphlet *Let's Make a Slave.* The principles are laid out the same for the slave breaking as they are for horse breaking: (1) both horse and nigger are no good to the economy in the wild or natural state, (2) both must be broken and tied together for orderly production, (3) for the orderly futures, special attention must be paid to the female and the young offspring, (4) both must be crossbred to produce a variety and division of labor, (5) both must be taught to respond to a particular language, and (6) psychological and physical instruction of containment must be created for both. These principles are taken one at a time and discussed from the perspective of how they are used in today's society. I have found these two pamphlets to be extremely effective in the beginning process of freeing the mind. Students have responded with comments such as: "Ms. Nzingha teaches us about who we are and from where we come"; "We want to learn where we come from, we don't want our minds to be blank"; "She teaches from the African perspective. She taught us who we were, bringing our self-esteem from five percent to ninety-five percent"; "She is correcting all the lies taught in the past and I learn from her and go home and teach my family." Other books I recommend using in the classroom are *What They Never Told You in History Class* by Indus Khamit-Kush; *Notes for an African World Revolution* by John Henrik Clarke; *Psychic Trauma* by Sultan A. Latif and Naimah Latif; *The Miseducation of the Negro* by Carter G. Woodson; *Before the Mayflower* by Lerone Bennett Jr.; *Afrikan People and European Holidays: A Mental Genocide (Book 1 and 2)* by Ishakamusa Barashango; *Black Women in Antiquity* by Ivan

Van Sertima, and the novel *The Genocide Files* by N. Xavier Arnold, to name a few. A good educator should adapt the contents of the suggested literature to suit the age level, but never lighten the subject matter.

Given that most literature used in educational institutions show enslaved Africans as docile beings with their heads lowered in "Oh why me?" position, students often raise questions about the complacency of those Africans and their refusal to rebel. It is essential for educators to include literature praising revolutionaries such as Nat Turner, Denmark Vesey, and Gabriel Prosser in their teachings, as well as modern-day warriors such as Kwame Toure and Khallid Abdul Muhammand.

HOW SHOULD REPARATIONS BE USED FOR EDUCATION?

Totally eliminating the damage done to African people during slavery and continuing into the twenty-first century is virtually impossible. Yet there are programs and opportunities that can and must be put in place to at least begin the process of repairing the damage. *I have concluded that the core of the problem in educating black children lies in the lack of connecting them to their African identity.* Most of them simply do not have a clue as to who they are culturally, historically, spiritually, or socially. *Until African people become clear that they are Africans with African connections they will not be able to move forward in a healthy way.* It is with that premise that I make the following recommendations on the use of reparations:

- Establishment of a permanent national panel of renowned African scholars (for example, Asa Hilliard, Molefi Asante, Ishakamusa Barashango) to develop the blueprint for the reafrikanization process of the mind. Cost over five years: $10 million.
- Development of African-centered schools with the curriculum based on the educational philosophy of psychologist Amos Wilson and like minds. There should be at least one such school per one thousand African K–12 students located in every school district in the United States. Cost over twenty-five-year period: $15 billion.
- Such schools should include interim, after-school, and Saturday programs designed to reinstate the spirituality, value systems, and socialization of African people, especially emphasizing the importance of extended families and communities. Cost over ten-year period: $5 billion.

- Establishment of language centers designed to teach African people an African language. One center should be established in any African community that exceeds 10 percent of the total population and instruction should be available for all persons regardless of age. Total cost over twenty-five-year period: $2 billion.
- Development of Ancestral Study Centers (ASC) to reconnect African people with the spiritual power of those who have come before them. Connected to the schools, each center would focus on helping African families link with their ancestral roots. The recent database established at Ellis Island can serve as a model for the ASCs. Total establishment cost: $1 billion.
- Development of training institutions for African teachers responsible for educating African students in the public school system and institutions of higher learning. Total cost over ten years: $10 billion.
- Development of workshops and seminars designed to teach the parents of African children parenting skills based on African traditions. Total cost over twenty-five years: $25 billion.
- Funds for the development and publication of textbooks to be used in private and public schools with accurate information on African people, their culture, contributions and ideologies. Total cost over fifteen years: $45 billion.
- Funds to institute trips for African people to visit Africa at least twice during their lifetimes. Cost over twenty-five years: $100 million.
- A trust fund available for African people who wish to permanently return to Africa. Cost over twenty-five years: $15 billion.

Total Education Reparations Budget = $128,100,000,000.

CONCLUSION

Reparations will never *totally* repair the horrific damage done to African people, but it is a beginning, a beginning long overdue. As Queen Mother Audley Moore states in the last verse of her poem "What's the Hour of the Night?":

> *Time is running out for us*
> *A deadline we must meet*
> *To file reparation petitions*
> *And make politicians earn their seats*

If we're to win this battle
Every organization must join the fight
Then we'll tell our watchman
What's the hour of the night.[13]

YAA ASANTEWA NZINGHA *is an educator, activist, and actress living in Brooklyn, New York. She is widely known for using theater to raise the academic and consciousness levels of African youth. She has received numerous awards from community and national organizations for instructional techniques with children of African descent.*

RAYMOND A. WINBUSH

Interview with Chester and Timothy Hurdle, Barbara Ratliff, and Ina Hurdle-McGee, October 29, 2002

EDITOR'S NOTE: *Several children of American slaves are still alive and form a direct link between what is often viewed as a historical institution with no relevance to the issue of reparations. These interviews, conducted on October 29, 2002, with the Hurdle brothers, their grand-niece, and their attorney, Barbara Ratliff, are compelling reminders of how slavery, like Toni Morrison's ghost child Beloved, still haunts the nation psychologically and economically.*

RAW: *Tell me something about your father, Andrew Jackson Hurdle. What kind of man was he? What do you remember most about him?*

TIMOTHY HURDLE: He was a dedicated man to his family and also a religious man. He was a minister. He brought the children up in the fear of the Lord and [was] quite concerned about ethics, and believed in a man being a man. He didn't like slothful people or people who were not useful. He just believed in people being of a good character and knowing how to look after and provide for themselves. He went through a lot as a child, but he was just a dedicated man. He was a father in charge.

RAW: *You mentioned in one interview that your father didn't talk too much about slavery; it wasn't something that you talked about at the breakfast table.*

CHESTER HURDLE: I was five years old when he became feeble and seven years old when he died. At the age of five I didn't even know how to spell *slave*.

So we talked about Christian things, how to live a wholesome life, how to treat your friends and neighbors in a positive way. He talked about getting along with my friends, trying to develop my education. I hadn't even started kindergarten and he was talking about education. He believed very firmly in education and Christianity.

BARBARA: I'm going to throw out some things I've heard Mr. Chester and Mr. Timothy say about their father. He was sold about 1855. Five sisters and brother were sold to Bennett Hazel in North Carolina.

CHESTER HURDLE: Near Burlington, North Carolina, near Alamance County. I've got a ninety-three-year-old cousin that lives there.

BARBARA: Then their father, Andrew, was sold to Turner—T. H. Turner—

CHESTER HURDLE: T. H. Turner in Texas.

BARBARA: In Dangerfield, Texas. I want to point out some things about their father, his personality, like his strength, his courage, his defiance, and his success. He was there [in Dangerfield] until he ran away in 1861, when he was confronted by a new overseer; Andrew had been living as a companion to the son of the slave owner. But when he was sixteen and this new overseer came, he had to prove himself by whipping everybody. Andrew resisted and he was not whipped. But he knew it was a price to pay for that . . . and that night he ran. When he ran, he was chased by bloodhounds into the woods, living off whatever he could find until he ran into the Union army. When the Civil War ended, they escorted him back to his hometown, where he married his childhood sweetheart, Bonnie, who was the mother of his first seventeen children; when she died and he remarried, he had eight more children. Timothy and Chester are the youngest of those eight.

There was another incident where he had to run away because there was some conflict with some whites, and he had to hide out for about a year. This was when he was trying to make his own way. Significantly, he was able to accumulate about five hundred acres of land there [in Texas]. By growing and having his own land, growing his own cotton and raising his own chickens, cattle, hogs, he had a certain amount of independence. Also, because he was a minister, there was a school, a college, like a Bible college, named after him. None of his children ever had any experiences with the law—being convicted or jailed or anything like that. A number of them went to college. Mr. Timothy went to three years of college, and Mr. Chester went three years.

RAW: *What colleges did you attend?*

CHESTER HURDLE: Timothy went to Jarvis Christian. My mother was born in that area, Big Sandy. I went to American River College, here in Sacramento, after I got out of the military.

RAW: *I read that in addition to running a farm in Greenville, Texas, not far from*

Dallas, Andrew Hurdle became a preacher and condemned slavery in his ser-
mons. So it wasn't as if your father "loved slavery" and afterward supported
it—a stereotype that is portrayed in some history books.

CHESTER HURDLE: No, he did not embrace it at all.

RAW: *Let me ask you this, Barbara. Why do you think it is important now to file*
these lawsuits? Why do this now? I think that one of you said in one of these
articles, "I don't want a penny, but the legacy is to leave something for the
coming generations." Why do you think reparations for living Africans in
America are important?

BARBARA: Well, why do it now? Because it is justice, it is fair, and because it is
something that should have been done when slaves were emancipated. In
fact, it was discussed in the Congress and by President Lincoln, but through
neglect nothing was ever done. And Black folk did not have anyone to stand
up for them and they weren't able to stand up for themselves. We're just
getting to the point that we can stand up for ourselves on these issues.

RAW: *Tell me why slavery still lingers in this society.*

CHESTER: The residual effect of slavery has affected us in so many different
ways. Monetarily, in our education system, in the military. In my thirty
years in the military, I ran into fifteen different incidents that happened to
me that would not have happened to a person not of color. I'll give you
one example: I worked in the post office in San Francisco before I went
into the service. So when I went in, I got a postal specialist AFSC. I was a
non-commissioned officer. I had only three stripes at the time. I was in
charge of the post office at Lakeland Air Force Base in Texas. I trained a
white guy under my thumb. He was my assistant. He had two stripes, I had
three. I trained him, he got his third stripe and then he got the fourth
stripe before I got my fourth stripe, and he became my supervisor. Here I
was, the guy that trained him. I knew all about the postal laws and regulations. I had to know it. Somebody who came with a postal question, I had
to answer it. But then for me to train somebody who had a stripe less than
I have and then they get the same stripes that I have, three. I was a sergeant
at the time. Then he makes staff sergeant and I'm still struggling. He became my supervisor. Now, he played softball with a white captain, who
was my supervisor. That had a lot to do with it, I'm quite sure.

RAW: *What would you tell white Americans and what do they need to under-*
stand about the impact of slavery?

CHESTER: Well, I would not generalize in the first place. Because if I generalize,
it would be very wrong. We belong to a church that is predominantly
white. It's only about two percent black there. We did a commercial here
on TV when the suit first was filed. I bet there were eighty members of our

church that came up to me and said "Chet, I'm so proud that you are finally bringing this to light." I can't look at it as white against black. On the other side of the coin, one of the TV channels showed a video of me talking on reparations, and then right behind it they showed one on [Ward] Connerly where he thinks its going to cause "division."

RAW: *Do you think slavery was a crime against humanity?*

TIMOTHY: Well, really, I don't.

CHESTER: I do.

BARBARA: Why do you say, Mr. Timothy, that it wasn't?

TIMOTHY: Well, it got me to America. I like to live in America. I was born here just like the other people. I never think of myself as being black. I think of myself as being an American. I just kinda like being an American and have noticed that a lot of other people like being American, too. I was born here.

BARBARA: You know what, I know a woman. Personally know a woman who was married to a man who was rich, wealthy. But he also beat her periodically. So the question for her was, Do I stay here and let him beat me, sometimes rape me, mistreat me, embarrass me, so that I can live in a beautiful home, and wear beautiful clothes? So you see what I mean, just because slavery brought us to America. We have to be careful about thinking that we are not in Africa. We look at Africa today, and see how bad it is. We see poverty over there, AIDS, they're killing each other. And so here we are in 2002, and we black folks can look at Africa and say I'm glad they made me a slave and brought me over here and put me through three hundred years of abuse, so that today, I am not over there, poor, diseased, and at war. But we also have to remember that small, elite group of Europeans not only took people from Africa to control those natural resources—that same mind-set also went throughout the whole continent of Africa and colonized it and forced those men to go off to their wars—but they had a sort of slavery right there, throughout the continent. So today, when you have all this poverty and warring and horrible conditions in Africa, it's not a result of the Africans, it's the result of what the Europeans did in Africa. So even though we might be in a better situation, we have to look at that.

RAW: *Mr. Chester, tell me why do you think it is a crime against humanity? I know what Ms. Ratliff said, but what do you think?*

CHESTER: I'll give you an example. I love America. I spent thirty years in the service showing that I'm proud to be an American. But my father, at the age of eight, was sold away, almost three thousand miles away from his five siblings. So those things like that . . . when I think about that and read about some of the other things that happened in this. . . . You know. I vote for this country. I love this country, but I don't love the actions that were taken back in those days.

RAW: *Let me ask one last question: If you were standing in front of an audience of young Black children, what would you tell them about reparations?*

CHESTER: Well, I would start out with something like this. Most of you are familiar with a dike. If you have a hole in a dike and don't repair that hole, it destroys the whole dike. So I look at reparations as repair; you're repairing something that should have been repaired many years ago. So if you leave that hole—that small hole—that started out in the dike there, eventually the water and everything are going to destroy the whole dike. It will flood on the other side. That's the example I would give.

RAW: *Now, Ms. McGee* [grand-niece of Timothy and Chester Hurdle], *what do you feel are the most devastating effects slavery has had on African Americans?*

INA: I think its greatest effects have been economical and psychological. Of all people in this country, we are "on the bottom." Whenever we move up one step, we move back a step and a half. The last hired and the first fired. I think part of this is that the bonding between black men and women is weak. This is a legacy of slavery that is still with us. When we came out of slavery we had strong family bonds and also a spiritual component in our lives. The church was the nucleus and the school. We had a thirst for education. You must have these three components: family, church, and education. Our minds were strong, given what we had suffered. We encouraged our children to read, write, and think for themselves to survive. These were survival skills, because if you didn't, you were cheated out of your assets, and the church would enhance what you were doing. All of us were striving for a better quality of life.

We also looked forward to being in the political process, even though we were excluded by it. We were strangled by the Black Codes of that era; they helped us get closer, but also kept us back.

RAW: *How should reparations be given to African Americans?*

INA: I think the older people should get them first. They are the hardest hit by poverty, medical issues, and abuse by health care systems. I believe a nest egg should be provided for elderly African Americans so that they will have something to leave their descendants when they die. Start with them and work down to the younger ones.

RAW: *Do you think that reparations will create an even greater racial divide between black and white Americans?*

INA: We're already divided! Human behavior has not changed that much between us for centuries.

RAW: *What prompted you to encourage your uncles to file a suit for reparations?*

INA: One night I was looking at David Horowitz on CNN, and he said the reason why reparations shouldn't be paid to African Americans is because no slaves *or their children* are alive to collect them! What really got me was that the two black guests on the show, one from N'COBRA, didn't contradict him, even though this information was absolutely false. Horowitz said that all of the slaves' children were dead and there was simply no one to give reparations to! He said that he would be the first to offer reparations if there were survivors.

I had uncles and aunts whose parents were slaves, and I also knew that there were hundreds of such people left. I was so depressed that no civil rights organizations came forth to challenge him. Our relatives are still alive. They had not vanished or been "raptured up." There was no excuse for people not knowing this. This is vital information and we had just had the 2000 census.

No one would challenge what Horowitz was saying, even though it was false, and this included civil rights organizations which had census information readily available. I think the interview showed how ignorant many of us are—both black and white—about slavery. I just wasn't going to let something like that go unchallenged. I knew about National Coalition of Blacks for Reparations in America in Dallas and their work on behalf of reparations. I went to the meeting and talked with them about [the descendants of slaves who are still alive], and they invited me to their national conference in Baton Rouge in 2001. I went to the conference and met several people involved in the reparations movement, including Dorothy Lewis and Conrad Worrill. Deadria Farmer-Paellmann was on a panel, and she talked about the research she was doing on corporations that had amassed millions of dollars on the backs of slavery. I talked to her the next morning and told her that I had family members who were the children of slaves. That began my legal pursuit of reparations for slavery.

TIMOTHY AND CHESTER HURDLE *are the only surviving sons of Andrew Jackson Hurdle, who was born in North Carolina as a slave in 1845. Timothy, eighty-three, lives in San Francisco, while his brother, Chester, seventy-five, lives about 100 miles east in Elk Grove, California. Their sole surviving sister, Hannah, seventy, lives in Illinois. In 2002, they filed a multimillion-dollar lawsuit in federal and state courts in San Francisco against corporations that profited from the enslavement of African Americans. This suit was encouraged by their grand-niece,* INA BELL DANIELS HURDLE MCGEE, *sixty-eight, of Dallas, who was also interviewed.*

PART VI

Historical Documents

Thirteenth, Fourteenth, and Fifteenth Amendments to the United States Constitution

AMENDMENT XIII

(1865)

Section 1. Neither slavery nor involuntary servitude, except as a punishment for crime whereof the party shall have been duly convicted, shall exist within the United States, or any place subject to their jurisdiction.

Section 2. Congress shall have power to enforce this article by appropriate legislation.

AMENDMENT XIV

(1868)

Section 1. All persons born or naturalized in the United States, and subject to the jurisdiction thereof, are citizens of the United States and of the state wherein they reside. No state shall make or enforce any law which shall abridge the privileges or immunities of citizens of the United States; nor shall any state deprive any person of life, liberty, or property, without due process of law; nor deny to any person within its jurisdiction the equal protection of the laws.

Section 2. Representatives shall be apportioned among the several states according to their respective numbers, counting the whole number of persons in each state, excluding Indians not taxed. But when the right to vote at any election for the choice of electors for President and Vice President of the United States, Representatives in Congress, the executive and judicial officers of a state, or the members of the legislature thereof, is denied to any of the male inhabitants of such state, being twenty-one years of age, and citizens of the United States, or in any way abridged, except for participation in rebellion, or other crime, the basis of representation therein shall be reduced in the proportion which the number of such male citizens shall bear to the whole number of male citizens twenty-one years of age in such state.

Section 3. No person shall be a Senator or Representative in Congress, or elector of President and Vice President, or hold any office, civil or military, under the United States, or under any state, who, having previously taken an oath, as a member of Congress, or as an officer of the United States, or as a member of any state legislature, or as an executive or judicial officer of any state, to support the Constitution of the United States, shall have engaged in insurrection or rebellion against the same, or given aid or comfort to the enemies thereof. But Congress may by a vote of two-thirds of each House, remove such disability.

Section 4. The validity of the public debt of the United States, authorized by law, including debts incurred for payment of pensions and bounties for services in suppressing insurrection or rebellion, shall not be questioned. But neither the United States nor any state shall assume or pay any debt or obligation incurred in aid of insurrection or rebellion against the United States, or any claim for the loss or emancipation of any slave; but all such debts, obligations and claims shall be held illegal and void.

Section 5. The Congress shall have power to enforce, by appropriate legislation, the provisions of this article.

AMENDMENT XV

(1870)

Section 1. The right of citizens of the United States to vote shall not be denied or abridged by the United States or by any state on account of race, color, or previous condition of servitude.

Section 2. The Congress shall have power to enforce this article by appropriate legislation.

MAJOR GENERAL WILLIAM T. SHERMAN

IN THE FIELD:
SAVANNAH, GEORGIA,
SPECIAL FIELD ORDERS, No. 15,
January 16, 1865

I. The islands from Charleston, south, the abandoned rice fields along the rivers for thirty miles back from the sea, and the country bordering the St. Johns River, Florida, are reserved and set apart for the settlement of the negroes now made free by the acts of war and the proclamation of the President of the United States.

II. At Beaufort, Hilton Head, Savannah, Fernandina, St. Augustine and Jacksonville, the blacks may remain in their chosen or accustomed vocations—but on the islands, and in the settlements hereafter to be established, no white person whatever, unless military officers and soldiers detailed for duty, will be permitted to reside; and the sole and exclusive management of affairs will be left to the freed people themselves, subject only to the United States military authority and the acts of Congress. By the laws of war, and orders of the President of the United States, the negro is free and must be dealt with as such. He cannot be subjected to conscription or forced military service, save by the written orders of the highest military authority of the Department, under such regulations as the President or Congress may prescribe. Domestic servants, blacksmiths, carpenters and other mechanics, will be free to select their own work and residence, but the young and able-bodied negroes must be encouraged to enlist as soldiers in the service of the United States, to contribute their share towards maintaining their own freedom, and securing their rights as citizens of the United States.

Negroes so enlisted will be organized into companies, battalions and regiments, under the orders of the United States military authorities, and will be paid, fed and clothed according to law. The bounties paid on enlistment may, with the consent of the recruit, go to assist his family and settlement in procuring agricultural implements, seed, tools, boots, clothing, and other articles necessary for their livelihood.

III. Whenever three respectable negroes, heads of families, shall desire to settle on land, and shall have selected for that purpose an island or a locality clearly defined, within the limits above designated, the Inspector of Settlements and Plantations will himself, or by such subordinate officer as he may appoint, give them a license to settle such island or district, and afford them such assistance as he can to enable them to establish a peaceable agricultural settlement. The three parties named will subdivide the land, under the supervision of the Inspector, among themselves and such others as may choose to settle near them, so that each family shall have a plot of not more than (40) forty acres of tillable ground, and when it borders on some water channel, with not more than 800 feet water front, in the possession of which land the military authorities will afford them protection, until such time as they can protect themselves, or until Congress shall regulate their title. The Quartermaster may, on the requisition of the Inspector of Settlements and Plantations, place at the disposal of the Inspector, one or more of the captured steamers, to ply between the settlements and one or more of the commercial points heretofore named in orders, to afford the settlers the opportunity to supply their necessary wants, and to sell the products of their land and labor.

IV. Whenever a negro has enlisted in the military service of the United States, he may locate his family in any one of the settlements at pleasure, and acquire a homestead, and all other rights and privileges of a settler, as though present in person. In like manner, negroes may settle their families and engage on board the gunboats, or in fishing, or in the navigation of the inland waters, without losing any claim to land or other advantages derived from this system. But no one, unless an actual settler as above defined, or unless absent on Government service, will be entitled to claim any right to land or property in any settlement by virtue of these orders.

V. In order to carry out this system of settlement, a general officer will be detailed as Inspector of Settlements and Plantations, whose duty it shall be to visit the settlements, to regulate their police and general management, and who will furnish personally to each head of a family, subject to the approval of the President of the United States, a possessory title in writing, giving as near as possible the description of boundaries; and who shall adjust all claims or conflicts that may arise under the same, subject to the like approval, treating such titles altogether as possessory. The same general officer will also be charged with the enlistment and organization of the negro recruits, and protecting their interests while absent from their settlements; and will be governed by the rules and regulations prescribed by the War Department for such purposes.

VI. Brigadier General R. SAXTON is hereby appointed Inspector of Settlements and Plantations, and will at once enter on the performance of his duties. No change is intended or desired in the settlement now on Beaufort *[Port Royal]* Island, nor will any rights to property heretofore acquired be affected thereby.

BY ORDER OF MAJOR GENERAL W. T. SHERMAN:

Special Field Orders, No. 15, Headquarters Military Division of the Mississippi, 16 Jan. 1865, Orders & Circulars, ser. 44, Adjutant General's Office, Record Group 94, National Archives.

A Bill Introduced by
Thaddeus Stevens of Pennsylvania,
H.R. 29, 40th Congress, 1st Session,
March 11, 1867:
A Plan for Confiscation

Whereas it is due to justice, as an example of future times, that some proper punishment should be inflicted on the people who constituted the "confederate States of America," both because they, declaring an unjust war against the United States for the purpose of destroying republican liberty and permanently establishing slavery, as well as for the cruel and barbarous manner in which they conducted said war, in violation of all the laws of civilized warfare, and also to compel them to makes some compensation for the damages and expenditures caused by the war: Therefore,

Be it enacted . . . That all the public lands belonging to the ten States that formed the government of the so-called "confederate States of America" shall be forfeited by said States and become forthwith vested in the United States.

Sec. 2. The President shall forthwith proceed to cause the seizure of such of the property belonging to the belligerent enemy as is deemed forfeited by the act of July 17, A.D. 1862, and hold and appropriate the same as enemy's property, and to proceed to condemnation with that already seized.

Sec. 3. In lieu of the proceeding to condemn the property thus seized as enemy's property, as is provided by the act of July 17, A.D. 1862, two commissions or more, as by him may be deemed necessary, shall be appointed by the President for each of the said "confederate States," to consist of three persons each, one of whom shall be an officer of the late or present Army, and two shall be civilians, neither of whom shall be citizens of the State for which he shall be appointed; the said commission shall proceed to adjudicate and condemn the property aforesaid, under such forms and proceedings as shall be prescribed by the Attorney General of the United States, whereupon the title to said property shall become vested in the United States.

Sec. 4. Out of the lands thus seized and confiscated, the slaves who have been liberated by the operations of the war and the amendment of the Constitution or otherwise, who resided in said "confederate States" on the 4th day of March, A.D. 1861, or since, shall have distributed to them as follows, namely: to each male person who is the head of a family, forty acres; to each adult male, whether the head of a family or not, forty acres; to each widow who is the head of a family, forty acres; to be held by them in fee simple, but to be inalienable for the next ten years after they become seized thereof. For the purpose of distributing and allotting said land, the Secretary of War shall appoint in each State as many commissions as he may deem necessary, to consist of three members each, two of whom at least shall not be citizens of the State for which he is appointed. At the end of ten years the absolute title to said homesteads shall be conveyed to said owners or to the heirs of such as are then dead.

Sec. 5. Out of the balance of the property thus seized and confiscated there shall be raised, in the manner hereinafter provided, a sum equal to fifty dollars, for each homestead, to be applied by the trustees hereinafter mentioned toward the erection of buildings on the said homesteads for the use of said slaves; and the further sum of $500,000,000, which shall be appropriated as follows, to wit: $200,000,000 shall be invested in the United States six per cent securities; and the interest thereof shall be semi-annually added to the pensions allowed by law to the pensioners who have become so by reason of the late war; $300,000,000, or so much thereof as may be needed, shall be appropriated to pay damages done to loyal citizens by the civil or military operations of the government lately called the "Confederate States of America."

Sec. 6. In order that just discrimination may be made, the property of no one shall be seized whose whole estate on the 4th day of March, A.D. 1865, was not worth more than $5,000, to be valued by the said commission, unless he shall have voluntarily become an officer or employee in the military or civil service of the "Confederate States of America," or in the civil or military service of some one of said States, and in enforcing all confiscations the sum or value of $5,000 in real or personal property shall be left or assigned to the delinquent.

Sec. 8. If the owners of said seized and forfeited estates shall, within ninety days after the first of said publications, pay into the Treasury of the United States the sum assessed on their estates respectively, all of their estates and lands not actually appropriated to the liberated slaves shall be released and restored to their owners.

Sec. 9. All the land, estates and property, of whatever kind, which shall not be redeemed as aforesaid within ninety days, shall be sold and converted into money, in such time and manner as may be deemed by the said commissioners the most advantageous to the United States: *Provided*, That no arable land shall be sold in tracts larger than 500 acres.

Congressional Globe, *March 19, 1867, p. 203.*

Congressman John Conyers's (D-MI) Bill for a Study on the Impact of Slavery on African Americans HR40IH, 105th Congress, 1st Session, H.R. 40

To acknowledge the fundamental injustice, cruelty, brutality, and inhumanity of slavery in the United States and the 13 American colonies between 1619 and 1865 and to establish a commission to examine the institution of slavery, subsequently de jure and de facto racial and economic discrimination against African-Americans, and the impact of these forces on living African-Americans, to make recommendations to the Congress on appropriate remedies, and for other purposes.

IN THE HOUSE OF REPRESENTATIVES

January 7, 1997

Mr. CONYERS (for himself, Mr. FATTAH, Mr. FOGLIETTA, Mr. HASTINGS of Florida, Mr. HILLIARD, Mr. JEFFERSON, Ms. EDDIE BERNICE JOHNSON of Texas, Mrs. MEEK of Florida, Mr. OWENS, Mr. RUSH, and Mr. TOWNS) introduced the following bill; which was referred to the Committee on the Judiciary

A BILL

To acknowledge the fundamental injustice, cruelty, brutality, and inhumanity of slavery in the United States and the 13 American colonies between 1619 and 1865 and to establish a commission to examine the institution of slavery, subsequently de jure and de facto racial and economic discrimination against African-Americans, and the impact of these forces on living African-Americans, to make recommendations to the Congress on appropriate remedies, and for other purposes.

Be it enacted by the Senate and House of Representatives of the United States of America in Congress assembled,

SECTION 1. SHORT TITLE.

This Act may be cited as the 'Commission to Study Reparation Proposals for African-Americans Act'.

SEC. 2. FINDINGS AND PURPOSE.

(a) FINDINGS– The Congress finds that—

(1) approximately 4,000,000 Africans and their descendants were enslaved in the United States and the colonies that became the United States from 1619 to 1865;

(2) the institution of slavery was constitutionally and statutorily sanctioned by the Government of the United States from 1789 through 1865;

(3) the slavery that flourished in the United States constituted an immoral and inhumane deprivation of Africans' life, liberty, African citizenship rights, and cultural heritage, and denied them the fruits of their own labor; and

(4) sufficient inquiry has not been made into the effects of the institution of slavery on living African-Americans and society in the United States.

(b) PURPOSE– The purpose of this Act is to establish a commission to—

(1) examine the institution of slavery which existed from 1619 through 1865 within the United States and the colonies that became the United States, including the extent to which the Federal and State Governments constitutionally and statutorily supported the institution of slavery;

(2) examine de jure and de facto discrimination against freed slaves and their descendants from the end of the Civil War to the present, including economic, political, and social discrimination;

(3) examine the lingering negative effects of the institution of slavery and

the discrimination described in paragraph (2) on living African-Americans and on society in the United States;

(4) recommend appropriate ways to educate the American public of the Commission's findings;

(5) recommend appropriate remedies in consideration of the Commission's findings on the matters described in paragraphs (1) and (2); and

(6) submit to the Congress the results of such examination, together with such recommendations.

SEC. 3. ESTABLISHMENT AND DUTIES.

(a) ESTABLISHMENT– There is established the Commission to Study Reparation Proposals for African Americans (hereinafter in this Act referred to as the 'Commission').

(b) DUTIES– The Commission shall perform the following duties:

(1) Examine the institution of slavery which existed within the United States and the colonies that became the United States from 1619 through 1865. The Commission's examination shall include an examination of—

(A) the capture and procurement of Africans;

(B) the transport of Africans to the United States and the colonies that became the United States for the purpose of enslavement, including their treatment during transport;

(C) the sale and acquisition of Africans as chattel property in interstate and intrastate commerce; and

(D) the treatment of African slaves in the colonies and the United States, including the deprivation of their freedom, exploitation of their labor, and destruction of their culture, language, religion, and families.

(2) Examine the extent to which the Federal and State governments of the United States supported the institution of slavery in constitutional and statutory provisions, including the extent to which such governments prevented, opposed, or restricted efforts of freed African slaves to repatriate to their home land.

(3) Examine Federal and State laws that discriminated against freed African slaves and their descendants during the period between the end of the Civil War and the present.

(4) Examine other forms of discrimination in the public and private sectors against freed African slaves and their descendants during the period between the end of the Civil War and the present.

(5) Examine the lingering negative effects of the institution of slavery and

the matters described in paragraphs (1), (2), (3), and (4) on living African-Americans and on society in the United States.

(6) Recommend appropriate ways to educate the American public of the Commission's findings.

(7) Recommend appropriate remedies in consideration of the Commission's findings on the matters described in paragraphs (1), (2), (3), and (4). In making such recommendations, the Commission shall address, among other issues, the following questions:

> (A) Whether the Government of the United States should offer a formal apology on behalf of the people of the United States for the perpetration of gross human rights violations on African slaves and their descendants.

> (B) Whether African-Americans still suffer from the lingering effects of the matters described in paragraphs (1), (2), (3), and (4).

> (C) Whether, in consideration of the Commission's findings, any form of compensation to the descendants of African slaves is warranted.

> (D) If the Commission finds that such compensation is warranted, what should be the amount of compensation, what form of compensation should be awarded, and who should be eligible for such compensation.

(c) REPORT TO CONGRESS– The Commission shall submit a written report of its findings and recommendations to the Congress not later than the date which is one year after the date of the first meeting of the Commission held pursuant to section 4(c).

SEC. 4. MEMBERSHIP.

(a) NUMBER AND APPOINTMENT– (1) The Commission shall be composed of 7 members, who shall be appointed, within 90 days after the date of enactment of this Act, as follows:

> (A) Three members shall be appointed by the President.

> (B) Three members shall be appointed by the Speaker of the House of Representatives.

> (C) One member shall be appointed by the President pro tempore of the Senate.

(2) All members of the Commission shall be persons who are especially qualified to serve on the Commission by virtue of their education, training, or experience, particularly in the field of African-American studies.

(b) TERMS– The term of office for members shall be for the life of the

Commission. A vacancy in the Commission shall not affect the powers of the Commission, and shall be filled in the same manner in which the original appointment was made.

(c) FIRST MEETING– The President shall call the first meeting of the Commission within 120 days after the date of the enactment of this Act, or within 30 days after the date on which legislation is enacted making appropriations to carry out this Act, whichever date is later.

(d) QUORUM– Four members of the Commission shall constitute a quorum, but a lesser number may hold hearings.

(e) CHAIR AND VICE CHAIR– The Commission shall elect a Chair and Vice Chair from among its members. The term of office of each shall be for the life of the Commission.

(f) COMPENSATION– (1) Except as provided in paragraph (2), each member of the Commission shall receive compensation at the daily equivalent of the annual rate of basic pay payable for GS-18 of the General Schedule under section 5332 of title 5, United States Code, for each day, including travel time, during which he or she is engaged in the actual performance of duties vested in the Commission.

(2) A member of the Commission who is a full-time officer or employee of the United States or a Member of Congress shall receive no additional pay, allowances, or benefits by reason of his or her service on the Commission.

(3) All members of the Commission shall be reimbursed for travel, subsistence, and other necessary expenses incurred by them in the performance of their duties to the extent authorized by chapter 57 of title 5, United States Code.

SEC. 5. POWERS OF THE COMMISSION.

(a) HEARINGS AND SESSIONS– The Commission may, for the purpose of carrying out the provisions of this Act, hold such hearings and sit and act at such times and at such places in the United States, and request the attendance and testimony of such witnesses and the production of such books, records, correspondence, memoranda, papers, and documents, as the Commission considers appropriate. The Commission may request the Attorney General to invoke the aid of an appropriate United States district court to require, by subpoena or otherwise, such attendance, testimony, or production.

(b) POWERS OF SUBCOMMITTEES AND MEMBERS– Any subcommittee or member of the Commission may, if authorized by the Commission, take any action which the Commission is authorized to take by this section.

(c) OBTAINING OFFICIAL DATA– The Commission may acquire directly from the head of any department, agency, or instrumentality of the executive branch of the Government, available information which the Commission considers useful in the discharge of its duties. All departments, agencies, and instrumentalities of the executive branch of the Government shall cooperate with the Commission with respect to such information and shall furnish all information requested by the Commission to the extent permitted by law.

SEC. 6. ADMINISTRATIVE PROVISIONS.

(a) STAFF– The Commission may, without regard to section 5311(b) of title 5, United States Code, appoint and fix the compensation of such personnel as the Commission considers appropriate.

(b) APPLICABILITY OF CERTAIN CIVIL SERVICE LAWS– The staff of the Commission may be appointed without regard to the provisions of title 5, United States Code, governing appointments in the competitive service, and without regard to the provisions of chapter 51 and subchapter III of chapter 53 of such title relating to classification and General Schedule pay rates, except that the compensation of any employee of the Commission may not exceed a rate equal to the annual rate of basic pay payable for GS-18 of the General Schedule under section 5332 of title 5, United States Code.

(c) EXPERTS AND CONSULTANTS– The Commission may procure the services of experts and consultants in accordance with the provisions of section 3109(b) of title 5, United States Code, but at rates for individuals not to exceed the daily equivalent of the highest rate payable under section 5332 of such title.

(d) ADMINISTRATIVE SUPPORT SERVICES– The Commission may enter into agreements with the Administrator of General Services for procurement of financial and administrative services necessary for the discharge of the duties of the Commission. Payment for such services shall be made by reimbursement from funds of the Commission in such amounts as may be agreed upon by the Chairman of the Commission and the Administrator.

(e) CONTRACTS– The Commission may—

(1) procure supplies, services, and property by contract in accordance with applicable laws and regulations and to the extent or in such amounts as are provided in appropriations Acts; and

(2) enter into contracts with departments, agencies, and instrumentalities of the Federal Government, State agencies, and private firms, institutions, and agencies, for the conduct of research or surveys, the preparation of reports,

and other activities necessary for the discharge of the duties of the Commission, to the extent or in such amounts as are provided in appropriations Acts.

SEC. 7. TERMINATION.

The Commission shall terminate 90 days after the date on which the Commission submits its report to the Congress under section 3(c).

SEC. 8. AUTHORIZATION OF APPROPRIATIONS.

To carry out the provisions of this Act, there are authorized to be appropriated $8,000,000.

The Dakar Declaration: African Regional Preparatory Conference for the World Conference against Racism, Racial Discrimination, Xenophobia and Related Intolerance, Dakar, 22–24 January 2001

I. DECLARATION AND RECOMMENDATIONS FOR A PROGRAMME OF ACTION

A. Declaration

Preamble

We, African Ministers, meeting at Dakar from 22 to 24 January 2001, within the framework of the World Conference Against Racism, Racial Discrimination, Xenophobia and Related Intolerance, in accordance with United Nations General Assembly resolution 52/111 of 12 December 1997,

Recalling the values and principles of human dignity and equality enshrined in the Charter of the United Nations, the Universal Declaration of Human Rights, the International Convention on the Elimination of All Forms of Racial Discrimination, the Convention on the Elimination of All Forms of Discrimination Against Women, the African Charter on Human and Peoples Rights and all other related international instruments,

Recalling also the great importance African peoples attach to the values of solidarity, tolerance and multiculturalism, which constitute the moral ground and the inspiration for our struggle against racism, racial discrimination, xenophobia and related intolerance, inhuman tragedies which Africa has been suffering for too long,

Realizing the urgent need to resuscitate and reinvigorate those cherished values, that the World Conference Against Racism, Racial Discrimination, Xenophobia and Related Intolerance is a historical opportunity to achieve this objective, and that its outcome should therefore be result oriented and bring added value to existing mechanisms,

Recalling the principles established by positive international law, including the non-applicability of statutory limitations to crimes against humanity,

Stressing that the fight against racism, racial discrimination, xenophobia and related intolerance is an arduous task, the proof from the most recent African experience being the enormous African sacrifices made and the unabated struggle waged for decades before Africa could convince the rest of the world of the imperative and urgent necessity of dismantling the abhorrent institutionalized racist system of apartheid,

Recognizing that racism, racial and ethnic discrimination, xenophobia and related intolerance affect women differently from men, aggravate their living conditions and generate multiple forms of violence, thus limiting or denying their enjoyment of their human rights,

Bearing in mind that availing ourselves of this historical opportunity requires political will, intellectual integrity and the analytical capacity to draw lessons from past experiences with a view to avoiding their recurrence,

Expressing in this regard our sincere appreciation and paying tribute to countries and personalities all over the world who lent their valuable support to Africa during its struggle against institutionalized racism, colonialism and apartheid,

Acknowledging the important role of African and international nongovernmental organizations, the media, national institutions and civil society in the fight against racism and encouraging them to intensify their endeavors in this respect,

HEREBY:

1. Salute the memory of all victims of racism and racial discrimination, colonialism and apartheid all over the world;

2. Note with grave concern that, despite the efforts of the international community, the principal objectives of the two Decades for Action to Combat Racism and Racial Discrimination have not been attained and that millions of human beings continue to this day to be victims of varied, evolving and sophisticated contemporary forms of racism and racial discrimination, in particular nationals of different origins, migrant workers, asylum-seekers, refugees and foreigners;

3. Express our concern that beyond the material progress of racism is the disturbing fact that contemporary forms and manifestations of racism are striving to regain political, moral and even legal recognition in many ways, including through legislative prescriptions such as those relating to the freedom of expression, the platforms of some political parties and organizations, and the dissemination through modern communication technologies of ideas based on racial superiority;

4. Regret the flagrant contradiction that, in an era when globalization and technology have contributed considerably to bringing people closer together, the international community is evidently receding from the notion of a "human family" based on equality, dignity and solidarity;

5. Bear in mind that although the African continent has regrettably suffered ethnic violence, including instances of genocidal acts, it is not an exclusively racial phenomenon but has many deeply rooted national and international dimensions;

6. Express our deep concern that the socio-economic development of our continent is being hampered by widespread internal conflicts which are due, among other causes, to violations of human rights, including discrimination based on ethnic or national origin and lack of democratic, inclusive and participatory governance;

7. Also express our concern in this regard that external interference, mainly linked to the exploitation of minerals and the arms trade, an unfavorable international economic environment and foreign debt are the main contributing factors in the spread of conflicts and instability in Africa;

8. Express our firm conviction that the development of democratic systems of government in Africa, that guarantee full access and representation of all sectors of our societies, respect for and protection of all human rights and fundamental freedoms, equitable distribution of wealth and access to economic advancement, active promotion of peace, preventive diplomacy and conflict resolution and an equitable international economic environment, is an essential prerequisite for the prevention of conflicts and instability in Africa;

9. Recall the historical fact that among the most hideous manifestations of racial discrimination the African continent and Diaspora have suffered, namely the slave trade, all forms of exploitation, colonialism and apartheid, were essentially motivated by economic objectives and competition between

colonial powers for strategic territorial gains, appropriation, and control over and pillage of natural and cultural resources;

10. Affirm that the slave trade, particularly of Africans, is a unique tragedy in the history of humanity, a crime against humanity which is unparalleled, not only because of its abhorrent barbarism but also in terms of its enormous magnitude, its institutionalized nature, its transnational dimension, and especially its negation of the essence of the human nature of the victims;

11. Also affirm that the consequences of this tragedy, accentuated by those of colonialism and apartheid, have resulted in substantial and lasting economic, political and cultural damage to African peoples and are still present in the form of damage caused to the descendants of the victims by the perpetuation of prejudice against Africans in the continent and people of African descent in the Diaspora;

12. Stress the negative economic consequences of racism, racial discrimination, xenophobia and related intolerance, conscious that the economic difficulties of Africa cannot be explained exclusively by foreign factors and historical events and aware that it is nevertheless a reality that those factors and events have had profound crippling effects on the economic development of Africa and that justice now requires that substantial national and international efforts be made to repair the damage;

13. Reaffirm that the discriminatory treatment of foreigners and migrant workers, established or practiced in certain countries, inter alia concerning granting visas, work permits, conditions of family members, housing and access to justice, based on race, colour, descent or national or ethnic origin, are human rights violations which seriously contradict the Universal Declaration of Human Rights, the International Convention on the Elimination of All Forms of Racial Discrimination and the International Convention on the Protection of the Rights of All Migrant Workers and Members of Their Families;

14. Also reaffirm that the stigmatization of people of different origins by acts or omissions of public authorities, institutions, the media, political parties or national or local organizations, is not only an act of racial discrimination but also an incitement to the recurrence of such acts, thereby resulting in the creation of a vicious circle which reinforces racist attitudes and prejudices; such acts should be declared offences and crimes punishable by law;

15. Express our concern that the complicating dimension of this vicious circle contributes to and intensifies racially discriminatory social attitudes which cannot be criminalized by law;

16. Recall that without the necessary political will to recognize and assume responsibility for historical injustices and their contemporary forms

and repercussions, programmes of action against racism, racial discrimination, xenophobia and related intolerance, as well as the anti-racist slogans and measures taken at the World Conference and at the regional and national levels, will not change deeply ingrained prejudices or reach the noble goal of a genuine human family based on equal dignity and equal opportunities;

17. Affirm that the first logical and credible step to be taken at this juncture of our collective struggle is for the World Conference to declare solemnly that the international community as a whole fully recognizes the historical injustices of the slave trade and that colonialism and apartheid are among the most serious and massive institutionalized forms of human rights violations;

18. Also affirm that this recognition would be meaningless without an explicit apology by the former colonial powers or their successors for those human rights violations, and that this apology should be duly reflected in the final outcome of the World Conference Against Racism, Racial Discrimination, Xenophobia and Related Intolerance;

19. Recall that article 6 of the International Convention on the Elimination of All Forms of Racial Discrimination already contains the obligation to provide effective protection and remedies for everyone against any acts of racial discrimination which violate human rights and fundamental freedoms, a principle reaffirmed by numerous subsequent human rights instruments including the basic principles and guidelines on the right to a remedy and reparation for victims of violations of international human rights and humanitarian law;

20. Strongly reaffirm that States which pursued racist policies or acts of racial discriminations such as slavery and colonialism should assume their moral, economic, political and legal responsibilities within their national jurisdiction and before other appropriate international mechanisms or jurisdictions and provide adequate reparation to those communities or individuals who, individually or collectively, are victims of such racist policies or acts, regardless of when or by whom they were committed;

21. Also strongly reaffirm that as a pressing requirement of justice, victims of human rights violations as a result of racism, racial discrimination, xenophobia and related intolerance should be assured effective protection and remedies as well as legal assistance, including the right to seek and receive just and adequate reparation or satisfaction for material and moral damage as a result of violations in the implementation of human rights standards;

22. Express the deep conviction that the right of everyone to an effective remedy by the competent national tribunals for acts violating the fundamental human rights stipulated in article 8 of the Universal Declaration of Human Rights, article 6 of the International Convention on the Elimination of All

Forms of Racial Discrimination and article 7 of the African Charter on Human and Peoples' Rights undoubtedly applies to victims of racial discrimination;

23. Commend the work of national institutions for the promotion and protection of human rights in Africa established in compliance with the Paris Principles, especially their role in raising awareness through human rights education and training at the national level and in facilitating protection from and prevention of human rights violations, especially racism, racial discrimination, xenophobia and all forms of racially motivated violence;

24. Encourage the Office of the United Nations High Commissioner for Human Rights to continue its support for those African States that are in the process of establishing national institutions by providing training and resources, and strongly advise African States that have not yet done so to consider establishing effective and independent national institutions for the promotion and protection of human rights;

25. Reaffirm that all individual human rights violations and collective violations such as racial discrimination should be condemned and appropriate remedies must be provided;

26. Express our conviction that applying a victim-oriented approach to victims of racial discrimination at both the national and the international level reaffirms the human values of tolerance and solidarity and thus strengthens the foundations of human rights law;

27. Affirm that, by enhancing the victims' right to benefit from international recognition and protection of their right to remedies and reparation, the international community strengthens its credibility in the cause of human rights, shows faith and human solidarity with victims, survivors and future human generations and reaffirms the principles of the equality and dignity of all human beings, accountability, justice and the rule of law;

28. Also affirm in particular that the victims' right to have access to justice is of special importance to victims of racial discrimination in the light of their vulnerable situation, socially, culturally and economically, and that the principle of equality of victims in legal systems is meaningless unless it is accompanied by affirmative action;

29. Note that other groups which were subjected to other scourges and injustices have received repeated apologies from different countries, as well as ample reparations, on a bilateral basis, from both public and private sources and lately through certain international organizations;

30. Affirm in that spirit that all human beings are equal and that all scourges and injustices should, therefore, be addressed with the same emphasis and that such fairness is a fundamental prerequisite for the creation of the

peace of mind of all parties involved, which gives future efforts better chances of success;

31. Also affirm the commitment by States to comply with their obligations relating to the promotion and protection of the human rights of refugees, asylum-seekers and internally displaced persons;

32. Bear in mind the situation of vulnerability in which migrants frequently find themselves, owing, inter alia, to their absence from their State of origin and to the difficulties they encounter because of differences in language, customs and culture, as well as economic and social difficulties and obstacles to the return of migrants who are undocumented or in an irregular situation;

33. Recognize that the number of recent and on going conflicts around the world reveals that racism, racial discrimination, xenophobia and related intolerance of peoples, groups and individuals are among the root causes of conflict and are very often also among its consequences, and in this regard recall that non-discrimination is a fundamental principle of international humanitarian and human rights law;

34. Note with grave concern the negative effects on health and the environment of environmental racism suffered, in particular, by countries in Africa, including the illicit dumping of toxic wastes and substances, hazardous working and living conditions and dangerous methods of extracting natural resources.

B. Recommendations for a Programme of Action

Recognizing the urgent need to translate the objectives of the Declaration into a practical and workable Plan of Action, we therefore recommend to the World Conference the following:

1. A follow-up mechanism, headed by the President of the World Conference and composed of five eminent persons from the different regions, appointed by the United Nations Secretary-General after due consultation with all regions, should be established. This mechanism will function in consultation with the High Commissioner for Human Rights, the Committee on the Elimination of Racial Discrimination and the Special Rapporteur on contemporary forms of racism, racial discrimination, xenophobia and related intolerance of the Commission on Human Rights. This mechanism would be entrusted with the supervision of the implementation of the Declaration and Programme of Action to be adopted by the World Conference and with submitting an annual report to the United Nations General Assembly.

2. An International Compensation Scheme should be set up for victims of the slave trade, as well as victims of any other transnational racist policies and acts, in addition to the national funds or any equivalent national mechanisms aimed at fulfilling the right to compensation.

3. A Development Reparation Fund should be set up to provide resources for the development process in countries affected by colonialism.

4. The modalities of such reparation and compensation should be defined by the World Conference in a practical and result-oriented manner.

5. On a collective basis, such reparation should be in the form of enhanced policies, programmes and measures to be adopted by States which benefit materially from these practices in order to rectify, through affirmative action, the economic, cultural and political [sic] which has been inflicted on the affected communities and peoples to the full implementation of their right to development.

6. The International Compensation Scheme and the Development Reparation Fund should be financed not only from governmental sources but also by private contributions emanating in particular from those elements of the private sector which benefited, directly or indirectly, from transnational racist policies or acts.

7. An international mechanism should be established to monitor racially discriminatory attitudes and acts, individual or collective, private or public, including by non-State actors, charged with the following tasks:

(a) To gather information about racial acts and related developments;

(b) To create and maintain, together with a coalition of non-Governmental organizations working in the field of combating racism and in collaboration with the Office of the High Commissioner for Human Rights, a website to receive and disseminate such information to the widest possible extent;

(c) To provide legal and administrative support and advice to victims of racial acts;

(d) To submit an annual report on its activities to the United Nations Secretary-General.

8. The Office of the High Commissioner for Human Rights should disseminate, in the most accessible manner, through its website and other appropriate means, all the remedies available through international mechanisms to victims of racial discrimination, as well as the national remedies, hopefully enhanced and developed in implementation of the Programme of Action to be adopted by the World Conference.

9. The international media, through their relevant associations and organizations at both regional and international levels, should consider the elaboration of an ethical code of conduct with a view to prohibiting the proliferation of ideas of superiority, justification of racial hatred and discrimination in any form and promoting mutual respect and tolerance among all peoples and human beings.

10. States should ensure the enactment of legislation declaring illegal and prohibiting all political platforms, organizations and propaganda activities which promote and incite racial discrimination and recognizing that participation in such organizations is an offence punishable by law.

11. States should give the utmost importance to the observations and recommendations of the Committee on the Elimination of Racial Discrimination. To that effect, States should consider setting up appropriate national monitoring and evaluation mechanisms to ensure that these observations and recommendations are duly addressed and that the relevant legislation is effectively implemented and that all necessary steps are taken to promote national harmony, equality of opportunity and good inter-ethnic and inter-racial relations.

12. States should review and reconsider their reservations to the International Convention on the Elimination of All Forms of Racial Discrimination with a view to withdrawing reservations that are incompatible with the purpose and objectives of the Convention.

13. States should facilitate access to all appropriate methods of justice and provide legal assistance to victims of racial discrimination in a manner adapted to their specific needs and vulnerability, including exemption from fees, simplification of procedures, legal representation and establishment as appropriate of special adapted jurisdictions to deal with such cases.

14. States that have not yet done so should adhere, as a matter of urgency and without reservations, to the Geneva Conventions on the protection of victims of international armed conflicts of 12 August 1949 and the two Protocols Additional thereto of 1977, as well as to other treaties of international humanitarian law. All States should carry out their obligation to respect and ensure respect for these fundamental norms.

15. States should enact, as a matter of the highest priority, appropriate legislation and take other measures required to give full effect to their obligations under international humanitarian law, in particular in relation to the rules prohibiting discrimination.

16. States should adopt legislation providing, in particular, for the prosecution and punishment of persons suspected of having committed or having ordered to be committed grave breaches of the Geneva Conventions and Additional Protocol I and of other serious violations of the laws and customs of war, in particular in relation to the rules prohibiting discrimination.

17. States should also take seriously their humanitarian obligations, without discriminating between the different regions of the world, with regard to the principles of international cooperation, burden-sharing and the resettlement of refugees in their countries and, in this regard, provide additional support to those African countries hosting refugees to enable them better to discharge their humanitarian obligations.

18. States should intensify their efforts in the field of education to promote the awareness of the evils of racism, racial discrimination, xenophobia and related intolerance, in order to ensure respect for the dignity and worth of all human beings. In this context, States should develop, where appropriate, and implement specific sensitization and training programmes, formulated in local languages and for all categories of society, in particular young people, to combat racism.

19. States should take measures to facilitate the full and active participation of youth in the preparatory process of the World Conference at the national and international levels.

20. The Office of the High Commissioner [for] Human Rights is invited, in cooperation with United Nations Economic and Social Cultural Organization, the concerned specialized and regional organizations, national institutions and non-governmental organizations active in the field of promotion and protection of human rights, to undertake periodic consultations and to encourage research activities aimed at collecting, maintaining and adapting the technical, scientific, educational and information materials produced by all cultures around the world to fight racism.

21. States should incorporate a gender perspective into all programmes of action against racism, racial discrimination, xenophobia and related intolerance and involve women in decision-making to ensure their full and equal participation in the entire process of development of the economy and the output of their communities.

22. All States which have not yet done so should sign and ratify the International Convention on the Protection of the Rights of All Migrant Workers and Members of Their Families.

23. Countries receiving migrants should strengthen training and awareness-raising activities designed for State personnel, especially the police and other civil servants in charge of enforcing laws, as well as teachers and local authorities, in order to prevent racial conflicts.

All States should take specific measures for the promotion and protection of the fundamental rights and freedoms of vulnerable groups, especially children, youth, the disabled, people with HIV/AIDS, refugees and indigenous populations.

Farmer-Paellmann v. FleetBoston, Aetna Inc., CSX

UNITED STATES DISTRICT COURT
for the
EASTERN DISTRICT OF NEW YORK

_____X

DEADRIA FARMER-PAELLMANN,

On behalf of herself
and all other persons
similarly situated,

PLAINTIFF

vs.

FLEETBOSTON FINANCIAL CORPORATION,
AETNA INC., CSX, and Their predecessors, successors
and/or assigns, and CORPORATE DOES NOS. 1-1000,
DEFENDANTS.

_____X

CIVIL ACTION

CLASS ACTION

COMPLAINT
AND JURY TRIAL
DEMAND

Plaintiffs, on behalf of themselves and all other persons similarly situated,
state, upon information and belief, as follows:

INTRODUCTION, JURISDICTION AND VENUE

Introduction

1. Over 8,000,000 Africans and their descendants were enslaved in the United States from 1619 to 1865. The practice of slavery constituted an immoral and inhumane deprivation of Africans' life, liberty, African citizenship rights, cultural heritage and it further deprived them of the fruits of their own labor.

2. The first slave ship that sailed into Jamestown Harbor in Virginia in 1619 contained a handful of captive Africans, but by the end of the Atlantic slave trade, more than two centuries later, somewhere between 8 million and 12 million Africans had arrived in the New World in chains.[1]

3. Historians estimate that one slave perished for every one who survived capture in the African interior and made it alive to the New World, meaning as many as 12 million perished along the way.[2]

4. Although, it is a common perception that the South alone received the enslaved Africans, many of them arrived in the Dutch colonial city of New Amsterdam that later became New York City. Integral to the colony from the start, slaves helped build Trinity Church, the streets of the city and the wall, from which Wall Street takes its name, that protected the colony from military strikes.[3]

5. These slaves in New York lived in attics, hallways and beneath porches, cheek to jowl with their masters and mistresses. In death, these same slaves were banished to the Negro Burial Ground, which lay a mile outside the city limits and contained between 10,000 and 20,000 bodies by the time it was closed in 1794.[4]

6. Further research conducted by Howard University of 400 skeletons of these buried slaves revealed that 40 percent were children under the age of 15 and the most common cause of death was malnutrition. Most of the children had rickets, scurvy, anemia or related diseases. The adult skeletons show that many people died of unrelenting hard labor. Strain on the muscles and ligaments was so extreme that muscle attachments were commonly ripped away from the skeleton taking chunks of bone with them—leaving the body in perpetual pain. The highest mortality rate is found among women ages 15 to 20. Investigators have concluded that some died of illnesses acquired in the holds of slave ships or from a first exposure to the cold or from the trauma of being torn from their families and shipped in chains halfway around the globe. Moreover, the research has concluded that these women were worked to death by owners who could simply go out and buy a new slave.[5]

7. But New Yorkers were not alone in the utilization of slaves, in fact, more recent research has revealed that many of our esteemed and celebrated institutions of learning had their origins in the profits derived from the slave trade. For instance, money from the slave trade financed Yale University's first endowed professorship, its first endowed scholarships and its first endowed library fund. Moreover, in the 1830s, Yale officials led the opposition that prevented the building of the first African American college, on the grounds that such an institution would have been incompatible with the existence of Yale. Nicholas and John Brown, two of the founders of what became Brown University were slave traders. Likewise, Harvard Law School was endowed by money its founder earned selling slaves in Antigua's cane fields.[6]

8. Many early American industries were based on the cotton, sugar, rice, tobacco, and other products African labor produced. Railroads and shipping companies, the banking industry and many other businesses made huge profits from the commerce generated by the output of enslaved labor.

9. Slaves built the U.S. Capitol, cast and hoisted the statue of freedom on top of its dome, and cleared the forest between the Capitol and the White House.[7]

10. Slavery fueled the prosperity of the young nation. From 1790 to 1860 alone, the U.S. economy reaped the benefits of as much as $40 million in unpaid labor.[8] Some estimate the current value of this unpaid labor at 1.4 trillion dollars.[9]

11. Not only did the institution of slavery result in the extinguishment of millions of Africans, it eviscerated whole cultures: languages, religions, mores, and customs, it psychologically destroyed its victims. It wrenched from them their history, their memories, and their families on a scale never previously witnessed.

12. When the institution finally ended, the vestiges, racial inequalities and cultural psychic scars left a disproportionate number of American slave descendants injured and heretofore without remedy.

13. Although the institution of slavery in the United States was officially outlawed in 1865, it continued, de facto, until as recently as the 1950s. National archive records reveal that in the 1920s and 1930s, the NAACP still received letters from African-Americans claiming to still be on plantations and forced to work without pay. Several claims were investigated and were found to be legitimate. Moreover, as late as 1954, the Justice Department prosecuted the Dial brothers in Sumpter County, Alabama, because they held blacks in involuntary servitude.[10]

14. Even for those who were freed, their lives remained locked in quasi-servitude, due to legal, economic and psychic restraints that effectively blocked their economic, political and social advancement.[11]

15. Hence, new measures called Black Codes guaranteed control of Blacks by white employers. As John Hope Franklin noted in *From Slavery to Freedom*:

the control of blacks by white employers was about as great as that which slaveholders had exercised. Blacks who quit their job could be arrested and imprisoned for breach of contract. They were not allowed to testify in court except in cases involving members of their own race; numerous fines were imposed for seditious speeches, insulting gestures or acts, absence from work, violating curfews and the possession of firearms. There was of course no enfranchisement of blacks and no indication that in the future they could look forward to full citizenship and participation in democracy.[12]

16. The post-Reconstruction Southern practices of peonage and share-cropping which continued well into the twentieth century were direct outgrowths of slavery that continued a system of complete control by the dominant culture. Peonage was a complex system where a black man would be arrested for vagrancy, ordered to pay a fine that he could not afford, and then incarcerated. A plantation owner would then pay the fine and then hire him until he could afford to pay off the fine. The peon was forced to work, locked up at night and if he escaped, was chased by bloodhounds until recaptured.[13]

17. Likewise, during the 1920s, fortunate African-Americans became share-croppers on land leased from whites whose grandparents had owned their fore-bears. These African Americans were not allowed to vote, and were socially and economically relegated to the left-overs in education, earnings, and freedoms.

18. More recently, a 1998 census report shows that 26 percent of African American people in the United States live in poverty compared to 8 percent of whites. It also showed that 14.7 percent of African Americans have four-year college degrees, compared with 25 percent of whites. The same year, African American infant-mortality rates were more than twice as high as those among whites. Federal figures also show that a Black person born in 1996 can expect to live, on average, 6.6 fewer years than a white person born the same year.

19. African-Americans are more likely to go to jail, to be there longer, and if their crime is eligible, to receive the death penalty. They lag behind whites according to every social yardstick: literacy, life expectancy, income and education. They are more likely to be murdered and less likely to have a father at home.

20. Defendants, including, but not limited to **FLEETBOSTON FINAN-CIAL CORPORATION, AETNA INC., CSX, through their predecessors-in-interest,** conspired with slave traders, with each other and other entities and institutions (whose identities are not yet specifically identified, but which are described herein as **CORPORATE DOES # 1–100**) and other un-named entities and/or financial institutions to commit and/or knowingly facilitate crimes against humanity, and to further illicitly profit from slave labor.

21. Plaintiffs and the plaintiff class are slave descendents whose ancestors were forced into slavery from which the defendants unjustly profited. Plaintiffs seek an accounting, constructive trust, restitution, disgorgement and compensatory and punitive damages arising out of Defendants' past and continued wrongful conduct.

JURISDICTION AND VENUE

22. This Court has jurisdiction over this matter pursuant to 28 U.S.C. 1332(a) since the amount in controversy exceeds $75,000 per plaintiff exclusive of interests and costs and there is diversity of citizenship.

23. The Court has personal jurisdiction over the parties in that the defendants conduct systematic and continuous business within the State of New York.

24. Venue is proper in this Court since the Defendants do business and may be found in the District within the meaning of 28 U.S.C. 1391(a).

25. Plaintiffs and the plaintiff class are African-American slave descendants.

26. Plaintiff is a New York resident whose ancestors were enslaved in the agricultural industry.

DEFENDANTS

27. Defendants and the other known and unknown defendants used and/or profited from slave labor and have retained the benefits and use of those profits and products derived from that slave labor. Defendants knew that the plaintiff class was subject to physical and mental abuse and inhuman treatment.

28. Defendants conspired with each other with intentions to violate Plaintiffs' ancestors' basic human rights from slavery in that and by so doing to profit from these violations.

29. Defendant **FLEETBOSTON** is a Delaware corporation with its principal place of business located at 100 Federal Street, Boston, Massachusetts 02110. It does continuous and systematic business in New York. **FLEETBOSTON** is the successor in interest to Providence Bank which was founded by Rhode Island businessman John Brown. Brown owned ships that embarked on several slaving voyages and Brown was prosecuted in federal court for participating in the international slave trade after it had become illegal under federal law. Upon information and belief, Providence Bank lent substantial sums to Brown, thus financing and profiting from the founder's illegal slave trading. Upon information and belief, **FLEETBOSTON** also collected custom fees due from ships transporting slaves, thus, further profiting from the slave trade.

30. Defendant **CSX** is a Virginia corporation with its principal place of business located at 901 E. Cary Street, Richmond, VA 23219. It is a successor-in-interest to numerous predecessor railroad lines that were constructed or run, at least in part, by slave labor.[14]

31. Defendant **AETNA INC.** (AETNA) is a corporation with its principal place of business located at 151 Farmington Avenue, Hartford, Connecticut 06156. Upon information and belief, **AETNA's** predecessor in interest, actually insured slave owners against the loss of their human chattel. **AETNA** knew the horrors of slave life as is evident in a rider through which the company declined to pay the premiums for slaves who were lynched or worked to death or who committed suicide. Additionally, **AETNA** insured enslaved Africans who worked in the agricultural industry of which Plaintiff's enslaved. **AETNA,** therefore, unjustly profited from the institution of slavery.

32. Defendants **CORPORATE DOES NOS. 1–100** are other companies, industrial, manufacturing, financial and other enterprises that, like the named Defendants, its/their predecessors, affiliates and/or assigns, unjustly profited from slave labor. The designation CORPORATE DOES NOS. # 1–100 is used until such time as the specific identity of such additional companies, as they relate to this action, is ascertained through discovery and/or other means.

CLASS ALLEGATIONS

33. This action is brought and may properly be maintained as a class action pursuant to the provision of the Federal Rules of Civil Procedure 23(a), 23(b)(2) and 23(b)(3). Plaintiffs seek certification of the following class: all African-American slave descendants.

34. The exact number of Plaintiff class members is not known. Plaintiffs estimate that the class includes millions of African-American slave descendants and the Plaintiffs estimate that the class is so numerous that joinders of individual members is impracticable. The number and identities of the class members can only be ascertained through appropriate investigation and discovery.

35. Questions of fact and law are common with respect to each class member. Common questions of fact and law include:

  a. Whether Defendants knowingly, intentionally and systematically benefited from the use of enslaved laborers;

  b. Whether Defendants wrongly converted to their own use and for their own benefit, the slave labor and services of the Plaintiffs' forebears, as well as the products and profits from such slave labor;

    c. Whether the Defendants knew or should have known that they were assisting and acting as accomplices in immoral and inhuman deprivation of life and liberty;

    d. Whether Defendants have been unjustly enriched by their wrongful conduct; and

    e. Whether, as a result of this horrific and wrongful conduct by the Defendants, the Plaintiff class is entitled to restitution or other equitable relief, or to compensatory or punitive damages.

36. The claims of the individually named Plaintiffs are typical of the claims of the Plaintiff Class Members. Plaintiffs and all members of the Plaintiff Class have been similarly affected by the Defendants' common course of conduct and the members of each class have similar claims against the Defendants. The claims of all class members depend on a showing of the Defendants' common course of conduct, as described herein, which gives Plaintiffs, individually and as class representative, the right to the relief sought herein.

37. There is no conflict as between Plaintiffs and the other members of the class with respect to this action or the claims for relief. Plaintiffs know and understand their asserted rights and their roles as class representatives.

38. Plaintiffs and their attorneys are able to and will fairly, and adequately, protect the interest of the Class. Several of Plaintiffs' attorneys are experienced class action litigators who are or will be able to conduct the proposed litigation. Plaintiffs' attorneys can vigorously prosecute the rights of the proposed class members.

39. Prosecution of separate actions by individual Plaintiffs will create the risk of inconsistent and varying adjudications and will establish incompatible standards of conduct for Defendants in that different Courts may order Defendants to provide different types of accounting or take other inconsistent actions.

40. Prosecutions of separate actions by individual plaintiffs of other proposed class members not party to the adjudications will substantially impair or impede their ability to protect their interest in that, for example, Defendants may exhaust their available funds in satisfying the claims of earlier plaintiffs to the detriment of later plaintiffs.

41. Defendants have acted and/or refused to act on grounds generally applicable to the proposed class, making final injunctive relief and correspondent declaratory relief appropriate with respect to the class as a whole in that Defendants have been unjustly enriched by participation in acts that were known to be immoral and inhumane, and Defendants: (a) prevented and or refused restitution to the proposed class members, (b) prevented and/or refused to disgorge wrongfully gained and/or earned profits and benefits, or (c) refused to provide a full and complete accounting and disclosure of the extent of their aforesaid actions.

42. Common questions of law and fact predominate in the claims of all class members, including the named Plaintiff. These claims depend on proving Defendants are liable for their acts and/or omissions based, in part, on evidence of a common scheme. Plaintiffs' and the plaintiff class members' proposed evidentiary showings would be based on the same documents and testimony concerning the Defendants' actions.

43. A class action is superior to the other available methods for the fair, just and efficient adjudication of the controversy. Plaintiffs and the Plaintiff class members have no interest in individually controlling the prosecution of separate actions and, instead, are on the whole incapable as a practical matter of pursuing individual claims. Even if individual class members had the resources to pursue individual litigation, it would be unduly burdensome to the Courts in which the individual litigation would proceed. Individual litigation magnifies the delay and expenses to all parties in that the Court system of resolving the controversies engendered by Defendants/individual and/or common course of conduct. The class action device allows a single court to provide the benefits of unitary adjudication, judicial economy and the fair and equitable handling of all plaintiffs' claims in a single forum. The conduct of this action as a class action conserves the resources of the parties and of the judicial system, and reserves the rights of each class member. Furthermore, for most class members, a class action is the only feasible mechanism that allows them an opportunity for legal redress and justice. A large concentration of proposed class members is estimated to reside in this District and nearby states. The management of the litigation as a class would pose few problems for this Court.

44. Certification of the Plaintiff class is appropriate under Fed. R. Civ. P. 23(a) and also under 23(b)(2), 23(b)(3).

EQUITABLE TOLLING

45. The plaintiffs have been unable to secure records with regards to their ancestors due to the failure of most to be able to reliably access ship manifestos, or human cargo lists that directly connect them to their descendants. Moreover, family names were changed once the Africans arrived in America making it nearly impossible to accurately trace records. Recent advances in Internet and computer databases have made these records more accessible to the average African-American.

46. Likewise, corporate histories and records have also been extremely difficult and inaccessible to most people. Hence, research tracing the monetary benefit derived by American corporations from the slave trade has only been accessible and discussed by prominent researches within the last year.

47. Moreover, efforts to attempt to raise the issue of reparations for African-Americans in an attempt to secure easier access to information have stalled in Congress. Representative John Conyers from Michigan has for the last 11 years attempted to propose a resolution, No. 40, seeking to set aside $8 million to study the effects of slavery and come up with a formula for reparations. His resolution has died in committee for each of these past eleven years.

48. Moreover, with the advent of litigation related to reparations for holocaust victims from government entities and corporations, more emphasis has been placed on the viability of lawsuits for reparations for human rights violations.

49. Finally, the action of each of the Defendants by their failure to provide an accounting to the plaintiff constitutes a continuing tort that tolls the statute.

COUNT I—CONSPIRACY

50. Each of the Defendants acted individually and in concert with their industry group and with each other, either expressly or tacitly, to participate in a plan that was designed in part to commit the tortious acts referred to herein.

51. For instance, each industry group was co-dependant on each other and operated as joint enterprise, designed in part, to maintain and continue a system of inhumane servitude. The shipping and railroad industry benefited and profited from the transportation of the slaves. The railroad industry utilized slave labor in the construction of rail lines. These transportation industries were dependent upon the manufacturing and raw materials industry to utilize the slaves they shipped. The cotton, tobacco, rice and sugar industries thrived on profits generated from their use of slave labor, and relied upon financial and insurance industries to finance and insure the slaves that they utilized and owned. All industries: raw market, retail, financial, insurance, and transportation, benefited from the reduced costs of slave-produced goods.

COUNT II—DEMAND FOR AN ACCOUNTING

52. Plaintiffs on behalf of themselves and all other descendants who are similarly situated, re-allege as if fully set forth, each and every allegation contained into the preceding paragraphs.

53. The Defendants knew or should have known of the existence of corporate records that indicate their profiting from slave labor. Plaintiffs and the public have demanded that the Defendants reveal their complete corporate records regarding same and that a just and fair accounting be made for profits derived from the slave trade.

54. Defendants have failed to provide said records and have failed to comply with plaintiffs' demand.

WHEREFORE, Plaintiffs demand judgment: (a) requiring defendants make a full disclosure of all of their corporate records that reveal any evidence of slave labor or their profiting from same; (2) seeking the appointment of an independent historic commission to serve as a depository for corporate records related to slavery; and (3) directing defendants to account to plaintiffs for any profits they derived from slavery.

COUNT III—HUMAN RIGHTS VIOLATIONS

57. Plaintiffs on behalf of themselves and all other descendants who are similarly situated, re-allege as if fully set forth, each and every allegation contained into the preceding paragraphs.

58. The Defendants participated in the activities of the institution of slavery and in so doing furthered the commission of crimes against humanity, crimes against peace, slavery and forced labor, torture, rape, starvation, physical and mental abuse, summary execution. Specifically, the Defendants profited from these wrongs.

59. Defendants knowingly benefited from a system that enslaved, tortured, starved and exploited human beings, so as to personally benefit them. In the process, the Defendants directly or indirectly subjected the plaintiffs' ancestors to inhumane treatment, physical abuse, torture, starvation, execution and subjected the plaintiffs to the continued effects of the original acts, including but not limited to: race discrimination, unequal opportunity, poverty, substandard health care, substandard treatment, substandard housing, substandard education, unjust incarceration, racial profiling, and inequitable pay.

60. The above referenced actions by the Defendants were in violation of international law.

61. As a result of the above referenced violations of international law, Plaintiffs and members of the Plaintiff class have suffered injury and are entitled to compensatory damages in an amount to be determined at trial.

COUNT IV—CONVERSION

62. Plaintiffs on behalf of themselves and all other slave ancestors who are similarly situated, re-allege as if fully set forth, each and every allegation contained in the preceding paragraphs.

63. As a result of Defendants' failure and refusal to account for, acknowledge and return to Plaintiffs and the members of the Plaintiff class, the value of their slave labor, Defendants have willfully and wrongfully misappropriated and converted the value of that labor and its derivative profits into Defendants' own property.

64. Defendants have never accounted for or returned the value of Plaintiffs' ancestors' slave labor and the profits Defendants derived from said slave labor.

65. As a result of Defendants' wrongful acts and omissions, Plaintiffs and members of the Plaintiffs class have been injured and demand judgment against the Defendants jointly, severally and/or in the alternative on this cause of action for, amongst other things: (a) an accounting of the slave labor monies, profits and/or benefits derived by Defendants; (b) a constructive trust in the value of said monies, profits and/or benefits derived by Defendants' use of slave labor; (c) full restitution in the value of all monies, profits, and/or benefits derived by Defendants' use of slave labor; (d) equitable disgorgement of all said monies, profits, and/or benefits derived by Defendants' exploitation of slave labor; and (e) other damages in an amount in excess of the jurisdictional limits of this Court and to be determined at the trial herein, together with interest, exemplary or punitive damages, attorneys' fees and costs of this action.

COUNT V—UNJUST ENRICHMENT

66. Plaintiffs on behalf of themselves and all other slave descendants who are similarly situated, re-allege as if fully set forth, each and every allegation contained into the preceding paragraphs.

67. Defendants have improperly benefited from the immoral and inhumane institution of Slavery in the United States.

68. Defendants have failed to account for and/or return to Plaintiffs and the Plaintiff class the value of their ancestors' slave labor and/or the profits and benefits the Defendants derived therefrom and Defendants have concealed the nature and scope of their participation in the Institution.

69. As a result of the Defendants' wrongful acts and omissions as described above, Defendants have been unjustly enriched.

70. Defendants have been unjustly enriched at the expense of Plaintiffs and the Plaintiffs class. Plaintiffs and the Plaintiffs class therefore demand restitution and judgment against the Defendants jointly, severally and/or in the alternative, in an amount in excess of the jurisdictional limits of this Court and to be determined at the trial herein, together with interest, exemplary or punitive damages, attorneys' fees and the costs of this action.

PRAYER FOR RELIEF

WHEREFORE Plaintiffs and the Plaintiffs class demand a jury trial and judgment and damages against the Defendants, jointly, severally and/or in the alternative, as follows:

(1) For an order certifying the Plaintiff class alleged herein;

(2) For an accounting;

(3) For the appointment of an independent historic commission;

(4) For the imposition of a constructive trust;

(5) For restitution of the value of their descendants' slave labor;

(6) For restitution of the value of their unjust enrichment based upon slave labor;

(7) For disgorgement of illicit profits;

(8) For compensatory damages in an amount to be determined by trial together with interest;

(9) For exemplary or punitive damages in an amount to be determined at trial;

(10) For attorneys' fees; and

(11) For the cost of this action.

Dated: March 26, 2002
New York, New York

By: _____

EDWARD D. FAGAN, ESQ. (EF–4125)
FAGAN & ASSOCIATES

By: _____

ROGER S. WAREHAM, ESQ.
(RW 4751)
JOMO SANGA THOMAS, ESQ.
(JT 7544)
301 South Livingston Avenue
Livingston, New Jersey 07039
(973) 994–2908

THOMAS WAREHAM & RICHARDS
572 Flatbush Avenue, Suite 2
Brooklyn, New York 11225
(718) 941–6407

-and-

237 Park Avenue, 21st Floor
New York, New York 10017
Please Respond to New Jersey Address

BRYAN R. WILLIAMS, ESQ. (BW–)
46 Trinity Place
New York, New York 10006

BRUCE H. NAGEL, ESQ. (BN–6765)
JAY J. RICE, ESQ. (JR–9171)
DIANE E. SAMMONS, ESQ. (DS–9029)
NAGEL RICE DREIFUSS &
MAZIE, LLP
301 South Livingston Avenue,
Suite 201
Livingston, New Jersey 07039
(973) 535–3100

-and-

237 Park Avenue, 21st Floor
New York, New York 10017
Please Respond to New Jersey Address

MORSE GELLER, ESQ.
116–10 Queens Boulevard
Forest Hills, New York 11375

PART VII

*Current Status and
Future Directions of the
Reparations Movement*

BABA HANNIBAL AFRIK, JILL SOFFIYAH ELIJAH,
DOROTHY BENTON LEWIS, AND CONRAD W.
WORRILL

Current Legal Status of Reparations, Strategies of the National Black United Front, the Mississippi Model, What's Next in the Reparations Movement

EDITOR'S NOTE: *In 1967, Martin Luther King Jr. penned his last book,* Where Do We Go from Here: Chaos or Community?. *The book served as a prophetic voice for the Civil Rights struggle and was written to locate the current status of the movement for human rights in the United States. This chapter of the book gives opinions, strategies, and directions by some of the most respected members of the reparations movement, including a summary of the legal aspects of the struggle.*

THE LEGAL STATUS OF REPARATIONS

Introduction

The end of the twentieth century was marked by a dramatic change in the popular thinking in America with respect to reparations for African Americans. A significant catalyst for this change was the filing of several lawsuits seeking monetary damages for the various abuses of the African slave trade, known as the *Maafa;*[1] however, no one had yet prevailed in *pro se* efforts to gain redress against the United States. At the dawning of the twenty-first century, various lawyers began lining up potential defendants from the private and public sectors with an aim of exacting from them both an acknowledgment of the unspeakable harm they had caused and monetary damages as a means of compensation.[2]

The genesis of the litigation efforts also saw the beginning of a wave of local, state, and national legislative initiatives involving reparations. The best known of these was H.R. 40 (see Part VI, page 331) which had been reintroduced every year since 1989 by Representative John Conyers Jr. (D-Mich.). Although H.R. 40 had not been successful, many other legislative efforts were attempted on the federal, state, and local levels—some with positive results.

Litigation Efforts

The case of *Cato v. United States*[3] decided in 1995 by the Ninth Circuit Court of Appeals is generally recognized by reparations proponents as significant in the development of litigation strategy.[4] The plaintiffs, representing themselves, unsuccessfully sought damages against the United States for the enslavement of African Americans and subsequent discrimination against them, for acknowledgment of discrimination, and for an apology. They sued pursuant to the Federal Tort Claims Act (FTCA)[5] and the Ninth Circuit identified several major obstacles to the plaintiffs' claims. First, it noted that the United States government enjoys sovereign immunity. Thus, the United States can be sued only to the extent that it has waived its sovereign immunity. However, the United States has not rendered itself liable under the FTCA for constitutional tort claims.

Second, the FTCA has a six-year statute of limitations. Suits that are not brought within six years of the harm being complained of are time-barred. The plaintiffs argued that the harm they sought to redress was a continuing one. They claimed that the vestiges of slavery, as manifested in the various ways in which racial discrimination had continued to work to the detriment of African Americans, persisted to the present, and, therefore, their suit had been timely filed. The court held that the claims arising out of slavery, kidnapping, and other offenses to the plaintiffs' ancestors fell outside of the FTCA's limited waiver of sovereign immunity.[6] Therefore, the shorter two-year statute of limitations applied, and those actions were not pursued within the two-year timeframe. The court further found that the plaintiffs could not rely upon the continuing-violations doctrine to create jurisdiction.

Third, the court held that an individual cannot base a claim on a generalized class-based grievance. In other words, the court found that the plaintiffs had failed to allege any conduct on the part of any specific government official or any specific government program that violated their constitutional or statutory rights and caused them a specific injury. Without a concrete "personal injury that is not abstract and is fairly traceable to the government," the plaintiffs lacked standing to pursue their claims.[7]

Cato, as a test case, delineated the maze that reparations litigants would have to navigate in order to prevail. The Ninth Circuit, although willing to acknowledge that "the enslavement of Africans by this Country is inexcusable," was unwilling to grant jurisdiction over the issue. Rather, the court suggested that plaintiffs take their claims to the legislature.

Using a different tact, seven years after *Cato*, plaintiffs in *Obadele v. United States*,[8] through their lawyer, filed claims under the Civil Liberties Act (CLA)[9] based on the enslavement of their ancestors and the continuing failure of the United States to recognize African Americans' right to self-determination following the abolition of slavery. The Office of Redress Administration denied the claims, and judicial review was sought before the Federal Court of Claims.[10] The CLA was enacted to provide a "formal apology and benefits, including redress payments of $20,000, to certain individuals affected by the Federal Government's evacuation, relocation, or internment of United States citizens and permanent resident aliens of Japanese ancestry during World War II."[11] The plaintiffs in *Obadele* were of African descent, and the court denied their claims for failure to meet eligibility criteria. Here, too, the court urged the plaintiffs to seek legislative remedies.

The same year that *Obadele* was decided a series of lawsuits were filed against various private-sector corporations. Based on the relentless research of Deadria Farmer-Paellmann, an attorney in New York City, major corporations, such as Aetna, FleetBoston Financial, New York Life Insurance, Brown Brothers Harriman & Company, Norfolk Southern, and the railroad giant CSX were sued for their "crimes against humanity." In March 2002, Farmer-Paellmann as lead plaintiff along with Andre Carrington and Mary Lacy Madison filed the first class-action lawsuit in the Eastern District of New York seeking redress for crimes of enslavement, kidnapping, and discrimination against their ancestors. Defendants in the case were Aetna, FleetBoston, and CSX.

Then, two months later, a similar action was filed in the New Jersey District Court against New York Life, Brown Brothers, and Norfolk Southern. Richard E. Barber Sr., former deputy director of the National NAACP and a former Regional Administrator of the Small Business Administration under President Jimmy Carter, was the lead plaintiff. The relief sought included an accounting by the corporations for money derived from the slave trade, production of corporate documents, and establishment of a humanitarian fund to benefit institutions in the African American community.

Subsequent to the initial New York and New Jersey filings another suit seeking reparations was brought in California in the Northern District Court located in San Francisco. In September 2002, brothers Chester and Timothy Hurdle filed suit as direct descendants of slave parents. The plaintiffs sought $1.4 trillion dollars in damages against FleetBoston, Aetna, Lloyd's of London, New York Life Insurance Company, Westpoint Stevens, R. J. Reynolds Tobacco Company, Brown and Williamson Tobacco Corporation, Loews Corporation, and the Canadian National Railway Company. The complaint sought monetary damages as lost wages. They also contemplated suing the descendants of their father's slave owner.

In September 2002, 200 residents of Louisiana filed suit in federal court against companies that allegedly profited from slave labor. The defendants were Lloyd's of London; Brown Brothers Harriman and Co.; R. J. Reynolds; Liggett Group; Brown and Williamson; and three railroad companies, the Canadian National Railway Company, Norfolk Southern, and Union Pacific.

In the fall of 2002, another reparations claim was filed in the Eastern District Court of New York by EddLee Bankhead of Mississippi. Mr. Bankhead, at 119 years of age, was the oldest known resident of the United States.[12] He was a direct descendant of slave parents and sought redress for their enslavement.

In January 2003, Julie Wyatt-Kervin filed suit in the United States District Court for the Southern District of Texas against J. P. Morgan Chase, Union Pacific Railroad Company, and Westpoint Stevens.[13] Wyatt-Kervin's parents were enslaved Africans who worked on a Texas plantation producing corn, cotton, and cane. Another plaintiff in the suit, Ina Daniels Hurdle McGee, was the great-granddaughter of a slave who was sold and taken away from his family at a young age to be a playmate to a Texas plantation owner's son. The plaintiffs filed a demand for the corporations to provide records of any involvement they may have had with slave labor. They also sought an accounting of the corporations' slave-based profits, the creation of a constructive trust, restitution, disgorgement, and compensatory and punitive damages. Their complaint sought certification of a class made up of "all African-American slave descendants."

Most of the federal reparations suits filed during 2002 were consolidated into one federal case and transferred to the U.S. District Court in Chicago, Illinois. Judge Charles R. Norgle was selected to preside. Judge Norgle is reputed to be conservative and not necessarily viewed as a provider of a "friendly" forum by reparations activists.

Nonetheless, these recent reparations suits brought on behalf of direct slave descendants presented a stronger position than the cases that preceded them. As direct descendants of enslaved Africans, the plaintiffs all had standing to sue. They were represented by counsel and therefore qualified to represent a similarly situated class of persons. None of the suits sought damages against the United States, and thus the plaintiffs were spared the problem of overcoming a sovereign-immunity defense. Yet two thorny problems remained: whether the suits were time-barred and whether there was a remedy available when the actions of the defendants were legal at the time they were practiced.

The first of these two issues was addressed in *Cato*. The plaintiff there asserted that "she need not allege discrimination within any particular time period because discrimination is a continuing act." She argued, and the court agreed, that the continuing-violations doctrine applied because African Americans were still subjected to the badges and indicia of slavery. The court stated that the doctrine was applicable to constitutional and statutory violations. *Cato*'s problem was that the doctrine could not overcome the sovereign-immunity protection of the United States. Presumably nongovernmental defendants will not prevail if they assert the statute of limitations as a defense for their discriminatory practices and unjust enrichment.

The plaintiffs in the recent spate of cases assert these "crimes against humanity" have been ongoing. Since there is no statute of limitations on human rights violations, these actions should withstand a time-bar challenge.

The second issue is far more problematic. The *ex post facto* doctrine precludes the retroactive imposition of sanctions for actions that were legal at the time they were practiced. Therefore, a person cannot be charged with a crime or held liable for his past activities that were legal at the time but subsequently deemed illegal by legislative enactment. United States' jurisprudence does not permit retroactive punishment. Although some of the defendants in the various reparations suits have asserted that their actions were legal at the time they were practiced, only a thorough accounting of their records will indicate if, in fact, their claims were true. For example, the Transatlantic Slave Trade was deemed illegal in 1808. Any companies that continued to participate in and profit from that trade after 1808 may be subject to sanctions.

In a move that threatened radically to alter the discourse on reparations and global considerations, a major apartheid reparations complaint was filed in February 2003 against international corporations and banks that assisted apartheid South Africa. The suit, filed in the Eastern District Court of New York, named seven banks and twelve international corporations from Britain, France, Germany, the Netherlands, Switzerland, and the United States. It was the largest and most complex suit filed in the history of the reparations movement. In connection with the suit, the South African government expressed its support for the victims of apartheid and vowed to protect their futures under international human rights law. Several of the defendants had threatened political and economic pressure on South Africa to stop the call for apartheid reparations, but their efforts were futile.

The apartheid suit sought damages for South African victims of apartheid, not African American victims of slavery and continuing discrimination. Significantly, it served to highlight the antagonistic and hostile attitude of the United States government and its continued refusal to acknowledge the immoral character and horrors of slavery and the Transatlantic Slave Trade.[14] It also served to expose the lackluster efforts on the part of the United States to rectify and eliminate the vestiges of slavery and compensate its victims and their families.

Legislative Efforts

The courts in *Obadele* and *Cato* both suggested that the proper remedy for the plaintiffs' allegations rested with Congress rather than the courts. The various obstacles confronting the litigation strategists make the legislative approach more attractive to some reparations activists.

There has been a groundswell of legislative efforts in the reparations arena at the end of the twentieth century. Perhaps the most well-known effort in Congress is that of Representative John Conyers Jr. (D-Mich.), who has reintroduced H.R. 40 every year since 1989. The bill seeks to acknowledge the inhumanity and injustice of slavery between 1619 and 1865, and to establish a committee to study the institution of slavery, post–Emancipation Proclamation racial and economic discrimination, and the current-day impact. Congressman Conyers has been able to garner only Democratic support for the bill. Congress has never seen fit to vote H.R. 40 out of committee, and it has died in the House Subcommittee on the Constitution since its debut.[15]

Representative Donald Payne (D-N.J.) has sponsored H. Con. Res. 103, which calls on the States to explore the establishment of a commission to encourage and promote the implementation of education and awareness programs concerned with slavery and the contributions of African Americans. The resolution was referred to the Committee on Education and the Workforce.

Representative Cliff Stearns (R-Fla.) sponsored H.R. 196, which proposes the establishment of a North American Slavery Memorial Council to establish a memorial and museum on American slavery in Washington, D.C. This initiative would come under the auspices of the Department of the Interior.

In 1994, the Florida Legislature awarded $150,000 to each of nine survivors of the 1923 attack on blacks in Rosewood, Florida, and $500,000 to heirs of Rosewood property, and created a $100,000 scholarship fund. It had taken more than seventy years to gain an apology from the State of Florida and compensation for the victims.[16]

In 2000, a special commission in Oklahoma recommended that reparations be paid for one of the nation's deadliest racial clashes: a little-known 1921 rampage by a white Tulsa mob that killed as many as three hundred people, most of them black. Sadly, the families of the victims of the Tulsa, Oklahoma, race riots of 1921 were each compensated approximately $300 for the murder of their loved ones through a fund established by an interfaith religious coalition, the Tulsa Metropolitan Ministry Reparations Gift Fund. Despite an extensive report by the Tulsa Reparations Coalition commissioned by the State of Oklahoma finding that the state was liable for the deaths of the riot victims, the state legislature refused to appropriate funds to make the required payments. As of March 2002 the Tulsa victims and their families still had not been compensated by the state. The previous fall, Governor Frank Keating had crafted an explanation for the state's refusal to pay. In a letter addressed to the Tulsa Reparations Coalition in October 2001, Governor Keating stated: "I have carefully reviewed the finding of the Tulsa Race Riot Commission and, contrary to the statement in your letter, I do not believe that it assigns culpability for the riot to the state." He went on to say that "a state law prohibits Oklahoma from making reparations for any past mass crime committed by its officials or on the state's behalf." Now with only an empty promise, these aging potential plaintiffs were left to reconsider litigation.

Symbolically, during Black History Month 2003, lawyers for the Reparations Coordinating Committee headed by Professor Charles Ogletree of Harvard Law School filed suit in the Northern District Court of Oklahoma located in Tulsa on behalf of more than one hundred survivors of the Tulsa riots and another one hundred victims' descendants. The suit was not brought as a class action but as a civil claim on behalf of those who were personally harmed and their heirs. Named as defendants in the suit were the governor of Oklahoma, the mayor of Tulsa, and the Tulsa police department.

In 2000, California insurance companies were subject to newly enacted legislation that mandated that they provide records of insurance policies they held on enslaved Africans during United States slavery. The legislation was introduced by the state's then senator Tom Hayden and was signed into law by Governor Gray Davis.[17] Governor Davis also signed into law the University of California Slavery Colloquium Bill, now Chapter 1038 of the Education Code. It asks the University of California to hold a research conference to explore and identify issues concerning the economic legacy of slavery in America.

State Representative Ed Vaughn (D-Mich.) introduced a reparations bill in

Michigan "to acknowledge the fundamental injustices, cruelty, brutality, and inhumanity of slavery in the United States and the thirteen American colonies between 1619 and 1865 and to establish a commission to examine the institution of slavery, subsequent *de jure* and *de facto* racial and economic discrimination against African-Americans, and the impact of that discrimination on living African-Americans; and to make recommendations to the legislature on appropriate remedies."[18] State Representative Derrick Hale (D-Mich.) introduced a bill in 2000 to allow all African Americans who have lived in Michigan for the past ten years to claim up to $16,500 in tax credits each year for the next twenty years. Generally, Republicans in Michigan opposed these measures, but a few, like Representative Paul DeWeese (R-Mich.), said that they might support measures that would assist low-income African Americans to overcome economic disadvantages.

Numerous municipalities have entertained and/or passed resolutions in support of reparations. Chicago was the first major city in the nation to approve an ordinance that required all businesses competing for city contracts to search their records and disclose whether they profited from slavery. The Slavery Era Disclosure Ordinance was modeled after California legislation passed in 2000 that required insurers doing business with that state to disclose their records on slave policies. The Chicago ordinance went further and required all companies that did business with the city to make such disclosures. Alderman Dorothy Tillman (3rd) led the effort to pass the resolution.

In New York City, Councilman Charles Barron (D-N.Y.) led the effort in 2002 to have the city council adopt a resolution in support of reparations. A group of city council members introduced a resolution calling for a commission to study reparations for the descendants of enslaved Africans.[19] The resolution was the focus of an unusual public hearing by the council on reparations. Barron emphasized that slavery existed in New York City for two hundred years, from 1627 to 1827. According to Barron, the resolution's purpose was to show that "a crime has been committed, a people have been injured and compensation is due."

Where Litigation and Legislation Intersect

Native Americans, Jewish Holocaust victims, and Japanese victims of internment have all negotiated reparations payments with the support and assistance of the United States government.

The Holocaust reparations movement began with the efforts of Hugo Princz. Princz was the son of a naturalized American father who was living with his family in Slovakia in 1942. At that time, the Germans sent the family to Maidanek concentration camp because they were Jewish. The family was killed by the Nazis and Princz was the lone survivor. He was then forced to work as a slave laborer, stacking corpses at Auschwitz. Princz was rescued three and a half years years later by United States soldiers when the train he was traveling on to Dachau was intercepted.

Princz fought for compensation for the murder of his family and the abuses he suffered on several fronts simultaneously: in the courts, the Congress, and the White House. He tried unsuccessfully to sue the German government but was able to sue German companies: one, a successor to Messerschmidt, a German maker of war machinery, and two, a pharmaceutical and chemical company, I. G. Farben. Although Princz failed in his attempt to get a United States court to allow him to sue Germany, he almost convinced Congress to pass a bill that would permit Unites States citizens to sue foreign countries in federal courts.

Three United States presidential administrations attempted to work behind the scenes with Germany to secure a diplomatic settlement. Princz's lobbying efforts were very successful. Through his lawyer, he was able to get congressional figures to lobby the German government. The State Department got involved as did key congressional members Joe Lieberman, Bill Bradley, Bill Clinton, and Al Gore.

In 1995, Princz and ten other United States citizens settled their reparations claim against Germany for more than $2 million. The agreement was between the governments of the United States and Germany. The United States then transferred the money to the survivors. Princz also settled against four German companies whose predecessors had used him as a slave laborer and collected an undisclosed amount.

Following Princz's success new negotiations between Germany and the United States began. In 1996, Janet Reno announced that the U.S. Foreign Claims Settlement Commission had been authorized by Congress to determine the validity of claims. A three-judge panel was assigned to evaluate the claims before passing them on to the State Department, which would then negotiate a settlement with the German government. More than three hundred Holocaust survivors came forward. Eligibility included United States citizenship at the time of persecution at the hands of the Nazis and internment in a concentration camp or under comparable conditions for at least six months. Claimants had to be of minimal means.

By 1996, the United States Senate passed a resolution supporting the Holocaust survivors. Then, in 1998, Congress passed a law establishing the Jewish Holocaust Commission to assist Jews in claiming assets from European governments. Any money collected was deemed tax exempt. Between 1999 and 2002, Germany bent to pressure from the United States government and Jewish groups, and agreed to pay $110 million to Holocaust survivors in Eastern Europe.

Several treaties between Native Americans and the United States have been in existence for more than one hundred years. Many of these treaties were designed to protect land interests of Native Americans. Despite these treaty protections, millions of acres of Native American land were stolen by the United States. Billions of dollars have been paid out and millions of acres of land returned by the United States to compensate for stolen land and mineral rights in Alaska, Oregon, South Dakota, Florida, Wisconsin, and Utah. As recently as 2000 the United States promised to return 80,000 acres of land in Utah taken from the Ute Indian Tribal Nation to help fuel Navy ships during World War I. And at least as far back as 1971 the United States was settling land claims pursuant to the Alaska Native Land Settlement Agreement. During the purchase of Alaska from Russia, the human, land, and mineral rights of Alaska's Native and Indian peoples were totally ignored. As a result of the settlement, 44 million acres of land were returned and $1 billion was paid.

Japanese Americans were forced to resort to the courts and Congress in order to secure reparations for their illegal internment during World War II. In 1942, President Franklin D. Roosevelt issued Executive Order 9066, which delegated "authority to military commanders to exclude 'any and all persons' from designated areas in order to provide security against espionage and sabotage. The entire Pacific Coast was determined to be susceptible to espionage, and all persons of Japanese descent were excluded from the area.[20] Those persons of Japanese descent who either violated curfew or refused to relocate were prosecuted and forcibly removed from their homes. The leading case challenging the internment was *Korematsu*.[21] Mr. Korematsu challenged his internment and the fact that he was forced to leave his home in California. The court of claims applied a strict-scrutiny standard, found no unconstitutional deprivation of liberty, and denied his claim.

Although *Korematsu* had not been overruled, in 1948 Congress passed the American-Japanese Evacuation Claims Act, which permitted claims for particular losses or damage to property, and for injury to physical or mental health.[22] Due to the Act's burdensome requirements, it afforded little relief for those whom it was intended to vindicate.[23]

The Commission on Wartime Relocation and Internment of Civilians encouraged Congress and the president to give recognition to the grave injustice that had occurred and to offer "the apologies of the nation for acts of exclusion and detention."[24] The commission recommended that a special foundation be funded but acknowledged that "no fund would be sufficient to make

whole again the lives damaged by the exclusion and detention."[25] The result-
ant Civil Liberties Act of 1988 awarded a symbolic $20,000 restitution to each
"eligible individual." "Eligible individual" was defined as

> any individual of Japanese ancestry who is living on the date of the en-
> actment of this Act . . . and who, during the evacuation, relocation,
> and internment period [December 7, 1941, through June 30, 1946] . . .
> was a United States citizen . . . and . . . was confined, held in custody,
> relocated, or otherwise deprived of liberty or property as a result of . . .
> [government actions].[26]

The Civil Liberties Act (CLA) has withstood Equal Protection challenges.[27]
Arthur D. Jacobs, an American citizen, was interned at Ellis Island in February of 1945 along with his German father. Mr. Jacobs's father was interned after an individualized hearing. The Jacobs family was transferred in April 1945
from Ellis Island to Texas, where they were interned with other Japanese-
American families. They remained interned until the beginning of December
1945.

Jacobs challenged the constitutionality of CLA because he was denied eligibility for compensation. The Court of Appeals in *Jacobs v. Barr* gave great
deference to the fact that Congress had found that Japanese Americans were
the victims of widespread racial prejudice, while other racial groups, such as
Italians and Germans, were not. According to the court, the decision by Congress to provide restitution to Japanese Americans was "narrowly tailored to
fit the compelling government interest of compensating those victims of
racial prejudice."[28]

The support and involvement of Congress has been essential in the acquisition of reparations for Native Americans, Japanese Americans, and Jewish
Holocaust victims. In the case of Jewish Holocaust victims, the highest eche-
lons of the United States government supported their quest for reparations
against Germany.

Conclusion

For African Americans, the legislative route has not yielded as many victories
as one might have hoped. The battle to overcome long-entrenched racist views
against African Americans in the United States Congress, state legislatures,
and local governing councils has been frustrating. Similarly, efforts to prevail
in the courts have met one roadblock after another.

More lawsuits are anticipated. Groups such as N'COBRA and the Reparations Coordinating Committee have been preparing detailed comprehensive litigation strategies. Their targets will likely be both the public and private sector. It can be expected that their claims will be informed and shaped by the Chicago and Tulsa litigation.[29] United States history is replete with examples of white mob attacks on black individuals and towns that were ignored by local and federal law enforcement. Places such as Wilmington, North Carolina, where in 1898 white supremacists murdered African Americans, exiled their government officials, and drove thousands of their businessmen out of town, may provide excellent forums for litigation of those past atrocities.

Ultimately mass pressure will be required to leverage even a partial victory in the courts or in the halls of Congress. It is essential for a successful reparations movement to build popular support throughout the United States and to educate the entire community on the issues. It seems that only then will Congress, state legislatures, and the courts respond.

CURRENT STRATEGIES OF THE NATIONAL BLACK UNITED FRONT

The momentum gained from the Millions For Reparations Mass Rally, held on August 17, 2002, in Washington, D.C., continued to solidify cooperative arrangements between grassroots organizational work and the legislative and legal components of the reparations movement. The people who attended the rally from sixty-six cities and thirty-eight states across the country comprise an ever-growing coalition of persons supportive of reparations even though media outlets often ignore the movement. The rally was viewed by thousands of African people in America unable to attend the rally and was joined by thousands who held simultaneous rallies in cities across the country.

As the century began, grassroots organizers are increasingly vocal about their support of reparations, and the movement is fast becoming the "glue" that unites Black human rights organizations across the board. Conrad Worrill, National Chair of the National Black United Front (NBUF), described the media's reluctance to give attention to the reparations struggle in these terms:

> We cannot let the media put us to sleep or sidetrack us with peripheral issues. To the contrary, we must intensify our organizational efforts and stay the course, demanding that we be repaired for the most heinous crime in the annals of human history, the European chattel

slave enterprise that enslaved millions of Africans and resulted in an incalculable number of deaths in the millions among Africans and their African descendants.

The December 12th Movement, one of the co-sponsors of the rally, distributed thousands of position statements that emphasized several points at the demonstration:

> Today's Rally occurs within the historical context of 400 years of struggling for justice and genuine freedom for our people. The legacy of the freedom fighters and organizations that fought and continue to fight this good fight have collectively contributed to our clarity on the line of march, the necessary tactics, and understanding of on who does this battle depend? . . . Therefore it is with clarity that we must return to the scene of the crime. Yes, because we are perfectly clear it was/is a crime, and the 2001 UN World Conference Against Racism, in Durban, South Africa, removed any doubt of the type and magnitude of the crime it was/is, by its declaration that the Trans Atlantic Slave Trade Was A Crime Against Humanity. And as such, crimes against humanity have no statutes of limitation. The crimes committed against our people have no parallel in the annals of human history. They have no historical precedent.

Three Specific Organizing Strategies

After the August rally, the NBUF Central Committee chose three specific organizing strategies to help fuel the momentum gained by the Millions For Reparations Mass Rally and to add to the increase in interest in the reparations movement among African people across the United States.

Strategy I: The Elected Official Reparations Survey Scorecard

This is a survey that will be distributed to elected officials in the U.S. Congress, in state government, and local government. The survey contains the following language:

> As you may be aware, there is a national debate taking place over the issue of reparations for the horrific crime against humanity, the enslavement of African people and its vestiges. We are conducting a survey to determine exactly where governmental representatives stand.

Questions:

1. Do you support the H.R. 40 Bill, sponsored by Congressman John Conyers, which proposes that a commission be formed to study the impact of the enslavement on African people and its vestiges, and to recommend various remedies? Yes ___ No ___

2. If this commission concluded that reparations were appropriate in some form (monetary compensation, land, educational opportunities, and so on), would you support the commission's recommendation? Yes ___ No ___

The result will be published and made available to the community so that we can be clear on where elected officials stand on the question of reparations for African people in America.

Strategy II: Petition/Declaration Campaign

NBUF will launch Phase IV of the NBUF Petition/Declaration Campaign. During Phase III, NBUF was successful in gathering 157,000 signatures from African people in America who agreed that the United States government committed acts of genocide against them. In May 1997, it officially filed the genocide petitions at UN headquarters in Geneva. The acts of genocide outlined by NBUF on the part of the United States government spurred NBUF to continue its Anti-Genocide Campaign. It is clear to NBUF that "Continued Genocide" by the United States government is intimately connected to the demand for reparations.

The NBUF Genocide/Reparations Petition will once again be submitted to the United Nations on September 13, 2003, in conjunction with the United Nations Reparations March, which is being organized to further expose and dramatize the connection between genocide and reparations.

Strategy III

NBUF has taken up as a part of reparations organizing work the mobilization of the African community in America to support reparations lawsuits filed by Deadria Farmer-Paellmann and others. This is also an outgrowth of the Millions For Reparations Mass Rally, where Farmer-Paellmann made an impressive appeal for the community to support the lawsuit by appearing in the courtroom during proceedings.

The Farmer-Paellmann lawsuit was filed on March 26, 2002, in the United States District Court for the Eastern District of New York. In this unprecedented lawsuit, which seeks payback from companies with historical ties to the slave trade over 137 years ago, eight major corporations were named. The suit named Aetna, CSX, and FleetBoston Financial, among others, as "unjustly profiting from the slave trade before the Civil War ended in 1865." Deadria C. Farmer-Paellmann, the lead plaintiff in the lawsuit, is considered the "Rosa Parks of the Reparations Litigation Movement." As an attorney-activist, she has studied the relationship of government and corporate policies that determined economic and social policies impacting Africans in America.

Since the filing of the March 26 lawsuit, momentum increased, and reparations lawsuits have been filed in New Jersey, Virginia, North and South Carolina, Mississippi, Louisiana, and Texas. The group of lawyers and plaintiffs led by two law firms—Thomas, Wareham, and Richards in New York, and Jean Baptiste and Associates in Chicago—are called the "Corporate Restitution Team." These cases were originally filed in several different states. Since then the cases have been consolidated before Judge Charles Norgle, U.S. District Court of Illinois, Eastern Division in Chicago, Illinois.

REPARATIONS: THE MISSISSIPPI MODEL

The contemporary demand for reparations is due to the United States–sponsored chattel enslavement of African men, women, and children that resulted in forced labor without compensation for nearly three hundred years. It is also a form of economic justice for the centuries of racial discrimination, Jim Crow laws, and vestiges of slavery that continue today.

Baba Hannibal Afrik served in several capacities within the reparations movement for nearly fifteen years as Life Member and past National Co-chair for the National Coalition of Blacks for Reparations in America (N'COBRA), 1995–2001, and Minister of the Interior, Provisional Government/Republic of New Afrika, 1999–2002. He was also a delegate to the United Nations World Conference Against Racism in Durban, South Africa, in August 2001. Our strong African in America delegation, the Durban 400, lobbied for a major resolution on reparations in concert with the African Heads of State.

Under the leadership of the December 12th Movement and the National Black United Front, the Durban 400 successfully influenced the final conference resolution, which proclaimed that the "Transatlantic slave trade was a crime against humanity" and compensation should be given to the victims and their descendants.

Plans were made to conduct forums to provide information and inspiration to our people in order to move to the next level, mass mobilization.

Hannibal was designated to be the Mississippi State Coordinator for the Millions for Reparations Mass Rally, which was held in Washington, D.C., on August 17, 2002.

With co-leader Asinia Lukata Chikuyu, a veteran organizer in the Jackson area, Hannibal formed the Mississippi Reparations Committee (MRC) and created a Love Caravan, which brought a busload of students, parents, educators, and community activists to that historic national event on August 17, 2002, where more than fifty thousand participants demanded "Reparations Now!"

The Mississippi delegation attracted national media coverage through the Associated Press wire services and C-SPAN television programs. The *Washington Post* and several newspapers in Alabama and Mississippi printed articles based on interviews of our delegates.

Community Awareness Campaign

With the successful completion of the Washington, D.C., rally, the Mississippi Reparations Committee expanded its organizing efforts by establishing a Task Force Leadership Structure. This included Jackson City Councilman Kenneth Stokes and veteran publisher Charles Tisdale of the *Jackson Advocate* as honorary co-chairs. There was also attorney Chokwe Lumumba, legal advisor, along with a steering committee of representatives for the Malcolm X Grassroots Movement, the Jackson chapter of N'COBRA, and Cultural X-Pressions entrepreneurs.

Local contact persons were identified in sixteen cities to help sponsor Reparations Town Hall Meetings throughout the state. Correspondence was sent to all of the states' African American senators, representatives, and mayors, requesting them to support the Congressional Reparations Bill, H.R. 40, sponsored by Congressman John Conyers every year since 1989.

Monthly press releases were sent to electronic and print media in order to provide updated information. Interviews have been conducted on various radio programs in Mississippi and nationally.

Mass Mobilization

The Reparations Town Hall meetings are designed to educate and organize grassroots support in several ways, including soliciting volunteers to secure signatures on our petitions and to challenge elected officials to sponsor reparations resolutions.

These gatherings have helped the community to understand the legal and legislative importance of our movement—nationally and internationally. A wide range of literature, books, and study materials has been made available as well as committee T-shirts and buttons.

The initial meetings in October 2002 were held at Jackson State University and Tougaloo College, where student committees are now involved in hosting classroom and campus discussions and organizing reparations study groups. Since that date, several mass meetings have been held throughout the Jackson area, including a "Reparations Sunday" event on Ujamaa (cooperative economics) during the Kwanzaa season.

A workshop was conducted at Rust College in Holly Springs during its Ninth Annual National Student Leadership Conference, January 2003. The presentation focused on the topic "Organizing Student Advocates for the Reparations Movement."

Town Hall meetings were sponsored in McComb, West Point, Starkville, Laurel, and Port Gipson, with plans for Natchez, Bolton, Belzoni, Philadelphia, Alcorn State University, and Mississippi Valley State University, among others.

An Action Agenda will be formulated to ensure operational unity between a diverse cross-section of Mississippi organizations and faith-based groups. This process will concretize our demands for economic justice and should become the litmus test to determine the moral integrity of Mississippi and America regarding social equality.

Conclusion

The question of reparations is the most unifying issue our people have today. However, education is an essential ingredient for mass mobilization. The Mississippi Reparations Committee considers this to be a sacred vocation for repairing, healing, and restoring African people, "those at home and abroad." It is a divine duty to obtain economic justice for our beloved ancestors.

WHAT'S NEXT IN THE REPARATIONS MOVEMENT?

As N'COBRA continues its work in the reparations movement, it has established three tracks around which it is organizing: education, economic development, and preparation for the victory.

Track I: Education

The average person or child on the street must be able to understand and explain briefly why reparations are due and needed. N'COBRA's goal is to demystify reparations such that "reparations" becomes a household word, and "of course" or "yes" is the first thought that follows it. Educational strategies include trials, tribunals, state and federal congressional hearings, United Nations interventions, town hall meetings, academic research, and well-informed speakers at national and international conferences. These venues allow the masses to participate in their own education and have a safe space to channel their hurt and grief into constructive action.

A. Information and Media

Educating and informing the public has been an uphill battle. It has been an uphill battle due to the lack of resources and control over the media and institutions that impact our lives. This is a vestige of slavery and current day practices.

N'COBRA members, reparations advocates, and others who produce books, records, compact disks, audio and video tapes, DVDs, newsletters, magazines, newspapers, use Web sites, list serves, and other tools of technology are vital to our efforts to overcome media biases and the suppression of the Black reparations movement.

N'COBRA is working with artists who want to become more involved with activists and academics in the reparations movement. It encourages our talented youth and adults to write scripts that include intelligent, serious, and comic reparations conversations in their movies, plays, and concerts. Creativity and culture are ideal mediums to spread the word to as broad an audience as possible.

Many years ago, N'COBRA declared February 25 as Reparations Awareness Day. This has become institutionalized in many colleges and communities, and provides a great opportunity for chapters and member organizations to host reparations events. Also, N'COBRA is working with members of the Faith Community to institute a Reparations Sunday to engage the spiritual aspect of the reparations movement.

B. Legislative

N'COBRA's Legislative Commission launched "A Year Of Black Presence" on Capitol Hill. On January 30, 2003, Philadelphia members filled the House and the Senate office buildings to educate members of Congress on the urgency of passing H.R. 40. This scene will be repeated each month by N'COBRA chapters from a different state until H.R. 40 is passed and a Congressional hearing is conducted. N'COBRA members are influencing representatives at state and local levels to pass resolutions in support of H.R. 40, set up state or local commissions similar to H.R. 40, and to enact specific reparations proposals.

C. International

N'COBRA established its International Commission recognizing that African descendants scattered throughout the Diaspora are, by birthright, connected to land, resources, and kinships in Africa that must be restored. Africans dispossessed of their land and their resources through European invasion, colonialism, and neo-colonialist terror are also due reparations and debt relief. Several N'COBRA members traveled to the United Nations in Geneva, Switzerland, to petition for our human rights and recognition of African descendants as a people. Without a status similar to that of indigenous nations, whose experience with Europeans was the same as that of Africans, N'COBRA felt that Africans have no human rights as a people, and are individual clones of our various oppressors, always at their mercy.

After N'COBRA's outstanding contributions to the successful outcome of the World Conference Against Racism (WCAR), members now participate in the development and work of the Global Afrikan Congress (GAC). The GAC grew out of a commitment of African and African Diasporans who participated in the WCAR process to stay connected and make real the UN Program of Action that we created. Dorothy Benton Lewis was elected as one of the seventeen representatives who formed the Interim Steering Committee of the GAC.

D. Legal

The Legal Strategies Commission now partners with the Reparations Coordinating Committee and is co-chaired by Charles Ogletree and Adjoa Aiyetoro. While N'COBRA does not diminish the legal efforts, it believes that the reparations movement does not turn, rest, or depend solely on a legal strategy. It is just one strategy among many. A case winding its way through the courts or lost in a court decision is simply fuel for another strategy: legislation, international, or direct action. N'COBRA encourages individuals to research their family, city, and state history. A number of N'COBRA members have individual lawsuits, and such research can provide evidence for state and local class-action cases. In order for their complicity to be fully known, every institution involved in the dehumanizing process of enslavement will ultimately be exposed by the reparations advocates.

Track II: Economic Development

N'COBRA supports a reparations settlement strategy that will benefit generations. This requires infrastructure development in preparation for the victory. N'COBRA is working with the business community to institutionalize a buy-Black project called "Black Fridays," and to design and develop a new community as a demonstration project.

N'COBRA's Economic Development Commission periodically canvases African people for views on what a reparations settlement should look like. The mass-based approach is designed to ensure community awareness of and involvement in reparations business conducted in their names on their behalf. The commission conducts elections of commissioners who must be nominated by an organization that can vouch for their integrity and competence to represent the African Community regarding any settlement. The Commission is committed to economic development for self-reliance and participation in projects that support that commitment.

Track III: Preparation for the Victory

There must be transformation within the context of the Reparations Movement. So often we think that our perceived differences can be bridged through rules and points of unity. Some of us, often the best among us, make the rules and are the first to break them. We have seen a number of brilliant organizations with brilliant people start and fail. We have seen grand projects with bright promise fizzle to a shadow of their potential. Through N'COBRA's Organizational Division and the Global Afrikan Congress, North American Regions, we are beginning the process of transformation work within the reparations movement.

N'COBRA is gratified to know that Africans have the skills among them to transform five hundred years of terror. It holds annual conferences in June to address the issues in Tracks I and II. Now it is time to institutionalize conference retreats that allow each of us to look within to see the kind of emotional and behavioral baggage we may be carrying as a consequence of our enslavement and the intergenerational trauma of ongoing oppression. This looking within together allows us to see ourselves in others and to remove any blockages that get in the way of our fulfilling on our commitments. This may be one of the most important tasks of the movement in that it will make all other work most effective.

BABA HANNIBAL AFRIK, *a former co-chair of National Coalition of Blacks for Reparations in America, is one of the founding members of the Republic of New Afrika. Known internationally as a community organizer, he is also one of the founders of the Council of Independent Black Institutions, a coalition of independent African educational organizations.*

JILL SOFFIYAH ELIJAH *is a Clinical Instructor at the Criminal Justice Institute of Harvard Law School. She is an expert on criminal justice and prison issues. Her research and involvement in the reparations movement spans more than two decades.*

DOROTHY BENTON LEWIS *came to the reparations movement in the early 1960s from Fairbanks, Alaska. She has participated in the founding and leadership of several reparations organizations, including the Black Reparation Commission, which lobbied other organizations to adopt the reparations agenda. She authored several reparations booklets that helped shape the early conversations and development of the reparations movement. Currently she is the National Co-chair of the National Coalition of Blacks for Reparations in America, and one of two North American representatives to the Global Afrikan Congress.*

CONRAD W. WORRILL *is a professor at Northeastern Illinois University in Chicago and National Chair of the National Black United Front. He is considered by many to be one of the best community organizers in the nation. The author of numerous articles on Black nationalism and political organizing, Worrill is one of the founding members of the National Black United Front.*

NOTES

FOREWORD

1. Jubilee South statement on Bush proposals—Thursday, July 19, 2001, Genoa, Italy.
2. The United States and Israel walked out of the United Nations World Conference Against Racism, Racial Discrimination, Xenophobia, and Related Intolerance four days after it began, reputedly because of the Zionism = Racism issue that was being discussed, as well as the harassment of Israeli delegates and distribution of anti-Semitic literature. While the Palestinian issue loomed large at the gathering, many observers, particularly South African journalists, felt that the real reason for the U.S.'s pullout reflected its trepidation and avoidance of the reparations issue, which had wide support by delegates to the conference.

THE AFRICAN AMERICAN WARRANT FOR REPARATIONS

1. Arnold Schuchter, *Reparations: The Black Manifesto and Its Challenge to White America* (Philadelphia: Lippincott, 1970), p. 15.

RESTITUTION AS A PRECONDITION OF RECONCILIATION

1. Wole Soyinka, *The Burden of Memory, the Muse of Forgiveness* (New York: Oxford University Press, 1999), pp. 28–29.
2. *The Star* (South African newspaper), *Independent* Online, December 10, 2000.
3. Universal Declaration of Human Rights, adopted by the U.N. General Assembly, December 10, 1948, U.N.G.A. res. 217 A (III), U.N. Doc. A/810, at 71 (1948), reprinted in Richard B. Lillich and Hurst Hannum, *International Human Rights: Documentary Supplement* (Aspen: Aspen Law & Business, 1995), pp. 17–22. American Convention on Human Rights, O.A.S. Treaty series No. 36, 1144 U.N.T.S. 123 centered into force July 18, 1978, reprinted in United Nations, *Treaty Series: Treaties and International Agreements Registered or Filed and Recorded with the Secretariat of the United Nations*, vol. 1144 (New York: United Nations, 1987), pp. 123–63, or in Richard B. Lillich and Hurst Hannum, *International Human Rights: Documentary Supplement* (Aspen: Aspen Law & Business, 1995), pp. 145–67.
4. Soyinka, *Burden of Memory*, p. 31.

5. *The Independent* (South African Newspaper), *Independent* Online, December 17, 2000.

6. Noam Chomsky and Edward Herman, *Manufacturing Consent: The Political Economy of the Mass Media* (New York: Pantheon, 1988), pp. 37–86.

7. For a comparative discussion of the Aleut and the Japanese internment cases, see "Report of the Commission on Wartime Relocation and Internment of Civilians" and "German Americans, Italian Americans, and the Constitutionality of Reparations: Jacobs v. Barr," in *When Sorry Isn't Enough: the Controversy Over Apologies and Reparations for Human Injustice*, edited by Roy L. Brooks (New York: New York University Press, 1999), pp. 171–76, 206–16.

8. David Stannard, *American Holocaust: Columbus and the Conquest of the New World* (New York: Oxford University Press, 1992), p. 11. "Contact" is used to mark the beginning of the European encounter with indigenous people.

9. See Sharon Venne, *Our Elders Understand Our Rights: Evolving International Law Regarding Indigenous Rights* (Penticton, British Columbia: Theytus Books, Ltd., 1998) for the best analysis of the history and formation of the Draft Declaration, and an explication of its legal and political significance for indigenous peoples.

10. Chomsky and Herman, *Manufacturing Consent*, p. 31.

11. Ian Hancock, "Romani Victims of the Holocaust and Swiss Complicity," in *When Sorry Isn't Enough: The Controversy over Apologies and Reparations for Human Injustice*, edited by Roy L. Brooks (New York: New York University Press, 1999), p. 68.

12. Ibid., pp. 69–70.

13. Ibid., p. 70.

14. See Queen Lili`uokalani's letter of protest reproduced in her statement to Commissioner James H. Blount, House Ex. Doc. No. 47, 53rd Congress, 2nd Session, *Foreign Relations of the United States, 1894: Affairs in Hawaii*, Appendix II (Washington: Government Printing Office, 1895), p. 866.

15. See U.S. President Grover Cleveland's message in "President's Message Relating to the Hawaiian Islands, December 18, 1893," House Ex. Doc. No. 47, 53rd Congress, 2nd Session, *Foreign Relations of the United States, 1894: Affairs in Hawaii*, Appendix II (Washington: Government Printing Office, 1895), p. 456.

16. Noenoe K. Silva, "Kanaka Maoli Resistance to Annexation," *'Oiwi: A Native Hawaiian Journal*, December 1998, p. 61. Also see Noenoe K. Silva, "Ke Ku`e Kupa`a Loa Nei Makou: Kanaka Maoli Resistance to Colonization" (Ph.D. diss., University of Hawai`i, 1999), pp. 184–201.

17. For a detailed analysis of continuing American violations of both universal and specific indigenous human rights of the Hawaiian people, see the discussion in my book *From a Native Daughter: Colonialism and Sovereignty in Hawai`i* (Honolulu: University of Hawai`i Press, 1999), pp. 27–40. For a historical discussion of the evolution of specific Hawaiian groups and their arguments regarding native self-determination, including land and water claims, see Melody MacKenzie, *Native Hawaiian Rights Handbook* (Honolulu: Native Hawaiian Legal Corporation, 1991), pp. 3–104. Also see the process for reconciliation between Hawaiians and the United States in the Ka Lahui Hawai`i Master Plan of 1995, reprinted in full in *From a Native Daughter*, pp. 211–36.

18. U.S. Public Law 103–150, S.J. Res. 19, 103d Congress, 1st Session, 107 Stat 1510 (November 23, 1993).

19. See Andrew Hacker's revealing comparisons between white and Black Americans regarding their state of health, economic and sociopolitical opportunities, and general living conditions in his pathbreaking book, *Two Nations: Black and White, Separate, Hostile, Unequal* (New York: Ballantine Books, 1995).

REPARATIONS FOR THE DESCENDANTS

1. The Fifth Amendment of the U.S. Constitution states that property cannot be taken except for a "public purpose" and only if "just compensation" is provided to the owner of the property. See

discussion of *Dames & Moore v. Regan*, in the text below at notes 36 and 37, for examples of judicial claims being recognized as protectable property interests.

2. See, for example, Vincene Verdun, "If the Shoe Fits, Wear It: An Analysis of Reparations to African Americans," *Tulane Law Review* 67 (1993): 597; Rhonda V. Magee, "The Master's Tools from the Bottom Up: Responses to African American Reparations Theory in Mainstream and Outsider Remedies Discourse," *Virginia Law Review* 79 (1993): 863; Eric K. Yamamoto, "Racial Reparations: Japanese American Redress and African American Claims," *Boston College Law Review* 60 (1998): 477; Adrienne D. Davis, "The Case for United States Reparations to African Americans," *Human Rights Brief* 7 (Spring 2000).

3. Randall Robinson, *The Debt: What America Owes to Blacks* (New York: Dutton, 2000). See also Boris Bittker, *The Case for Black Reparations* (1973); Robert Westley, "Many Billions Gone," *Boston College Law Review* 40 (1998): 429; and *Boston College Third World Law Journal* 19 (1998): 429, republished in this volume.

4. Robinson, *The Debt*, p. 216. Robinson argues that slavery involved the "loss of millions of lives," but also was worse than other acts of genocide because "its sadistic patience, asphyxiated memory, and smothered cultures, has hulled empty a whole race of people with inter-generational efficiency."

5. *Id.*, pp. 204–05, 211.

6. *Id.*, p. 208.

7. *Id.*, p. 201; see, for example, H.R. 3745 (1989) and H.R. 40 (1999).

8. See H.R. Res. 96, 105th Cong. (1997) and H. Con. Res. 346 (2000) (the Apology for Slavery Resolution of 2000).

9. Universal Declaration of Human Rights, art. 4, December 10, 1948, U.N.G.A. Res. 217A, 3 U.N. GAOR, U.N. Doc. A/810, at 71 (1948)("No one shall be held in slavery or servitude; slavery and the slave trade shall be prohibited in all their forms").

10. International Covenant on Civil and Political Rights, art. 8(1), December 16, 1966, 999 U.N.T.S. 171 (1966)("No one shall be held in slavery; slavery and the slave trade in all their forms shall be prohibited").

11. European Convention for the Protection of Human Rights and Fundamental Freedoms, art 4(1), November 4, 1950, 213 U.N.T.S. 221, Eur. Treaty Ser. No. 5 (1950), revised by Protocol 11 ("No one shall be held in slavery or servitude").

12. American Convention on Human Rights, art. 6(1), November 22, 1969, 1144 U.N.T.S. 123, O.A.S. Treaty Ser. No. 36 (1969), 9 I.L.M. 673 (1970)("No one shall be subject to slavery or to involuntary servitude, which are prohibited in all their forms, as are the slave trade and traffic in women").

13. African Charter on Human and Peoples' Rights, art. 5, O.A.U. Doc. CAB/LEG/67/3 Rev. 5 (1981)("Every individual shall have the right to the respect of the dignity inherent in a human being and to the recognition of his legal status. All forms of exploitation and degradation of man, particularly slavery, slave trade, torture, cruel inhuman or degrading punishment or treatment shall be prohibited").

14. Slavery Convention, September 25, 1926, 60 L.N.T.S. 253, 46 Stat. 2183; Protocol Amending Slavery Convention, December 7, 1953, 182 U.N.T.S. 51.

15. *Restatement (Third) of the Foreign Relations Law of the United States* sec. 702 (1987)("A state violates international law if, as a matter of state policy, it practices, encourages, or condones . . . (b) slavery or slave trade . . ."); *id.* sec. 703(3)(recognizing the rights of victims to pursue all applicable remedies against perpetrators of such grave human rights abuses). See also Steven R. Ratner and Jason S. Abrams, *Accountability for Human Rights Atrocities in International Law—Beyond the Nuremberg Legacy* (Oxford: Clarendon Press, 1997): p. 105 ("Slavery violates international law in all circumstances, its prohibition rising to the level of a *jus cogens* norm").

16. Rome Statute of the International Criminal Court, art. 7(1)(c), July 17, 1998. This treaty was rat-

ified by the sixtieth nation in the spring of 2002, and took effect for those nations that have ratified it on July 1, 2002.

17. See, for example, *United States v. The Schooner Amistad,* 40 U.S. (15 Pet.) 518 (1841) (releasing Africans from the schooner *Amistad* and recognizing the Africans' right to resist "unlawful" slavery); *Prigg v. Pennsylvania,* 41 U.S. (16 Pet.) 536 (1842)("By the general Law of nations, no nation is bound to recognise the state of slavery, as to foreign slaves found within its territorial dominions . . .").

18. See, for example, *Iwanowa v. Ford Motor Company,* 67 F. Supp. 2d 424, 440–41 (D.N.J. 1999) (citing Robert Jackson, *The Nuremberg Case* (1971), xiv–xv, and the Nuremberg Charter, art. 6, 59 Stat. 1544, 82 U.N.T.S. 279); see also *Princz v. Federal Republic of Germany,* 26 F. 3d 1166, 1180 (D.C. Cir. 1994)(acknowledging that forced labor of civilians in World War II violated international law).

19. See, for example, *United States v. Matta-Ballesteros,* 71 F. 3d 754, 764 n. 5 (9th Cir. 1995)(recognizing the prohibition on slavery as one of the *jus cogens* norms of international law, "which are nonderogable and peremptory, enjoy the highest status within customary international law, are binding on all nations, and can not be preempted by treaty"); *Siderman de Blake v. Republic of Argentina,* 965 F. 2d 699, 715 (9th Cir. 1992), *cert. denied,* 113 S. Ct. 1812 (1993)(concluding that the "universal and fundamental rights of human beings identified by Nuremberg" included "rights against genocide, enslavement, and other inhuman acts"); *Tel Oren v. Libyan Arab Republic,* 726 F. 2d 774, 781 (D.C. Cir. 1984)(Edwards, J., concurring)(stating that slavery is a "heinous act," which violates "definable, universal and obligatory norms" of international law); *John Doe I v. Unocal Corp.,* 110 F. Supp. 2d 1294, 1304, 1305 (C.D. Cal. 2000)(stating that "[i]t is well accepted that torture, murder, genocide and slavery all constitute violations of *jus cogens* norms," that "the law of nations has historically been applied to private actors for the crimes of piracy and slave trading, and for certain war crimes," and that "the term 'slavery' now encompasses forced labor"); *Iwanowa v. Ford Motor Co.,* 67 F. Supp. 2d 424, 441 (D.N.J. 1999)("Thus, the case law and statements of the Nuremberg Tribunals unequivocally establish that forced labor violates customary international law").

20. Robinson, *The Debt,* p. 216.

21. Ibid., p. 204.

22. Chris K. Iijima, Speech to Annual Meeting of American Association of Law Schools, January 4, 2001.

23. Dr. Loury's observations are quoted in Diane Cardwell, "Seeking Out a Just Way to Make Amends for Slavery," *New York Times,* August 12, 2000, http://www.nytimes.com/library/arts/081200 slavery-reparations. htm.

24. Davis, "The Case for United States Reparations."

25. Paul Brest, Sanford Levinson, J. M. Balkin, Akhil Reed Amar, *Processes of Constitutional Decisionmaking* 4th ed. (New York: Aspen Publishing, 2000), p. 352.

26. Ibid.

27. Robinson, *The Debt,* p. 27.

28. Universal Declaration of Human Rights, note 9 above, art. 8 (emphasis added).

29. International Covenant on Civil and Political Rights, note 10 above, art. 2(3)(a) (emphasis added).

30. European Human Rights Convention, note 11 above, art. 6(1).

31. American Human Rights Convention, note 12 above, art. 25(1).

32. 4 Inter-Am. Ct. H.R. (Ser. C), para. 174d (1988), reprinted in 28 I.L.M. 291 (1989).

33. Ibid., para. 166 (emphasis added).

34. *Mentes v. Turkey,* 37 I.L.M. 858, 882 para. 89 (1998). The European Court of Human Rights ruled similarly in the *Golder Case,* Ser. A, No. 18 at 17 (Eur. Ct. H.R., May 7, 1974), that the right to bring a civil claim to an independent judge "ranks as one of the universally 'recognized' fundamental principles of law."

35. General Comment No. 20 (on Article 7), in Compilation of General Comments and General Recommendations Adopted by Human Rights Treaty Bodies, U.N. Doc. HRI/GEN/1/Rev. 1 at 30 (1994)(emphasis added). For criticism of the amnesty in Haiti, see Michael P. Sharf, "Swapping Amnesty for Peace: Was There a Duty to Prosecute International Crimes in Haiti?" *Texas International Law Journal* 31 (1996): 1.

36. 453 U.S. 654 (1981).

37. Indeed, the Court's opinion noted that if plaintiffs could later establish "an unconstitutional taking by the suspension of the claims, we see no jurisdictional obstacle to an appropriate action in the United States Court of Claims under the Tucker Act." 453 U.S. at 689–99. See also *Coplin v. United States*, 6 Ct. Cl. 115 (1984), *rev'd on other grounds*, 761 F. 2d 688 (Fed. Cir. 1985)(stating that the *Dames & Moore* opinion "noted that the abrogation of existing rights might constitute a taking").

38. 3 U.S. 199 (1796).

39. See also 3 U.S. at 229 ("It is admitted, that Virginia could not confiscate private debts without a violation of the modern law of nations . . .") and *id.*, 242 ("If the treaty had been silent as to debts, and the law of Virginia had not been made, I have already proved that debts would, on peace, have revived by the law of nations").

40. *Id.*, at 279.

41. *Id.*

42. 63 Ky. (2 Duv.) 502 (1866).

43. *Id.*, at 505–6.

44. For a survey of the approaches countries have taken toward human rights abuses committed by authoritarian regimes after they return to democratic rule, see Jon M. Van Dyke and Gerald W. Berkley, "Redressing Human Rights Abuses," *Denver Journal of International Law and Policy* 20 (1992): 243.

45. See text below at notes 57–58.

46. Agence France-Presse, "Iranians Respond to Overture from the U.S. with Mixed Signals," *New York Times*, March 19, 2000, A13, col. 1 (nat'l ed.).

47. Alessandra Stanley, "Pope Asks Forgiveness for Errors of the Church Over 2,000 Years," *New York Times*, March 13, 2000, A1, col. 4 (nat'l ed.).

48. Joint Resolution to Acknowledge the 100th Anniversary of the January 17, 1893, Overthrow of the Kingdom of Hawaii, Pub. L. 103–150, 107 Stat. 1510 (1993).

49. Civil Liberties Act of 1988, 50 U.S. C. sec. 1989.

50. Reuters, "Australia Expressing Regret to Aborigines," *New York Times*, August 27, 1999. For a listing of additional apologies by governments, groups, businesses, and individuals for racial wrongs, see Eric K. Yamamoto, "Race Apologies," *Journal of Gender, Race & Justice* 1 (1997): 47, 68–88.

51. Jeffrey Ghannam, "Repairing the Past," *ABA Journal*, November 2000, p. 39.

52. See Jon M. Van Dyke and Gerald W. Berkley, "Redressing Human Rights Abuses," *Denver Journal of International Law and Policy* 20 (1992): 243, 249–51.

53. *Truth and Reconciliation Commission of South Africa Report* (5 volumes, 1999). A challenge to the legitimacy of granting amnesties was rejected in *Azanian Peoples Organization v. The President of the Republic of South Africa*, 1996(4) South Africa Law Reports 637 (Constitutional Court of South Africa, 1996). The court justified its conclusion by explaining that the amnesty was not "a uniform act of compulsory statutory amnesia," but was appropriately linked to promoting "a constructive transition towards a democratic order" and was "available only where there is a full disclosure of all facts" and for acts committed "with a political objective." *Id.*, para. [32]. See generally Eric K. Yamamoto, "Race Apologies," *Journal of Gender, Race & Justice* 1 (1997): 47, 55–64.

54. Truth and Reconciliation Commission, "Amnesty Hearings & Decisions," January 11, 2000, http://www.doj-gov.za/trc/amntrans/index.htm (site visited April 28, 2002).

55. Dean E. Murphy, "Ex-Apartheid Minister Offers Lone High-Ranking Voice of Remorse," *Los Angeles Times*, December 17, 1999, A2, cols. 4, 5 (nat'l ed.).

56. Steven Mann, "Govt Hedges on TRC Reparations," *Daily Mail & Guardian* (Cape Town), May 11, 2000.

57. Mireya Navarro, "Guatemalan Army Waged 'Genocide,' New Report Finds," *New York Times*, February 26, 1999, A1, col. 8 (nat'l ed.).

58. John M. Broder, "Clinton Offers His Apologies to Guatemala," *New York Times*, March 11, 1999, A1, col. 5 (nat'l ed.).

59. Larry Rohter, "Past Military Rule's Abuse Is Haunting Brazil Today," *New York Times*, July 11, 1999, A11, col. 1 (nat'l ed.); Reuters, "Brazil Probes Military Over Secret 'Condor' Murders," CNN, May 17, 2000, http://www.rose-hulman.edu/~delacova/terrorism/condor-murders.htm (site visited April 28, 2002).

60. *Chicago Tribune*, "Tulsa Seeks to Redress Victims of 1921 Race Riot," *Honolulu Advertiser*, February 5, 2000.

61. Abraham Lama, "Peru's Truth Commission Off to a Slow Start, Say Activists," Digital Freedom Network, January 4, 2002, http://www.dfn.org/focus/peru/truth-commission.htm (site visited April 28, 2002).

62. *Id.;* see also Tania Mellado, "Peruvians Tell of Years of Rape and Torture," *Boston Globe*, April 9, 2002, http://www.boston.com/dailyglobe2/099/nat..._tell_of_years_of_rape_and_torture+.shtml (site visited April 28, 2002).

63. *Factory at Chorzow*, Merits, Judgment No. 13, P.C.I.J., Series A, No. 17, 47 (1928).

64. *The M/V Saiga Case (Saint Vincent and the Grenadines v. Guinea)*, para. 170 (ITLOS July 1, 1999). The text of this opinion can be found at http://www.un.org/Depts/los/ITLOS/Saiga_cases.htm (site visited March 22, 2000).

65. Law Nr. 19,123 Creating the National Corporation for Reparation and Reconciliation (Chilean National Congress 1992).

66. Civil Liberties Act of 1988, 50 U.S.C. sec. 1989; see generally Chris K. Iijima, "Reparations and the 'Model Minority' Ideology of Acquiescence: the Necessity to Refuse the Return to Original Humiliation," *Boston College Law Review* 40 (1998): 385.

67. Settlement Agreement, *Mochizuki v. United States*, No. 97–294C (Fed. Cl. 1997); "WWII Internees Get 5,000 Dollars, Official Apology," *Yomiuri Shimbun*, January 11, 1999.

68. Examples of Canadian "reconciliations" that involve substantial financial transfers include Canada's "Statement of Reconciliation" issued January 7, 1998, establishing a $245 million "healing fund" to provide compensation for the thousands of indigenous children who were taken from their homes and forced to attend boarding schools where they were sometimes physically sexually abused, and Canada's transfer in August 1998 of 750,000 square miles in British Columbia, just south of Alaska, to the 5,000-member Nisga'a Tribe. Anthony DePalma, "Canada Pact Gives a Tribe Self-Rule for the First Time," *New York Times*, August 5, 1998, A1. The basis for the "Statement of Reconciliation" can be found in Benjamin C. Hoffman, "The Search for Healing, Reconciliation, and the Promise of Prevention" (presented to the Reconciliation Process Implementation Committee in 1995, and documenting physical and sexual abuse at St. Joseph's and St. John's Training Schools for Boys), and Douglas Roche and Ben Hoffman, *The Vision to Reconcile* (1993).

69. For an example of the settlement obtained by one Maori group, see Ngai Tahu—New Zealand Maori Tribe Web site, http://www.ngaitahu.iwi.nz/.

70. Mireya Navarro, "Freed Puerto Rican Militants Revel in Life on the Outside," *New York Times*, January 27, 2000, A14, col. 3 (nat'l ed.).

71. Robinson, *The Debt*, p. 225 (citing House Bill 591).

72. Chris K. Iijima, "Reparations and the 'Model Minority' Ideology of Acquiescence: The Necessity to Refuse the Return to Original Humiliation," *Boston College Law Review* 40 (1998): 385, 388 (citing Lori S. Robinson, "Righting a Wrong Among Black Americans, the Debate Is Escalating Over Whether an Apology for Slavery Is Enough," *Seattle Post-Intelligencer*, June 29, 1997, E1).

73. See *Trajano v. Marcos (In re: Estate of Ferdinand E. Marcos Litigation)("Estate I")*, 978 F. 2d 493 (9th Cir. 1992), *cert. denied*, 508 U.S. 972 (1993); *In re Estate of Ferdinand E. Marcos Human Rights*

Litigation, 910 F. Supp. 1460, 1462–63 (D. Haw. 1995); *In re Estate of Ferdinand Marcos, Human Rights Litigation—Hilao v. Estate of Ferdinand Marcos ("Estate II")*, 25 F. 3d 1467 (9th Cir. 1994), *cert. denied*, 513 U.S. 1126; *Hilao v. Estate of Ferdinand Marcos ("Estate III")*, 103 F. 3d 767 (9th Cir. 1996).

74. See Karen Parker and Jennifer F. Chew, "Compensation for Japan's World War II War-Rape Victims," *International and Comparative Law Review* 17 (1994): 497, 528–32.

75. See, for example, *Burger-Fischer v. Degussa AG*, 65 F. Supp. 2d 248 (D.N.J. 1999), and *Iwanowa v. Ford Motor Company*, 67 F. Supp. 2d 424 (D.N.J. 1999).

76. See *In re Austrian and German Holocaust Litigation*, 250 F. 3d 156 (2d Cir. 2001); Ron Grossman, "Germans OK Paying $5 Billion to War Slaves; Deal Puts Pressure on U.S. Corporations," *Chicago Tribune*, December 15, 1999, A1.

77. See, for example, *In re Austrian and German Holocaust Litigation*, 250 F. 3d 156 (2d Cir. 2001).

78. *Chicago Tribune*, "Hitler Deportees to Receive $400 Million," *Honolulu Advertiser*, October 7, 2000, A3.

79. See Statute of the International Criminal Tribunal for the Former Yugoslavia, S.C. Resolution 827 (1993). See generally Louis Henkin, Gerald L. Neuman, Diane F. Orentlicher, and David W. Leebron, *Human Rights* (1999): 618–30. The International Criminal Tribunal for Rwanda was established by the Security Council in November 1994, S.C. Res. 995, Annex (1994), in response to the more than 500,000 minority ethnic Tutsi and Hutu opposition members who were killed during three months of slaughter in 1994 led by the Hutu-dominated government.

80. International Criminal Tribunal for Rwanda, "ICTR Detainees—Status on 16 April 2002," http://www.ictr.org/wwwroot/ENGLISH/factsheets/detaine.htm (site visited April 28, 2002); Associated Press, "Two Rwandans Held in Europe in 1994 Deaths," *New York Times*, February 16, 2000, A6, col. 6 (nat'l ed.).

81. International Criminal Tribunal for the Former Yugoslavia, "Judgments," http://www.un.org/icty/judgement.htm and "Indictments and Proceedings," http://www.un.org/icty/ind-e.htm (sites visited April 28, 2002).

82. Chris McGreal, "Unique Court to Try Killers of Sierra Leone," *Guardian Unlimited*, January 17, 2002, http://www.guardian.co.uk/sierra/article/0,2763,634697,00.html (site visited April 28, 2002).

83. Evelyn Leopold, "Pentagon Lawyer Named Prosecutor in Sierra Leone," Reuters, April 19, 2002, http://story.news.yahoo.com/...u=/nm/20020419/wl_nm/leone_usa_court_dc_2 (site visited April 28, 2002).

84. See *Regina v. Bow Street Metro, Stipendary Magistrate, Ex Parte Pinochet*, 2 All E.R. 97, 98 (1999).

85. Associated Press, "Menem Rejects Spain's Bid to Try Ex-Leaders," *Los Angeles Times*, November 4, 1999, A6, col.1.

86. Associated Press, "Argentine Arrested in Spanish 'Dirty War' Inquiry," *New York Times*, October 8, 1997, A4, col. 3 (nat'l ed.).

87. *Los Angeles Times*, "Argentina Arrests Ex-Dictator," *San Francisco Chronicle*, June 10, 1998, A10, col. 1.

88. Clifford Krauss, "New Argentine President Orders Purge of Remnants of 'Dirty War,' " *New York Times*, February 16, 2000, A12, col. 5 (nat'l ed.).

89. *Case Concerning the Arrest Warrant of 11 April 2000 (Dem. Rep. of the Congo v. Belgium)*, International Court of Justice, Judgment, February 14, 2002, para. 60.

90. Ibid., para. 61.

91. See, for example, Thomas Sowell, "Slave Reparations—a Scam at Tax Time and Any Time," *Honolulu Star-Bulletin*, April 8, 2002, A11, col. 1.

92. The Reparations Coordinating Committee, consisting of lawyers, academics, public officials, and activists, has been focusing its efforts on bringing claims against the federal and state governments, but will also be targeting private corporations and universities that have benefited from slavery. See, for example, Charles J. Ogletree Jr., "Slavery Reparations About Truth, Not Money," *Seattle Post-Intelligencer*, April 7, 2002.

93. On March 26, 2002, a lawsuit was filed in the U.S. District Court for the Eastern District of New York on behalf of a class of 35 million African Americans seeking compensation for the value of stolen labor and unjust enrichment against Aetna, FleetBoston Financial, CSX, and other unnamed companies. The case is styled *Farmer-Paellmann v. FleetBoston Financial Corporation*, and the main plaintiff in the lawsuit is New York attorney Deadria Farmer-Paellmann. Peter Viles, "Suit Seeks Billions in Slave Reparations," CNN.com/LawCenter, March 27, 2002, http://www.cn.com/2002/LAW/03/26/slavery.reparations/ (site visited April 28, 2002).

94. Emily Newburger, "Breaking the Chain," *Harvard Law Bulletin* (Summer 2001): 18, 20.

95. Ibid., p. 21.

96. See text accompanying notes 74–77.

97. Jeffrey Ghannam, "Repairing the Past," *ABA Journal* (November 2000): 39.

98. Associated Press, "Reparations for Slavery Demanded from Aetna," *San Francisco Chronicle*, March 20, 2000, A2, col. 3.

99. Ghannam, note 97 above, p. 43.

100. See text above accompanying note 16.

101. See, for example, *Cato v. United States*, 70 F. 3d 1103 (9th Cir. 1995)(concluding, in a case seeking $100 million for the descendants of the victims of slavery, that the United States had not waived its sovereign immunity).

102. See, for example, *Bd. of Trustees of the University of Alabama v. Garrett*, 531 U.S. 356 (2000).

103. A notable exception are the Native Hawaiians, whose situation is discussed in Dr. Trask's chapter in this volume. They are one of the very largest groups of Native Americans and are almost unique in having never been given a claims commission to present their grievances to or a settlement package through which their lands and resources could be returned to them. See generally Jon M. Van Dyke, "The Political Status of the Native Hawaiian People," *Yale Law and Policy Review* 17 (1998): 95.

104. See text accompanying note 60.

SLAVE TAXES

1. *Compania General de Tabacos de Filipinas v. Collector of Internal Revenue*, 275 U.S. 87, 100 (1927) (Holmes, J., dissenting).

2. In this essay, the term "slave taxes" will refer to direct taxes on slaves (poll and capitation taxes; property taxes; import duties; and occupational taxes); indirect taxes on slaves (taxes on slave transfer and manumission; and discriminatory taxes in the slave era on free blacks); and taxes on slave commerce (export duties on slave commodities such as tobacco, cotton, and rice; import duties on articles purchased with slave commodities; and *ad valorem* taxes on slave improvements such as houses built by slaves and land improved and worked by slaves). The term as used herein is thus much broader than the federal direct slave tax of 1798 or state poll taxes on slaves. One important and controversial aspect of this definition is that it specifically includes that portion of federal customs revenue on imports made possible by the export of slave commodities.

3. I do not suggest filing a refund claim with the Internal Revenue Service just yet, for many important legal questions would first have to be resolved. Initially, it must be acknowledged that rarely did the slaves themselves pay any slave taxes. In the ordinary course, the masters paid the tax on their property. Other obstacles include sovereign immunity, the statute of limitations, the lack of a procedure for refunds of this nature, and the lack of adequate documentary records. IRS officials have warned against filing "slave reparations" refund claims, which are viewed as an illegal tax scam. J. J. Thompson, "IRS Officials Warn of 'Slave Reparations' Tax Scam," *Philadelphia Inquirer*, October 15, 2000, available at http://inq.philly.com/inquirer/2000/10/15/south_jersey/15jscam.htm (site visited July 10, 2001); Michelle Singletary, " 'Slave Reparation' Tax Scam Targets Blacks," *Seattle Post-Intelligencer*, March 5, 2001, available at http://seattlepi.nwsource.com/money/sing053.shtml (site visited July 10, 2001).

4. See, for example, Commission to Study Reparation Proposals for African-Americans Act, H.R. 40, 107th Cong., 1st Sess. (2001). This legislation has not been passed by Congress. For a successful prior effort to create a similar commission, see Commission on Wartime Relocation and Internment of Civilians Act, Pub. L. No. 96–317 (1980) (established study commission for persons of Japanese ancestry who were interned by the U.S. government during World War II) (for background, see *Korematsu v. United States*, 323 U.S. 214 (1944); *Hirabayashi v. United States*, 320 U.S. 81 (1943); and *Yasui v. United States*, 320 U.S. 115 (1943). The Report of the Commission on Wartime Relocation and Internment of Civilians led to the passage of the Civil Liberties Act of 1988, Pub. L. No. 100–383 (1988) (Congress accepted the findings of the Commission) and subsequent appropriations. Mitchell T. Maki et al., *Achieving the Impossible Dream: How Japanese Americans Obtained Redress* (Champaigne: University of Illinois Press, 1999).

5. See generally Elazar Barkan, *The Guilt of Nations* (New York: W. W. Norton, 2000).

6. Legal action against corporations avoids the difficult questions of sovereign immunity. My preference is for a political and factual discussion of black reparations rather than a series of class-action lawsuits. If the wound in American society is racial, legal approaches may polarize rather than encourage positive societal transformation.

7. James Walvin, *Black Ivory: A History of British Slavery* (London: HarperCollins, 1992), p. 25.

8. The Royal African Company, chartered in 1672 by Charles II, was the first to meet with success. It was preceded by the Company of Royal Adventurers trading to Africa in 1662 and a prior company chartered by Charles I in 1631.

9. Martina Elbl, "Portuguese Trade with West Africa, 1440–1520" (1986), pp. 340–41 (unpublished Ph.D. dissertation, University of Toronto) cited in John Thornton, *Africa and Africans in the Making of the Atlantic World, 1400–1800* (2d ed., 1998), p. 58, n. 48.

10. John Thornton, *Africa and Africans*, pp. 63–64.

11. John Thornton, *African Political Ethics and the Slave Trade: Central African Dimensions* (unpublished manuscript on file with the author, 2001) available at http://www.millersv.edu/~winthrop/Thornton.html (site visited July 11, 2001).

12. John Thornton, *Africa and Africans*, p. 71. John Thornton's main point was that no European nation could control the terms of trade at the expense of Africans, the examination of the tax revenue to Europe being peripheral to his inquiry.

13. John Thornton, *Africa and Africans*, pp. xiii (Map 4), 130–41; Walvin, *Black Ivory*, pp. 6–10. The legacy of slavery in Central and South America is largely unknown to the general American audience.

14. W. E. B. Du Bois, *The Suppression of the African Slave-Trade to the United States of America, 1638–1870* (1896) ed. Philip S. Foner (New York: Dover Publications, 1970), p. 3.

15. This discussion does not include the enslavement of indigenous people in the Americas, which was substantial, particularly in Central and South America.

16. In 1660, the bulk of the tobacco labor force was free or indentured rather than slave, but massive slave importations began in earnest with the tobacco boom, reaching 40,000 resident slaves by 1670 with an additional 100,000 imported into the region from 1690 to 1770. In addition, North American slaves were able to support internal population growth. Walvin, *Black Ivory*, pp. 8–9.

17. Tax Analysts, Inc., Tax History Project, 1660–1713 available at http://www.tax.org/museum/1660–1712.htm.

18. An Act for the laying on Imposition upon Negroes Slaves & White servants Imported into this Province, 38 Md. Arch. 51 (May 1695), in *Proceedings and Acts of the General Assembly of Maryland, 1694–1729*, edited by Bernard Christian Steiner.

19. Walvin, *Black Ivory*, p. 9.

20. George Ruble Woolfolk, "Taxes and Slavery in the Ante Bellum South," *Journal of Southern History* 26 (1960): 180, 182.

21. Robert A. Becker, *Revolution, Reform, and the Politics of American Taxation, 1763–1783* (Baton Rouge: University of Louisiana Press, 1980) pp. 243–44, Appendix Tables 77, 192.

22. South Carolina colonial currency.

23. Margaret G. Myers, *A Financial History of the United States* (New York: Columbia University Press, 1970) p. 19 (data from 1731).

24. Becker, *Revolution*, pp. 16–18 (citing data from September 1763 to December 1769).

25. Becker, *Revolution*, pp. 80, 208–09; Du Bois, *Suppression*, pp. 9–11; Darnold D. Wax, "Preferences for Slaves in Colonial America," *Journal of Negro History* 58 (1973): 371–401.

26. Wax, "Preferences," pp. 371–401.

27. Becker, *Revolution*, p. 207.

28. J. Mills Thornton, III, "Fiscal Policy and the Failure of Radical Reconstruction in the Lower South," in *Region, Race, and Reconstruction: Essays in Honor of C. Vann Woodward* edited by J. Morgan Kousser & James M. McPherson (New York: Oxford University Press, 1982), p. 351; Peter Wallenstein, *From Slave South to New South: Public Policy in Nineteenth-Century Georgia* (Chapel Hill: University of North Carolina Press, 1987), p. 41. In 1850, the tax was $0.56 per slave and accounted for 63.1% of South Carolina's revenue. *Id.*

29. Becker, *Revolution*, p. 245, Appendix Table 19.

30. Wallenstein, *From Slave South to New South*, pp. 40, 57.

31. Ibid., p. 14 (1860 data).

32. Ibid., p. 41.

33. Becker, *Revolution*, p. 77.

34. Ibid., p. 78.

35. Ibid., pp. 196–98.

36. Ibid., p. 199.

37. Walvin, *Black Ivory*, p. 9.

38. Becker, *Revolution*, pp. 79, 91.

39. J. Mills Thornton, "Fiscal Policy," p. 351.

40. Wallenstein, *From Slave South to New South*, p. 41, n. 12.

41. Woolfolk, "Taxes and Slavery," pp. 183–84 ($27,046.50 out of a total of $121,195.50 in 1838).

42. Myers, *Financial History*, p. 17.

43. "An Act to Lay a Duty on the Goods, & a Tax on the Slaves therein Mentioned During the time & for the Uses Mentioned in the Same," *The Colonial Laws of New York from the Year 1664 to the Revolution*, vol. 2, ch. 624, pp. 876–84 (Nov. 28, 1734).

44. Becker, *Revolution*, p. 47.

45. Gary K. Wolinetz, "New Jersey Slavery and the Law," *Rutgers Law Review* 50 (1998): 2227, 2228, 2238, 2256–57.

46. Ibid., p. 2235.

47. Richard T. Ely, *Taxation in American States and Cities* (1888), p. 118.

48. Davis Rich Dewey, *Financial History of the United States* (2d ed., 1903), p. 16.

49. Du Bois, *Suppression*, pp. 20–24; Myers, *Financial History*, pp. 15, 18.

50. Du Bois, *Suppression*, p. 34, n. 4.

51. "An Act for Granting unto Their Majesties a Tax of Twelvepence a Poll, and One Penny on the Pound for Estates 1694–95," *Acts and Resolves, Public and Private, of the Province of Massachusetts Bay*, vol. 1, chap. 2 (21 vols., Boston, 1869–1922).

52. Du Bois, *Suppression*, pp. 27–31.

53. James Walvin, "The Public Campaign in England Against Slavery, 1787–1834," in *The Abolition of the Atlantic Slave Trade, Origins and Effects in Europe, Africa, and the Americas* edited by David Eltis and James Walvin (1981), pp. 71–72.

54. David Eltis, *Economic Growth and the Ending of the Transatlantic Slave Trade* (New York: Oxford University Press, 1987), p. 49.

55. The price paid for slaves on the coast of Africa was a small percentage of the price for a healthy arrival on the auction block in Havana or Charleston; nevertheless, the African governments that

controlled the slave trade were paid for delivering slaves. The enormous social cost to Africa and Africans of this trade is not belittled by this observation.

56. Charles H. Fairbanks Jr. with Eli Nathans, "The British Campaign Against the Slave Trade," in *Human Rights in Our Time: Essays in Memory of Victor Baras,* edited by Marc F. Plattner (1983), pp. 53–59.

57. Becker, *Revolution,* pp. 80, 208–09; Du Bois, *Suppression,* pp. 9–11; Wax, "Preferences," pp. 371–401.

58. Hans Christian Johansen, "The Reality behind the Demographic Argument to Abolish the Danish Slave Trade," *The Abolition of the Atlantic Slave Trade,* edited by Eltis and Walvin, pp. 221–24.

59. Woolfolk, "Taxes and Slavery," pp. 188, 190–91. Georgia lowered taxes on whites and slaves in 1763 and again in 1770 and 1773, but the poll tax on free blacks remained high, increasing to £1 in 1773, up 33% from 1768. Becker, *Revolution,* pp. 77–78.

60. J. Mills Thornton, "Fiscal Policy," p. 353. Higher taxes would both encourage migration out of the state as well as hinder free black immigration.

61. Wallenstein, *From Slave South to New South,* p. 87.

62. Ibid., p. 90. The $5 tax was the same amount required from white professionals, such as lawyers. *Id.,* p. 41.

63. Wallenstein, *From Slave South to New South.*

64. Frederic D. Ogden, *The Poll Tax in the South* (1958), p. 281.

65. Ibid., p. 282.

66. Ibid., p. 284.

67. Woolfolk, "Taxes and Slavery," pp. 184–86.

68. Planters held a disproportionate share of political power, and although they paid significant taxes on slaves, the tax was generally lower than the tax on land when one compares actual market values. The slave owner paid taxes, but not excessively so, compared to the wealth held by the plantation owners. Wallenstein, *From Slave South to New South,* pp. 41–42, n. 12.

69. Robin L. Einhorn, "Slavery and the Politics of Taxation in the Early United States," *Studies in American Political Developments* 14 (2000): 156, 178–79 (Federal tariffs and direct taxes to 1800); J. Mills Thornton, "Fiscal Policy," pp. 356–57 (Lower South); Donald C. Butts, "The 'Irrepressible Conflict': Slave Taxation and North Carolina's Gubernatorial Election of 1860," *North Carolina Historical Review* 58 (1981): 44–66, and Donald Cleveland Butts, "A Challenge to Planter Rule: The Controversy Over the Ad Valorem Taxation of Slaves in North Carolina: 1858–1862" (1978) (unpublished Ph.D. dissertation, Duke University) (North Carolina); Dall W. Forsythe, *Taxation and Political Change in the Young Nation 1781–1833* (New York: Columbia University Press, 1977) (taxation and regime politics generally); and Woolfolk, "Taxes and Slavery," p. 197 (Georgia).

70. Einhorn concludes that the unique features of the apportioned direct federal tax on slaves adopted in 1798 muted class conflict in the South, but exacerbated it in the North. Einhorn, "Slavery and the Politics of Taxation," pp. 178–79.

71. Tenn. Const. of 1796, Art. I, Sec. 26 ("All lands liable to taxation in this state, held by deed, grant or entry, shall be taxed equal and uniform, in such manner that no one hundred acres shall be taxed higher than another, except town lots, which shall not be taxed higher than two hundred acres of land each; no free man shall be taxed higher than one hundred acres, and no slave higher than two hundred acres, on each poll."); Ely, *Taxation in American States and Cities,* p. 117.

72. J. Mills Thornton, "Fiscal Policy," p. 351.

73. Wolinetz, "New Jersey Slavery," p. 2234.

74. Wallenstein, *From Slave South to New South,* p. 89.

75. Ibid., pp. 10, 27, 40.

76. Ibid., p. 95.

77. W. Elliot Brownlee, *Federal Taxation in America* (1996), p. 16 ("Tariff revenues, in combination with the debt finance that the general [slave, houses and land] taxing power made possible, funded the Louisiana Purchase"). Note my use here of the broader term "slave taxes," as described above in note 2.

78. Speech given by Frederick Douglass to the Rochester Ladies' Anti-Slavery Society, Rochester, New York, July 5, 1852. James M. Gregory, *Frederick Douglass, the Orator* (1893), pp. 103–06.

79. John Adams, "Notes of Debates on the Articles of Confederation," July 30, 1776, in *Diary and Autobiography of John Adams*, ed. L. H. Butterfield, 4 fols. (Cambridge, MA: Harvard University Press, 1961), 2:245–46, and Thomas Jefferson, "Notes of Proceedings in the Continental Congress," July 12, 1776, in *The Papers of Thomas Jefferson*, ed. Julian Boyd et al., 27 vols. (Princeton: Princeton University Press, 1950–1997), 1:322.

80. John Adams, "Notes of Debates on the Articles of Confederation," July 30, 1776, in *Diary and Autobiography of John Adams*, ed. L. H. Butterfield, 4 fols. (Cambridge, MA: Harvard University Press, 1961), 2:245–46.

81. Brownlee, *Federal Taxation in America*, p. 11; Roger H. Brown, *Redeeming the Republic: Federalists, Taxation and the Origins of the Constitution* (Baltimore: Johns Hopkins University Press, 1993), pp. 11–12; Forsythe, *Taxation*, p. 14.

82. *See* above nn. 20–47 and text accompanying.

83. Brown, *Redeeming the Republic*, pp. 69–70. This amount declined sharply in 1784 and reached zero in 1785 as the debts were paid off.

84. Brown, *Redeeming the Republic*, p. 70.

85. Attributed to James Madison.

86. *McCulloch v. Maryland*, 17 U.S. 316, 431 (1819) (Marshall, C.J.).

87. Robin Einhorn points out that while counting slaves would increase total Southern taxes, it also increased class conflict in the North while assuaging it in the South. Einhorn, "Slavery and the Politics of Taxation," p. 178.

88. Einhorn, "Slavery and the Politics of Taxation," p. 168.

89. U.S. Const., Art. I, Sec. 2, cl. 3.

90. U.S. Const., Art. I, Sec. 8, cl. 1 and Art. I, Sec. 10, cl. 2.

91. U.S. Const., Art. I, Sec. 9, cl. 1. This provision also delayed the abolition of the slave trade until at least 1808. Congress passed the law abolishing the slave trade in the United States on March 2, 1807, effective January 1, 1808.

92. Annals of Congress, 1st. Cong., 1st sess., 349–56; 1st Cong., 2nd sess., 1223–33, 1239–47, 1464–66, 1500–14, 1516–25.

93. An Act to Prohibit the Importation of Slaves into any Port or Place Within the Jurisdiction of the United States, From and After the First Day of January, in the Year of our Lord One Thousand Eight Hundred and Eight, U.S. Stat. at Large, Sec. 2 (March 2, 1807); Du Bois, *Suppression*, pp. 117–18 (description of an 1818 capture near Mobile, Alabama).

94. Einhorn, "Slavery and the Politics of Taxation," p. 182.

95. Forsythe, *Taxation*, pp. 51–52.

96. 1 U.S. Stat. at Large (1798) 597–98.

97. Einhorn, "Slavery and the Politics of Taxation," p. 178.

98. Annals of Congress, 4th Cong., 2nd Sess. (Jan. 20, 1797); Einhorn, "Slavery and the Politics of Taxation," p. 180.

99. 1 U.S. Stat. at Large, Sec. 16 (1798) 597–98.

100. Forsythe, *Taxation*, p. 52; Dewey, *Financial History*, p. 109. Federal expenditures were modest in the period to 1797, used mainly to repay Revolutionary War debt, including the debts assumed from the states. 1796: $5,800,000; and 1797: $6,000,000. As tensions heated up with France, federal military spending increased: 1798: $7,600,000; 1799: $9,300,000; 1800: $10,800,000. Sidney Ratner, *Taxation and Democracy in America* (1942, rep. 1967), p. 29.

101. Dewey, *Financial History*, p. 109.

102. Ratner, *Taxation and Democracy*, p. 30.

103. The tax on slavery was not controversial. Public opposition centered on the Federalist use of any internal taxes at all. Forsythe, *Taxation*, pp. 54–57.

104. Dewey, *Financial History*, p. 121.

105. The Napoleonic Wars produced customs revenues for the United States that far exceeded expectations, but the surge in revenue was only temporary. Forsythe, *Taxation*, p. 57; Dewey, *Financial History*, p. 121.

106. Statutes at large, 3 (1813), 26, 53–71; Forsythe, *Taxation*, p. 58–59.

107. Einhorn, "Slavery and the Politics of Taxation," p. 183.

108. Statutes at Large, 3 (1815), 164–80; Dewey, *Financial History*, p. 139.

109. Statutes at Large; Ratner, *Taxation and Democracy*, p. 34.

110. Dewey, *Financial History*, p. 140.

111. Brownlee, *Federal Taxation in America*, p. 21; see also Dewey, *Financial History*, p. 140.

112. Statutes at Large, 3 (1817), 401–3; Dewey, *Financial History*, p. 141.

113. Einhorn, "Slavery and the Politics of Taxation," p. 160.

114. House Journal, 30th Cong., 1st sess., 347–48; Ratner, *Taxation and Democracy*, pp. 43–44.

115. Wallenstein, *From Slave South to New South*, p. 96.

116. Hopkins, *Works* II.615 (1854) quoted in De Bois, *Suppression*, pp. 34–35.

117. Wallenstein, *From Slave South to New South*, p. 8.

118. Forsythe, *Taxation*, p. 83.

119. Ibid.

120. Ronald Bailey, "The Slave(ry) Trade and the Development of Capitalism in the United States: The Textile Industry in New England," in *The Atlantic Slave Trade, Effects on Economies, Societies, and Peoples in Africa, the Americas, and Europe*, edited by Joseph E. Inikori and Stanley L. Engerman (Durham: Duke University Press, 1992), p. 205.

121. U.S. Const., Art. I, Sec. 9, cl. 5. In Madison's notes to the Constitutional Convention debates on this provision, he wrote: "General Pinkney . . . was alarmed at what was said yesterday, concerning the Negroes. He was now again alarmed at what had been thrown out concerning the taxing of exports. S. Carola. has in one year exported to the amount of £600,000 Sterling all which was the fruit of the labor of her blacks. Will she be represented in proportion to this amount? She will not. Neither ought she then to be subject to a tax on it. He hoped a clause would be inserted in the system restraining the Legislature from taxing Exports." [*1:592; Madison, 12 July*] Max Farrand, ed., *The Records of the Federal Convention of 1787*, rev. ed., 4 vols. (New Haven and London: Yale University Press, 1937).

122. Brownlee, *Federal Taxation in America*, pp. 15–16. Total federal customs duties until 1815 were $223 million; all other federal revenues were $24 million, or about 10%. Myers, *Financial History*, p. 59. After the repeal of the second direct federal tax in 1817, almost all of the federal government's revenue came from duties.

123. [*1:592; Madison, 12 July*] Farrand, ed., *The Records of the Federal Convention*.

124. See Section I above.

125. Similar research could be undertaken with regard to corporations.

126. The dominant *mythos* of American democracy is the subject of hearty critique, beyond the scope of this essay.

THE NATIONAL BLACK UNITED FRONT

1. Lisa N. Nealy and Pfizer Shellow. "Our Land is a'comin and Our Mule is on de Way." Research paper submitted to the National Black Caucus of State Legislators, Washington, D.C., 2002.

2. Roger Wareham, Esq., "U.S. Seeks to Sabotage Reparations Movement," *Final Call Newspaper*, April 3, 2001, p. 23.

3. United Nations Economic and Social Council, E/CN.4/1995/78/1, January 16, 1995, p. 3.

4. Lord Anthony Gifford, "For Whom the Bell Tolls: The Legal Basis for the African Reparations Claim," *New African*, December 1999, p. 3.

5. "Facts on Reparations to Africa and Africans in the Diaspora," prepared by the Group of Eminent Persons on Reparations, June 1992, p. 11.

6. John Hope Franklin, "Horowitz's Diatribe Contains Historical Inaccuracies." Letter to the editor in the *Duke Chronicle*, March 29, 2001.
7. John Henrik Clarke, *Notes for an African World Revolution—Africans at the Crossroads* (Trenton, NJ: Africa World Press, 1992), p. 420.

N'COBRA

1. Statement of I. H. Dickerson, General Manager, and Mrs. Callie House, Promoter, Ex-Slave Mutual Relief, Bounty and Pension Association, "To all Local Ex-Slave Associations in the United States," circa 1896.
2. Notice for convention of Ex-Slave Mutual Relief, Bounty and Pension Association held in Columbia, Tennessee, March 24–27, 1899.
3. *Reparations Yes!* 4th ed. (Baton Rouge: The House of Songhay Commission and The Malcolm Generation, Inc., 1995).
4. The Asante people living in Ghana and the Ivory Coast stamp *Adinkra* symbols on cloth. Each of the symbols has a name and meaning based on a proverb or historical legend. The shapes of humans, animals, plants, and objects inspired the geometric forms of the symbols.
5. Salim Muwakkil, "Black America: Blacks call for reparations to break the shackles of the past," *In These Times* (October 11–17, 1989), p. 6.
6. 101st Cong. Status Profile for H.R. 3745.

THE POPULARIZATION OF THE INTERNATIONAL DEMAND

1. Throughout this article the term "African people" is used to denote Black people, i.e., people of African descent whether born on the African continent or in the African Diaspora.
2. E.CN.4/Sub.2/1993/8, July 2, 1993.
3. Ibid, para. 24.
4. E/CN.4/1995/78/Add.1
5. E/CN.4//1998/68/Add.3

REPARATIONS AND HEALTH CARE

1. W. Michael Byrd and Linda A. Clayton, *An American Health Dilemma* (New York: Routledge, 2000), pp. 130–31.
2. Ibid., p. 181.
3. Marvin Perry, Alan H. Scholl, Daniel F. Davis, Jeanette G. Harris, Theodore H. Von Laue, *History of the World* (Evanston: McDougal Littell Inc., 1995), p. 79.
4. Ibid., pp. 180–81.
5. Byrd and Clayton, *American Health Dilemma*, p. 179.
6. Ibid., pp. 179, 195, 196.
7. Ibid., p. 179.
8. Ibid., p. 196.
9. John Hope Franklin and Alfred A. Moss, *From Slavery to Freedom* (New York: McGraw Hill, Inc., 1994), p. 288.
10. Ibid., p. 130.
11. Ibid., p. 142.
12. Byrd and Clayton, *American Health Dilemma*, p. 281.
13. Bruce Wiseman, *Psychiatry: The Ultimate Betrayal* (Los Angeles: Freedom Press, 1995).
14. See Psychiatry's Betrayal for the history of drug experimentation on African Americans, available at http://www.cchr.org/racism/pdf/racism.pdf.

15. Raymond Winbush, *The Warrior Method: A Program for Rearing Healthy Black Boys* (New York: Amistad/HarperCollins, 2001), pp. 11–12.

16. George Breitman, *Malcolm X Speaks: Selected Speeches and Statements* (New York: Grove Press, 1990).

17. Malcolm X, *The Final Speeches, February 1965*, ed. Steve Clark (New York: Pathfinder, 1992).

18. Alvin Poussaint, *Why Blacks Kill Blacks* (New York: Emerson Hall Publishers, Inc., 1972), p. 19.

19. David B. Gaspar and Darlene C. Hine, eds. *More Than Chattel: Black Women and Slavery in the Americas* (Bloomington: Indiana University Press, 1996), p. 121.

20. Office of Minority Health Web site, http://www.omhrc.gov.

21. "Promoting Children's Health: Progress Report of the Child Health Workgroup Board of Scientific Counselors, With New Guidance for Toxicological Profiles, 1998–1999," Agency for Toxic Substances and Disease Registry, Office of Children's Health, Atlanta, 1999.

22. See particularly on AIDS *http://www.redcross.org/services/hss/hivaids/AfrAmInfoSht.pdf.* For a comprehensive study of health care disparities between Black and white Americans, see *Achieving Equitable Access: Studies of Health Care Issues Affecting Hispanics and African Americans,* published by the Joint Center for Political and Economic Studies in 1996.

23. http://www.thehutchinsonreport.com/061101feature.html

24. Jennifer Brooks, "The Minority AIDS Crisis," in *Closing the Gap* (April 1999) p. 1.

25. Ronald L. Braithwaite and Sandra E. Taylor, eds., *Health Issues in the Black Community* (San Francisco: Jossey Bass Publishers, 1992), p. 97.

RIDING THE REPARATIONS BANDWAGON

1. See Edwin Black's *IBM and the Holocaust: The Strategic Alliance Between Nazi Germany and America's Most Powerful Corporation* (New York: Crown Publishing, 2001) as an example of understanding private industry's complicity in committing atrocities toward Jews and the issue of compensatory measures.

REPARATIONS + EDUCATION

1. John Henrik Clarke, *Africans at the Crossroads: Notes for an African World Revolution* (Trenton, NJ: African World Press, 1991), p. 419.

2. Lunsford Lane, *The Narrative of Lunsford Lane* (Boston: J. G. Torrey, Printer, 1842).

3. Harriet Beecher Stowe, *The Education of Freedmen*, part 1, 1999, p. 60.

4. Willie Lynch, *The Willie Lynch Letter and The Making of a Slave* (Chicago: Lushena Books, 1999).

5. Marimba Ani, *Yurugu* (Trenton, NJ: Africa World Press, Inc., 1994), p. 294.

6. Carter Woodson, *The Miseducation of the Negro* (Trenton, N.J.: Africa World Press), p. 84.

7. Michael Kelly, "Test results illustrate racial divide that can't be ignored," *Milwaukee Journal Sentinel* (April 2001).

8. Amos N. Wilson, *The Falsification of Afrikan Consciousness* (New York: Afrikan World InfoSystems, 1993), p. 21.

9. Gregory Wright, "A Romantic Take on Slavery?" *Emerge* (February 1999).

10. James W. C. Pennington, *The Fugitive Blacksmith*, 2nd ed. (London, 1849).

11. Kwame Agyei and Akua Nson Akoto, *The Sankofa Movement* (Washington, D.C.: Oyoko Nifco, Inc., 1999).

12. Wilson, *Falsification,* p. 40.

13. Moore, Audley, "What's the Hour of the Night?" Poem read at Gary, Indiana, Black Political Convention, 1972.

FARMER-PAELLMAN V. FLEETBOSTON, AETNA INC., CSX

1. Brent Staples, African Holocaust, *The Lessons of a Graveyard.*
2. Ira Berlin, "Many Thousands Gone."
3. Brent Staples, African Holocaust, *The Lessons of a Graveyard.*
4. *Id.*
5. Brent Staples, African Holocaust, *Lessons from a Graveyard,* quoting in part from Dr. Michael Blakey, Howard University.
6. Kate Zernike, "Slave Traders in Yale's Past Fuel Debate on Restitution," *New York Times* (August 13, 2001).
7. Randall Robinson, *Compensate the Forgotten Victims of America's Slavery Holocaust.*
8. Tim Wise, "Breaking the Cycle of White Dependency" (5/22/02).
9. Tamara Audi, "Payback for Slavery: Growing Push for Reparations Tries to Fulfill Broken Promise," quoting Randall Robinson (9/18/00).
10. Alistair Highet, "Will America Pay for the Sins of the Past, Slavery's Past," *The Hartford Advocate* (February 14, 2002), quoting Dr. Ronald Waters.
11. *Id.*
12. John Hope Franklin, *From Slavery to Freedom* (New York: Knopf, 1947).
13. Yuval Taylor, *I Was Born a Slave.*
14. Jim Cox, "Rail Networks Own Lines Bult with Slave Labor," *USA Today* (02/21/02).

CURRENT LEGAL STATUS OF REPARATIONS

1. *Maafa* is a Kiswahili term for "Disaster" or "Terrible Occurrence."
2. At least as early as 1990, lawyers from N'COBRA (National Coalition of Blacks for Reparations in America) had begun plans for a comprehensive series of lawsuits.
3. 70 F. 3d 1103 (9th Cir. 1995).
4. A year prior to *Cato,* the District Court for the Northern District of California dismissed two similar suits by *pro se* litigants seeking redress for the horrors of slavery. In *Bailey v. United States,* 1994 WL 374524 (N.D. Cal. 1994) the Court dismissed the plaintiff's *pro se* claim brought on behalf of himself and family members, finding that he had not asserted a claim that arose under the Constitution, laws, or treaties of the United States. Phillip Bailey asserted that the kidnapping and enslavement of his ancestors was sanctioned by the Constitution and laws of the United States. The Court found that he had not presented a federal question and therefore it was without jurisdiction to hear the matter. The Court also held that the government's sovereign immunity to the suit was an obstacle to the plaintiff's claim unless the United States had consented to being sued. There was no such consent and therefore the plaintiff could not sue the United States.

    In *Berry v. United States,* 1994 WL 374537 (N.D. Cal. 1994), the plaintiff brought a claim pursuant to the Freedmen's Bureau Act of 1865. *Pro se* litigant Mark Berry sought to quiet title to forty acres of land located in San Francisco, or, in the alternative, for three million dollars in damages. Multiple problems with the claim resulted in its dismissal. First, the Freedmen's Bureau Act of 1865 lapsed on July 17, 1869. Therefore, it no longer provided a viable remedy. Second, even if the Act was still enforceable, it only covered land "within the insurrectionary states." Berry's claim was for title to land located in California. Third, the Federal Court of Claims had exclusive jurisdiction over claims brought against the United States. The Quiet Title Act, 28 USC §2409a and §1346(f), read in conjunction with the Tucker Act, 28 U.S.C. §1346(a)(2) and §1491(a)(1), rested jurisdiction over property claims against the United States with the Federal Court of Claims. Even if the Quiet Title Act was applicable, it contained a twelve-year statute of limitations and therefore the claim was time-barred.
5. 28 USC §1346.
6. The waiver of sovereign immunity in tort actions against the United States in the FTCA is limited to claims arising on or after January 1, 1945. 28 USC §1346(b).

7. *Pro se* litigants are not allowed to bring suit on behalf of anyone else, therefore the case could not proceed as a class action. Even if it had been litigated as a class action, plaintiffs still needed to prove that they suffered a personal harm. The laws of the United States do not provide individuals with standing to complain generally that their government is violating the law.

8. 32 Fed. Ct. 32 (2002).

9. Civil Liberties Act of 1988 50 USC app. § 1989b-4(h), Pub.L. 100-383 (Aug. 10, 1988), 102 Stat. 903, (the Act, or CLA), as amended Pub.L. 102-371 (September 27, 1992), 106 Stat. 1167 (West 2001).

10. The plaintiffs were constrained to pursue their claim in the Federal Court of Claims. Pursuant to the Tucker Act, exclusive subject matter jurisdiction of any "civil action or claim against the United States . . . founded either upon the Constitution, or any Act of Congress" where money damages exceed $10,000 rests with the Federal Court of Claims. 28 U.S.C §1346(a)(2) and §1491(a)(1).

11. Civil Liberties Act of 1988 *supra.*

12. Mr. Bankhead passed away not long after the suit was filed.

13. *Wyatt-Kervin, et al., v. J. P. Morgan Chase & Co., et al.,* No. 03-CV-36 (S.D. Tex., complaint filed 01/21/03).

14. The tepid almost-apology of former president William Clinton in 1997 for slavery provided nothing of legal, political, social, or economic significance for African Americans. The United States further solidified its position against reparations and any acknowledgement of its violations against people of African descent with respect to the Transatlantic Slave Trade when it refused to participate in the historic UN-sponsored World Conference Against Racism held in Durban, South Africa, in 2002.

15. See H.R. 3745, 101st Cong. (1989); H.R. 1684, 102nd Cong. (1991); H.R. 40, 103rd Cong. (1993); H.R. 891, 104th Cong. (1995); H.R. 40, 105th Cong. (1997).

16. Over a period of seven days in 1923, whites attacked and burned the small black town of Rosewood, Florida. Local police did nothing to stop the attack or protect the victims.

17. Chapter 934 of the Insurance Code of California (formally SB 2199).

18. HB 5483 was introduced in February 2000.

19. The other sponsors of the measure were John Liu (Queens), Bill Perkins (Harlem), Helen Foster (Highbridge), James Sanders (Queens), Larry Seabrook (North Bronx), Miguel Martinez (Washington Heights) and Robert Jackson (West Harlem).

20. See *Tanihara v. United States,* 32 Fed. Cl. 805 (1995).

21. *Korematsu v. United States,* 323 U.S. 214, 89 L. Ed. 194, 65 S. Ct. 193 (1942).

22. 50 U.S.C. app. § 1981–1987 (1993).

23. Personal Justice Denied Part 2; Recommendations: Report of the Commission on Wartime Relocation and Internment of Civilians (1983), p.7.

24. Id. at p. 8.

25. Id. at p. 9.

26. 50 U.S.C. app. § 1989b-7(2).

27. See *Jacobs v. Barr,* 959 F. 2d 313 (D.C. Cir. 1992); *Obadele v. United States,* 52 Fed. Cl. 432 (2002).

28. 959 F. 2d, p. 320.

29. Some reparations proponents have maintained that the Tulsa situation really falls outside of the reparations arena because what was being sought there was compensation for criminal activities that occurred well after slavery had ended. The victims should be compensated, they argue, similar to the way restitution or damages are routinely ordered in many criminal and civil cases.

INDEX

≛ Amistad

Also by Raymond A. Winbush, Ph.D.:

THE WARRIOR METHOD
A Parents' Guide to Rearing Healthy Black Boys
ISBN 0-380-79275-3 (paperback)

Young black boys and men are becoming an endangered species in the United
States: imprisoned at a greater rate than their white counterparts and dying younger
and more violent deaths than white boys. *The Warrior Method* is the first book to
present a parenting program to combat these atrocities by teaching parents how to
raise their sons. In a program modeled after the Poro societies of West Africa, this
program seeks to imitate the rite of passage in which young men were taken by the
men of the village and taught how to live as responsible adults.

"Clear, concise, and insightful. This book should truly be close at hand to anyone
committed to raising black boys."
—Molly Secours, counselor to at-risk teens and incarcerated youth

"An important book appearing at a crucial time . . . as we struggle to raise healthy
black men in an unhealthy society."
—Jill Nelson, author of *Police Brutality: An Anthology*

"If you care about the future of young black sons and brothers, *The Warrior
Method* provides deliverance."
—Deborah Mathis, nationally syndicated columnist, Gannett